Lecture Notes in Computer Science 14600

Founding Editors

Gerhard Goos
Juris Hartmanis

Series Editors

Elisa Bertino, Purdue University, West Lafayette, IN, USA
Wen Gao, Peking University, Beijing, China
Bernhard Steffen, TU Dortmund University, Dortmund, Germany
Moti Yung, Columbia University, New York, NY, USA

The series Lecture Notes in Computer Science (LNCS), including its subseries Lecture Notes in Artificial Intelligence (LNAI) and Lecture Notes in Bioinformatics (LNBI), has established itself as a medium for the publication of new developments in computer science and information technology research, teaching, and education.

LNCS enjoys close cooperation with the computer science R & D community, the series counts many renowned academics among its volume editors and paper authors, and collaborates with prestigious societies. Its mission is to serve this international community by providing an invaluable service, mainly focused on the publication of conference and workshop proceedings and postproceedings. LNCS commenced publication in 1973.

Jesper Larsson Träff

Lectures on Parallel Computing

　Springer

Jesper Larsson Träff
Vienna University of Technology (TU Wien)
Vienna, Austria

ISSN 0302-9743　　　　　　ISSN 1611-3349　(electronic)
Lecture Notes in Computer Science
ISBN 978-3-031-86577-0　　ISBN 978-3-031-86578-7　(eBook)
https://doi.org/10.1007/978-3-031-86578-7

© The Editor(s) (if applicable) and The Author(s), under exclusive license to Springer Nature Switzerland AG 2026

This work is subject to copyright. All rights are solely and exclusively licensed by the Publisher, whether the whole or part of the material is concerned, specifically the rights of translation, reprinting, reuse of illustrations, recitation, broadcasting, reproduction on microfilms or in any other physical way, and transmission or information storage and retrieval, electronic adaptation, computer software, or by similar or dissimilar methodology now known or hereafter developed.
The use of general descriptive names, registered names, trademarks, service marks, etc. in this publication does not imply, even in the absence of a specific statement, that such names are exempt from the relevant protective laws and regulations and therefore free for general use.
The publisher, the authors and the editors are safe to assume that the advice and information in this book are believed to be true and accurate at the date of publication. Neither the publisher nor the authors or the editors give a warranty, expressed or implied, with respect to the material contained herein or for any errors or omissions that may have been made. The publisher remains neutral with regard to jurisdictional claims in published maps and institutional affiliations.

The cover image, taken on Amrum, is used with permission from Ingeburg Neisius-Knichel.

This Springer imprint is published by the registered company Springer Nature Switzerland AG
The registered company address is: Gewerbestrasse 11, 6330 Cham, Switzerland

If disposing of this product, please recycle the paper.

Preface

These lecture notes are designed to accompany an imaginary, virtual, undergraduate, one- or two-semester course on the fundamentals of Parallel Computing as well as to serve as background and reference for graduate courses on High-Performance Computing, parallel algorithms and shared-memory multiprocessor programming. They introduce theoretical concepts and tools for expressing, analyzing and judging parallel algorithms and, in detail, cover the two most widely used concrete frameworks OpenMP and MPI as well as the threading interface `pthreads` for writing parallel programs for either shared or distributed memory parallel computers with emphasis on general concepts and principles. Code examples are given in a C-like style and many are actual, correct C code. The lecture notes deliberately do not cover GPU architectures and GPU programming, but the general concerns, guidelines and principles (time, work, cost, efficiency, scalability, memory structure and bandwidth) will be just as relevant for efficiently utilizing various GPU architectures. Likewise, the lecture notes focus on deterministic algorithms only and do not use randomization. Slides or blackboard drawings are imagined to be worked out for the actual lectures by the lecturer, so the lecture notes deliberately do not provide such important visual aids: some may be available from the author on request. Also the student of this material will find it instructive to take the time to understand concepts and algorithms visually. The exercises can be used for self-study and as inspiration for small implementation projects in OpenMP and MPI that can and should accompany any serious course on Parallel Computing. The student will benefit from actually implementing and carefully benchmarking the suggested algorithms on the parallel computing system that may or should be made available as part of such a Parallel Computing course. In class, the exercises can be used as the basis for hand-ins and small programming projects for which sufficient, additional detail and precision should be provided by the instructor.

Acknowledgments

These lecture notes grew out of a bachelor's course given at TU Wien, Austria, since 2011, and have benefitted much from comments, often severe criticism, weariness and occasionally very good questions and suggestions by the students who have taken (and had to take) this course over the years. The starred material ($\star$) is usually not covered in the actual lecture. The lecture notes themselves were written starting from March 2020. The author sincerely thanks everyone who contributed in spirit and materially, perhaps unbeknownst to themselves. In particular, Sascha Hunold has over the years significantly influenced the shape of the Parallel Computing course and the thinking and presentation in these lecture notes. Leonhard Patoschka did an extremely careful proof-reading in Spring 2024 which much improved the presentation in the last phase. Enjoyable discussions with Thom Frühwirth, Christian Siebert and Martti Forsell have likewise been of value over the years. Springer has been supportive of the idea of bringing out these lecture notes in the series Lecture Notes in Computer Science. The winding stairs with a vanishing point on Amrum is from a photo by Ingeburg Neisius-Knichel. The responsibility for the selection and presentation of the material as well as any mistakes, errors or omissions in these lecture notes is solely the author's.

Jesper Larsson Träff
TU Wien, April 2024

Deutsches Vorwort

Dieses Skriptum ist als Lesehilfe für die Folien und den Vortrag der Bachelorvorlesung "Parallel Computing" an der TU Wien gedacht. Wir versuchen, auf die besonders wichtige Punkte aufmerksam zu machen und die jeweiligen Vorlesungseinheiten zusammenzufassen. Ergänzende Textbücher, die Material enthalten, das nicht in der Vorlesung besprochen wird, sind das Buch von Rauber und Rünger [90], das Buch von Grama et al. [49] sowie das Buch von Schmidt et al. [96]. Umgekehrt enthält die Vorlesung auch viel Material, das nicht in diesen Büchern zu finden ist. Das Skriptum ist auf Englisch verfasst.

Die mit ⋆ markierten Abschnitte sind nicht Teil des Stoffes für die Bachelorvorlesung.

Contents

1 Introduction to Parallel Computing: Architectures and Models, Algorithms and Measures 1
  1.1 First Block (1–2 Lectures) 1
      1.1.1 "Free Lunch" and Moore's Law 2
      1.1.2 Performance of Processors 2
      1.1.3 Parallel vs. Distributed vs. Concurrent Computing....... 4
      1.1.4 Sample Computational Problems 5
      1.1.5 Models for Sequential and Parallel Computing 6
      1.1.6 The PRAM Model.................................. 7
      1.1.7 Shared vs. Distributed Memory Models and Systems..... 11
      1.1.8 Flynn's Taxonomy.................................. 12
  1.2 Second Block (1–2 Lectures) 13
      1.2.1 Sequential and Parallel Time 13
      1.2.2 Speed-up 16
      1.2.3 "Linear Speed-up Is Best Possible" 17
      1.2.4 Cost and Work..................................... 18
      1.2.5 Relative Speed-up and Scalability 21
      1.2.6 Overhead and Load Balance 22
      1.2.7 Amdahl's Law 24
      1.2.8 Efficiency and Weak Scaling 26
      1.2.9 Scalability Analysis.................................. 28
      1.2.10 Relativized Speed-up and Efficiency 28
      1.2.11 Measuring Parallel Time and Speed-up Empirically 29
      1.2.12 Examples 30
  1.3 Third Block (1–2 Lectures) 39
      1.3.1 Directed Acyclic Task Graphs★....................... 39
      1.3.2 Loops of Independent Iterations 43
      1.3.3 Independence of Program Fragments★................. 44
      1.3.4 Pipeline★.. 46
      1.3.5 Stencil 47
      1.3.6 Work Pool... 49

-   1.3.7 Master-Worker/Master-Slave 50
-   1.3.8 Domain Decomposition* 50
-   1.3.9 Iteration Until Convergence, Bulk Synchronous Parallel .. 51
-   1.3.10 Data Distribution 52
-   1.3.11 Compaction, Gather and Scatter 53
-   1.3.12 Data Exchange, Collective Communication 54
-   1.3.13 Reduction, Map-Reduce and Scan 54
-   1.3.14 Barrier Synchronization 55
- 1.4 Fourth Block (1 Lecture) 56
  - 1.4.1 Merging Ordered Sequences in Arrays 56
  - 1.4.2 Merging by Ranking 57
  - 1.4.3 Merging by Co-ranking 59
  - 1.4.4 Bitonic Merge* 60
  - 1.4.5 The Prefix Sums Problem 62
  - 1.4.6 Load Balancing with Prefix Sums 63
  - 1.4.7 Recursive Prefix Sums 64
  - 1.4.8 Solving Recurrences with the Master Theorem 66
  - 1.4.9 Iterative Prefix Sums 67
  - 1.4.10 Non-work-optimal, Faster Prefix Sums 69
  - 1.4.11 Blocking 70
  - 1.4.12 Related Problems 71
  - 1.4.13 A Careful Application of Blocking* 72
  - 1.4.14 A Very Fast, Work-Optimal Maximum Algorithm 73
  - 1.4.15 Do Fast Parallel Algorithms Always Exist?* 75
- 1.5 Exercises 76

2 Shared Memory Parallel Systems and OpenMP 85
- 2.1 Fifth Block (1 Lecture) 85
  - 2.1.1 On Caches and Locality 86
  - 2.1.2 Cache System Recap 87
  - 2.1.3 Cache System and Performance: Matrix–Matrix Multiplication 89
  - 2.1.4 Recursive, Divide-and-Conquer Matrix–Matrix Multiplication 90
  - 2.1.5 Blocked Matrix–Matrix Multiplication 91
  - 2.1.6 Multi-core Caches 91
  - 2.1.7 The Memory System 94
  - 2.1.8 Super-linear Speed-up Caused by the Memory System ... 95
  - 2.1.9 Application Performance and the Memory Hierarchy 96
  - 2.1.10 Memory Consistency 96
- 2.2 Sixth Block (1–2 Lectures) 99
  - 2.2.1 pthreads Programming Model 99
  - 2.2.2 pthreads in C 100
  - 2.2.3 Creating Threads 100
  - 2.2.4 Loops of Independent Iterations in pthreads 103
  - 2.2.5 Race Conditions and Data Races 103

- 2.2.6 Critical Sections, Mutual Exclusion, Locks 105
- 2.2.7 Flexibility in Critical Sections with Condition Variables .. 108
- 2.2.8 Versatile Locks from Simpler Ones 110
- 2.2.9 Locks in Data Structures 112
- 2.2.10 Problems with Locks 113
- 2.2.11 Atomic Operations 114
- 2.3 Seventh Block (3 Lectures) 118
  - 2.3.1 The OpenMP Programming Model 119
  - 2.3.2 OpenMP in C 120
  - 2.3.3 Fork-Join Parallelism with the Parallel Region 120
  - 2.3.4 OpenMP Library Calls 122
  - 2.3.5 Sharing Variables 122
  - 2.3.6 Work Sharing: Master and Single 123
  - 2.3.7 The Explicit Barrier 125
  - 2.3.8 Work Sharing: Sections 125
  - 2.3.9 Work Sharing: Loops of Independent Iterations 126
  - 2.3.10 Loop Scheduling 128
  - 2.3.11 Collapsing Nested Loops 131
  - 2.3.12 Reductions 132
  - 2.3.13 Work Sharing: Tasks and Task Graphs 135
  - 2.3.14 Mutual Exclusion Constructs and Atomic Operations ... 138
  - 2.3.15 Locks 140
  - 2.3.16 Special Loops 141
  - 2.3.17 Parallelizing Loops with Hopeless Dependencies 142
  - 2.3.18 Example: Parallelizing a Sequential Algorithm with Dependencies 143
  - 2.3.19 Cilk: A Task Parallel C Extension 145
- 2.4 Exercises 149

3 Distributed Memory Parallel Systems and MPI 171
- 3.1 Eighth Block (1 Lecture) 171
  - 3.1.1 Network Properties: Structure and Topology 172
  - 3.1.2 Communication Algorithms in Networks 175
  - 3.1.3 Concrete Communication Costs 178
  - 3.1.4 Routing and Switching 179
  - 3.1.5 Hierarchical, Distributed Memory Systems 182
  - 3.1.6 Programming Models for Distributed Memory Systems ... 182
- 3.2 Ninth Block (3–4 Lectures) 183
  - 3.2.1 The Message-Passing Programming Model 184
  - 3.2.2 The MPI Standard 185
  - 3.2.3 MPI in C 185
  - 3.2.4 Compiling and Running MPI Programs 186
  - 3.2.5 Initializing the MPI Library 187
  - 3.2.6 Failures and Error Checking in MPI 188
  - 3.2.7 MPI Concept: Communicators 189
  - 3.2.8 Organizing Processes* 194

		3.2.9	MPI Concepts: Objects and Handles 199
		3.2.10	MPI Concept: Process Groups★ 200
		3.2.11	Point-to-Point Communication 203
		3.2.12	Determinate vs. Non-determinate Communication 208
		3.2.13	Point-to-Point Communication Complexity and Performance 213
		3.2.14	MPI Concepts: Semantic Terms 215
		3.2.15	MPI Concept: Specifying Data with (Derived) Datatypes . 218
		3.2.16	MPI Concept: Matching Communication Operations 222
		3.2.17	Nonblocking Point-to-Point Communication 223
		3.2.18	Exotic Send Operations★ 226
		3.2.19	MPI Concept: Persistence★ 228
		3.2.20	More on User-Defined, Derived Datatypes★ 229
		3.2.21	MPI Concept: Progress 235
		3.2.22	One-Sided Communication 236
		3.2.23	One-Sided Completion and Synchronization............ 240
		3.2.24	Example: One-Sided Stencil Updates................... 242
		3.2.25	Example: Distributed Memory Binary Search 244
		3.2.26	Additional One-Sided Communication Operations★ 245
		3.2.27	MPI Concept: Collective Semantics 246
		3.2.28	Collective Communication and Reduction Operations 248
		3.2.29	Complexity and Performance of Applications with Collective Operations 261
		3.2.30	Examples: Elementary Linear Algebra 262
		3.2.31	Examples: Sorting Algorithms........................ 266
		3.2.32	Nonblocking and Persistent Collective Operations★ 270
		3.2.33	Sparse Collective Communication: Neighborhood Collectives★ 273
		3.2.34	MPI and Threads★ 275
		3.2.35	MPI Outlook 276
	3.3	Exercises 277
A	Proofs and Supplementary Material 289
	A.1	A Frequently Occurring Sum............................... 289
	A.2	Logarithms Reminder 289
	A.3	The Master Theorem 290
References.. 295
Index 301

Chapter 1
Introduction to Parallel Computing: Architectures and Models, Algorithms and Measures

Abstract. The first third of the lectures on Parallel Computing deals with the fundamental facts of Parallel Computing which distinguish it from Distributed Computing and Concurrent Computing. Computational problems can be creatively and constructively explored and studied with the PRAM model and judged by comparing against the best (known) sequential baselines. This naturally leads to the fundamental concepts of (parallel) work, time, work- and cost-optimality, speed-up, efficiency, and various notions of scalability. These concepts are helpful and meaningful, theoretically as well as practically and empirically. Parallelization patterns that can help both in design and analysis of parallel algorithms and programs are described. As concrete examples, parallel algorithms for important problems with easy linear-time, sequential algorithms are discussed at some length.

1.1 First Block (1–2 Lectures)

Parallel computers, meaning computers and computer systems with more than one processing element, each capable of executing a program and collaborating with other processing elements, are everywhere. The number of processing elements, in modern terminology often called a *processor-core* or just *core*, range from a few (embedded systems, mobile devices) to tens and hundreds (desktops, servers), to thousands, ten-thousands, and even millions in the largest High-Performance Computing (HPC) systems (see http://www.top500.org for some such systems). Every computer scientist has to be aware of this fact and know something about Parallel Computing.

Despite being an active area of research and also of commercial developments of actual parallel computer systems in the mid-1980s to mid-1990s, parallel computing was largely absent from mainstream computer science during the 1990s to early 2000s. This has had and still has dire consequences. The area was largely missing from university curricula (e.g., parallel algorithms, programming and software development), leading to a lack of knowledgeable experts and professionals and to quite frequent rediscovery of already known results and techniques: It still makes much sense to read books and technical papers from the 1980s and 1990s.

1.1.1 "Free Lunch" and Moore's Law

One reason for all this was the "free lunch" phenomenon [108], also sometimes called *Moore's Law*: The performance of sequential computers was observed (and projected) to increase exponentially, with a doubling rate of 18 to 24 months. To many, this made the more modest performance improvements by the use of more processing elements seem uninteresting and (commercially) irrelevant. This popular version of this "Law" held from the 1970s until the early- to mid-2000s, but is not exactly what Gordon Moore actually speculated [81]. Nevertheless, the exponential increase in sequential computer performance made building and selling parallel computers commercially tough. Many ambitious and well-founded companies folded in the early 1990ies, and other companies changed their strategies: HPC was one niche where some companies could survive. Conversely, "Moore's Law" exerted enormous pressure on processor manufacturers; also this had consequences leading, for instance, to many fantastic and fantastically useless HPC systems being built.

In the early 2000s (say, 2005) the "free lunch" was largely over. The performance of sequential processors has not increased as dramatically since then, as has been documented by many popular studies (that may deserve a closer look)[1]. A way out to continue increasing nominal and possibly achieved performance is to employ *parallelism*.

1.1.2 Performance of Processors

For now, we define *nominal processor performance* strictly processor-centrically as the maximum (best-case) number of operations (of some type, often: *FLoating point OPerations per Second, FLOPS*) that can be carried out per unit of time (second) by the processor. The performance of a single processor-core is calculated as the product of the clock frequency, number of "ticks" (cycles) per second, usually measured in GHz, and the number of instructions that the processor can complete per clock cycle (FLOPs/cycle). The number of instructions per clock cycle is determined by the processor architecture: Number of pipelines, depth of pipelines, number of functional units, types of instructions (fused multiply-add, for instance, other complex instructions), super-scalar capabilities, vectorization (*SIMD*) capabilities, etc. [25, 58]. The nominal processor performance provides

---

[1] e.g., https://www.karlrupp.net/2018/02/42-years-of-microprocessor-trend-data

an optimistic upper bound on the performance that can actually be achieved by real-world applications by assuming that all capabilities of the processor can be utilized during the execution of the application. We note that the FLOPS abbreviation is ambiguous and quite unfortunate: sometimes the number of FLOPs are meant, sometimes the FLOPs/second.

Whether the nominal performance of a processor can be reached depends on at least two factors. First, whether the program being executed contains operations in the right mix and with the right dependencies to allow full utilization of the components and features of the processor-core. For instance, a program solving a graph problem may use integers and therefore executes 0 FLOPs. It does not exploit any of the floating point capabilities of the processor (likely a major part). A fused multiply-add instruction (and the related parts of the processor) may be good for matrix–vector multiplication, but not for many other tasks. Second, the memory system must be able to supply the data needed to keep all parts of the processor busy fast enough. This is often a or even the major reason for observed, "poor" performance.

The ratio between processor performance and memory access time has not improved at the pace processor performance has improved (Moore's Law). The main idea to narrow the gap has been the introduction of (larger and larger, hierarchically organized) caches [25, 89]. Caches and the memory system play an important role in Parallel Computing and later in these lectures (see Sect. 2.1.1 and onwards).

With current terminology, a *processor* (CPU) consists of multiple *(processor-)cores*, also called *processing elements* (PE) or processing units (PU): These are the entities that are capable of executing a program. What is now called cores used to be called processors. A processor with a smaller number of cores (a handful, e.g., 4, 8, 10, 16, 24, 32, 48, and 64 which is typical of current server processors) is termed a *multi-core processor*, and a processor with a large number of cores a *many-core processor*. The distinction is blurry and mostly connotative. The prototypical example of the latter is the *graphics processing unit* (GPU), which will play almost no role in these lectures. We will use only the term *multi-core* (where needed). The *nominal performance* of a multi-core processor is calculated by multiplying the nominal per-core performance with the number of cores. For several reasons, also the nominal multi-core processor performance is a very optimistic upper bound on the performance that applications can actually reach.

To make matters more complicated, many modern processor-cores are capable of executing a small number of independent instruction streams (programs, processes, threads) *simultaneously*, typically two to four, with the purpose of exploiting the core's various functional units associated with the core more efficiently and possibly also to be able to hide memory access latencies by switching between streams. Such techniques implemented in hardware are called *hardware multi-threading*, *hyperthreading* or *simultaneous multi-threading* (SMT). To the application programmer, they make the multi-core processor look as if it had two (or four) times the number of hardware cores. Hardware multi-threading effectively improves the number of instructions per clock and is thus accounted for in the nominal processor performance as calculated above. Hardware multi-threading can sometimes improve the measured performance on the order of 10%, but certainly not by the number of supported hardware threads. Therefore, hardware threads are not counted as cores.

Some recommended text books to check up on computer systems and computer architecture are [25, 58, 89, 109] (regularly updated).

1.1.3 Parallel vs. Distributed vs. Concurrent Computing

The focus of Parallel Computing is on using parallel resources (processors, processor-cores) *efficiently* for solving given *computational (algorithmic) problems*. Towards this end, Parallel Computing is concerned with *algorithms*, their *implementation* in suitable *programming languages* that realize more or less explicitly formulated *programming models* capturing the essentials for analyzing and reasoning about programs, and the structure and capabilities of the underlying actual or imagined *computer architecture*. We judge efficiency in all these respects, both theoretically and practically/experimentally. Parallel Computing is thus theoretical, practical, and experimental Computer Science and much broader in scope than just *parallel programming*, which will also be treated in these lectures with C and `pthreads` and OpenMP and MPI as concrete examples.

Parallel Computing is intimately related to the disciplines of distributed and concurrent computing, and distinguishing is a matter of what we are interested in (our focus). In these lectures we propose and use the following definitions.

Definition 1 (Parallel Computing). The discipline of efficiently utilizing dedicated parallel resources for solving given computational problems.

The focus of Parallel Computing is on *problem solving efficiency*, and a fundamental assumption is that the full computer system is at our disposal (dedicated). Interesting parallel computing problems are those that require significant interaction (communication, be it via memory reads/writes, or explicit communication over some interconnection network) between the parallel resources (cores), on systems that actually provide significant inter-communication and processing capabilities. Therefore, real parallel computers are commonly not thought of as spatially (widely) distributed (the internet) [16].

Parallel Computing is related to and can benefit from results in *distributed* and *concurrent* computing, by which the following is meant (our definitions, others may disagree).

Definition 2 (Distributed Computing). The discipline of making independent, non-dedicated resources available to cooperate toward solving specified problem complexes.

The focus of Distributed Computing is on availability of resources that are not readily at hand, may be spatially widely distributed, may change dynamically, and may *fail*. In Parallel Computing, processor-cores *do not fail* (at least not in this lecture!). Specific, individual problems or larger problem complexes may be studied. A central tenet in Distributed Computing is that there is no centralized control. Example: Acquiring resources from the cloud, subject to certain constraints and requirements, may, on the one hand, be a Distributed Computing problem. Using the resources as a (virtual) parallel machine for solving the problem we are interested in efficiently (for instance, within given time constraints) is, on the other hand, a Parallel Computing problem. Example: Routing data through a (dynamically changing) network while sustaining a high (guaranteed) throughput and low latency with no possibility of deadlock or lost data can be viewed as a Distributed Computing problem, solutions to which are obviously relevant for Parallel Computing.

Definition 3 (Concurrent Computing). The discipline of managing and reasoning about interacting processes that may or may not progress simultaneously.

The focus of Concurrent Computing is on *concurrency*, activities that may or may not happen at the same time, are usually not centrally coordinated, and therefore, on reasoning about and establishing correctness (in a broad sense) in such situations (e.g., by process calculi [63, 80]). In contrast, Parallel Computing is specifically concerned with bounds on the performance that can be also practically achieved, and typically make much more and stronger assumptions about progress and actual concurrency in the system.

1.1.4 Sample Computational Problems

Some computational problems that will be considered and used as examples throughout these lectures are:

- computing sums and maxima over objects stored in arrays,
- matrix–vector multiplication, matrix–matrix multiplication,
- merging of ordered sequences of numbers and objects,
- sorting numbers or objects from ordered sets by merging, by counting, by Quicksort, ... and other methods,
- performing reductions over sets of numbers and objects with given associative operators,
- computing prefix sums over arrays, compacting arrays,
- listing prime numbers,
- performing stencil computations on matrices, and
- graph search problems (e.g., Breadth-First Search or Depth-First Search).

Such computational problems that can be precisely and quantitatively defined are routinely considered and solved in algorithms courses [33]. Most of them, e.g., the matrix-computations from *basic linear algebra* and the stencil computations, are clearly important enough by themselves. Almost all of them are crucial as building blocks in more complex algorithms, e.g., sorting, prefix sums and the graph search problems. More importantly, the solutions illustrate general patterns, approaches, and techniques for analyzing and solving similar problems. We define the problems more precisely as we deal with them. Some of the problems exhibit *regular* computational patterns, e.g., the matrix problems, that may even be *oblivious* to (independent of) the actual input. Some of the problems have more *irregular* computational patterns that depend on the specific input, e.g., some sorting algorithms and many graph search algorithms. For the matrix-problems we will here consider only so-called *dense* variants that are solved by regular, oblivious algorithms. In other words, we will not in any way take the (algebraic) structure of the matrices and vectors into account (triangular matrices, diagonal matrices, block matrices, matrices with many zero and one elements, ...). Doing so and dealing with *sparse* matrix-problems is considerably more challenging, for sequential as well as for parallel algorithmics; but can sometimes lead to faster solution.

1.1.5 Models for Sequential and Parallel Computing

For designing and analyzing parallel algorithms, a suitable model of computation is needed. A good model is one which makes it possible to derive interesting algorithms and results, makes analysis tractable, and bears enough resemblance to actual machines and systems that the algorithms can be implemented and results predictive of, say, performance.

A model with the last property is sometimes called a *bridging model* (we use the term in this fashion), a term originally introduced by Les Valiant [118, 120] who proposed a specific model as bridge for Parallel Computing, the so-called *Bulk Synchronous Parallel (BSP)* model. A minimum requirement for a good bridging model is that if some algorithm A is shown to perform better than algorithm B in the model, then a (faithful) implementation of A should perform better than an (equally faithful) implementation of B on the real machine ("bridging"). The (vague) notion of *performance portability* is related to the bridging idea. It says that the good performance of a program can be preserved when moving from one system to another. This is clearly a desirable property.

While there are various "bridging models" in sequential computing with the *RAM, Random Access Machine* being the most important one, although it is not unproblematic and has many restrictions, the situation is completely different for Parallel Computing. There are many different parallel computer architectures (multi-core CPU vs. GPU; distributed memory system vs. shared memory system, etc.), at vastly different scales, and no model (so far) bridges them all to any useful extent. The BSP model has so far not been successful (in finding universal or even widespread use). Also, model assumptions that are desirable for the design of algorithms do, to an even lesser extent than for sequential models, hold for parallel computer systems. Many such assumptions are related to the memory behavior. For instance, the assumption of unit-time, uniform memory access of the RAM is already problematic for sequential computers, and even more so for large parallel systems with widely distributed memory.

1.1.6 The PRAM Model

One extremely useful, but unrealistic model of parallel computing is the *Parallel Random Access Machine (PRAM)* [64], a natural generalization of the equally useful and pervasive, sequential *Random Access Machine (RAM)*. Like the RAM, the PRAM assumes a large (as large as needed) memory where processors (as many as needed) can read and write words (addresses) in unit time. A more concrete PRAM, closer to physical reality, would have a certain, given number of processors. These processors all execute their own program, but do so in *lock-step*: strictly synchronized, all following the same, global clock and performing an instruction in each *time step*. This means that the machine is always in a well-defined *state* comprised of the program counter of the processors, contents of the memory and the processor registers. State transitions happen instantaneously by the synchronous clock ticks, and reasoning with state invariants, as done in sequential RAM algorithms, is a way to prove properties. A PRAM algorithm specifies what the processors are to do in each step.

With many processors operating in lock-step, it can potentially happen that more than one processor is accessing some memory word in the same time step. The PRAM model needs to define what happens in such cases. First, a memory word can, in a step, be either read or written; but not read by some processor(s) and written by another. For potentially *concurrent accesses* to a memory word in a step by two or more processors, there are three main variations of the PRAM that have been used in the literature:

- An *EREW* (Exclusive Read Exclusive Write) *PRAM* disallows accesses to the same memory word in the same step by more than one processor. It is the algorithm designer's responsibility to make sure that simultaneous accesses do not happen.
- A *CREW* (Concurrent Read Exclusive Write) *PRAM* allows simultaneous (parallel, concurrent) reads to any memory word by more than one processor in a time step, but not simultaneous writes to a memory word in a time step.
- A *CRCW* (Concurrent Read Concurrent Write) *PRAM* allows both simultaneous reads and simultaneous writes to the same memory word in the same step; but not reads and writes to one and the same word in the same time step (many or all processors may read the same word; many or all processors may write to the same word). What happens when two or more processors write to a word in a step? In a *Common CRCW PRAM*, it must be ensured that the writing processors all write the same value. In an *Arbitrary CRCW PRAM*, either of the written values will survive in the memory word. A *Priority CRCW PRAM* has some priority associated with the processors, and

the writing processor with the highest priority will successfully write its value to the memory word.

What happens in case the EREW/CREW/CRCW constraints are violated by our algorithm is just a matter of model design: perhaps the machine breaks down, explodes, halts, delivers incorrect results, or some other outcome. The important requirement is that the algorithm designer has to make sure (prove!) that the constraints of the PRAM variant at hand are never violated when the algorithm is executed. Per definition, any algorithm that can be executed correctly on an EREW PRAM, can execute on any of the, in that sense, stronger variants.

The PRAM is largely a purely theoretical construct; there have been several attempts to realize emulated PRAMs in real hardware, but so far none have been entirely or commercially successful [1,2,41,66,85]. We use it as an analytical tool to precisely describe and analyze (fast) parallel algorithms with high parallelism: many processors relative to the size or computational demands of the problem to be solved. We can therefore freely invent convenient pseudo-code to liberally express algorithms, as long as it is clear that the PRAM model assumptions are satisfied. The goal is to be able to characterize time (number of parallel steps) and effort (number of processors used in the parallel steps) of parallel computations. For this, we allow to freely choose, for each parallel step, the number of PRAM processors to be used in that step. This can be a fixed number (sometimes just one), a function of the input size or a free parameter. On a physical PRAM with some fixed number of processors, the allocated (virtual) processors would be emulated by the available, possibly fewer physical processors.

In order to be able to describe interesting algorithms more concretely, we introduce a pseudo-code construct for starting a set of processors, each being assigned an identity (some integer) to which it can refer. This is the `par`-construct that looks similar to a C pseudo-code `for`-loop. This construct allows us to declare a range or set of processors to start working. We will assume that starting a reasonably specified set of processors can be done even on an EREW PRAM in a constant number of operations, $O(1)$ per processor. This is reasonable for simple ranges where each processor identity can be computed by simple arithmetic. On a physical PRAM realized in hardware, it would be the task of the run-time system and compiler to provide constructs for starting or allocating well-defined sets of (virtual) processors with some well-defined (small) overhead. In order to fulfill the lock-step assumption, correct pseudo-code will make sure that the allocated processors in a `par`-construct all perform the exact same number of instructions. This means that open `while`-loops where the number of iterations may be different for different processor identities are not allowed. Also, `if`-statements have to be written in such a way that both branches will have the same number of instructions to execute; but we will here just leave it to the (virtual) compiler to pad branches with the needed no-op instructions to ensure this. If it is not obvious how this can be done, the algorithm-code should be rewritten.

Using the analytic PRAM, we can now give interesting algorithms for many of our computational problems, for instance, for finding the maximum among n

1.1 First Block (1–2 Lectures)

numbers, and for doing matrix–matrix multiplication of $m \times l$ and $l \times n$ matrices into an $m \times n$ result matrix.

Our first, non-obvious PRAM algorithm expressed in PRAM-pseudo code for finding the maximum in a set of numbers (stored in an array) is given below, and the results are summarized in the theorems that follow.

```
par (0<=i<n) b[i] = true; // a[i] could be maximum
par (0<=i<n, 0<=j<n) {
  if (a[i]<a[j]) b[i] = false; // this a[i] is not maximum
}
par (0<=i<n) if (b[i]) x = a[i];
```

Theorem 1. *The maximum of n numbers stored in an array can be found in $O(1)$ parallel time steps, using n^2 processors and performing $O(n^2)$ total operations on a Common CRCW PRAM.*

In the program, the input is stored in the n-element array a, indexed C-style from 0 to $n-1$. The idea of this fastest possible algorithm is to do all the n^2 pairwise element comparisons in one parallel step (actually, the $n(n-1)$ comparisons with different element indices would suffice), and use the outcome to knock out the elements that cannot possibly be the maximum. This is done with the Boolean array b, which is used to mark each of the n elements as a candidate for being a maximum. As outcome of the pairwise comparisons, elements that cannot be maximum by virtue of being smaller than some other element are unmarked by one or more of the n^2 assigned processors. The three par-constructs start n, n^2, and n processors, respectively, first for initializing the b-array, second for performing all the n^2 comparisons in parallel, and finally for writing out the maximum to the result variable x. Since in one step, several (up to n) processors can discover that some element a[i] cannot be a maximum since a[i]<a[j], concurrent writing to the same b[i] can happen. Whether and at which indices this happens is dependent on the input. When several processors write to a location b[i] or x in a step, they, however, write the same value (false, or the maximum value, respectively), and therefore a Common CRCW PRAM suffices for this algorithm. This is an interesting, maximally fast (there is nothing faster than constant time, and the constants here seem to be small) algorithm: The PRAM model is good for exposing the maximum amount of parallelism in a problem. The time taken by the algorithm is the number of parallel time steps (here three), and the number of processors used is the maximum number of processors assigned in a parallel step (here n^2).

Simultaneous, concurrent writing to the same memory location or memory module is a (too?) powerful capability of a parallel computer, which should presumably be avoided if possible. In order to avoid concurrent writing in the maximum finding problem, a different algorithmic idea is needed: Instead of doing all pairwise comparisons in a step, do only up to $n/2$ comparisons in parallel between disjoint pairs of elements. The pseudo-code below implements this idea.

```
nn = n;  // number of elements per while-loop iteration
while (nn>1) {
  k = (nn+1)>>1;  // ceil(nn/2) by shift
  par (0<=i<k) {
    if (i+k<nn) a[i] = max(a[i],a[i+k]);
  }
  nn = k;
}
```

Theorem 2. *The maximum of n numbers stored in an array can be found in $O(\log n)$ parallel time steps, using a maximum of $n/2$ processors (but performing only $O(n)$ operations) on a CREW PRAM.*

This algorithm goes through $\lceil \log_2 n \rceil$ iterations, in each one roughly halving the number of element pairs to compare. Elements are stored in the a-array which as can be seen is destructively updated. The resulting maximum ends up in location a[0]. In each iteration, the algorithm performs comparisons between $\lfloor n/2 \rfloor$ pairs only, in each of which the larger element is stored. This reduces the number of possible maximum elements for the next iteration to $\lceil n/2 \rceil$. The comparison step needs to be iterated $\lceil \log_2 n \rceil$ times, after which a maximum element is left in a[0]. As written, the algorithm requires concurrent reading, namely of k and nn, but it can be modified to run also on an EREW PRAM, and it is a good exercise to do so.

The last example turns the definition of matrix–matrix multiplication, into parallel PRAM code. The $m \times n$ matrix product C of $m \times l$ and $l \times n$ input matrices A and B is defined as

$$C[i,j] = \sum_{k=0}^{l-1} A[i,k]B[k,j]$$

for $0 \leq i < m, 0 \leq j < n$. Since we do not (yet) know how to compute the sum of l elements (the l element products in the sum), this part of the definition is implemented as a sequential loop, but all mn sums are computed in parallel as specified by the outer **par**-construct.

```
par (0<=i<m, 0<=j<n) {
  C[i,j] = 0;
  for (k=0; k<l; k++) {
    C[i,j] += A[i,k]*B[k,j];
  }
}
```

Theorem 3. *Two $m \times l$ and $l \times n$ matrices can be multiplied into an $m \times n$ matrix in $O(l)$ time steps and $O(mnl)$ operations on a CREW PRAM.*

The algorithm shown can also be improved to run on an EREW PRAM by using extra space for intermediate results. It can be made faster by employing a variant of the maximum finding algorithm to do the summations in parallel.

The complexity properties of the PRAM algorithms so far were stated in terms of the total number of parallel steps required for the given input, the maximum number of processors needed in some parallel step, the total number of operations carried out by all allocated processors during the course of execution, and the PRAM model required by the algorithm. The natural goal when studying the parallel complexity of specific problems is to minimize these requirements on all counts: as few parallel steps, as few total operations, and as weak a PRAM model as possible. As the observations and theorems above show, some of these goals seem contradictory and not achievable simultaneously. A strong Common CRCW PRAM model made it possible to find the maximum of n numbers optimally fast (constant time), but at the additional cost of a large number of (redundant) operations (Theorem 1). An algorithm for the weaker, possibly less expensive CREW PRAM using less operations and processors was given; but it uses more time (parallel steps) (Theorem 2). We elaborate on these measures and trade-offs which will be a main theme in the following parts of these lectures.

The PRAM model has been productive in finding highly parallel, fast algorithms for many interesting problems and also in establishing lower bounds on how fast and with how many resources (processors) they can be solved [64]. Whether the algorithms studied so far are good or useful, will be elaborated on in the following.

Other theoretical models for Parallel Computing that we may encounter but will not use here include comparator networks, systolic arrays, cellular automata, ... The theoretician (and computer architect, but with the constraints of the world we live in) is free to invent models that serve the purpose: such models have been productive in establishing important results on how to do and not to do things.

1.1.7 Shared vs. Distributed Memory Models and Systems

The PRAM model is an example of a Parallel Computing model with a shared memory from and into which processors can freely read and write data and thus exchange information with each other, subject only to the EREW, CREW or CRCW constraints of the particular PRAM. The PRAM model allows us to formulate algorithms using as many processors as needed. All processors can access all words of a common shared memory which is also as large as needed. Access to memory always takes unit time, namely a single clock cycle, and this is independent of which location is being accessed by which processor. The PRAM is the most extreme case of a *Uniform Memory Access* (UMA) model. Access times are uniformly the same. Furthermore, memory operations are fine-grained and done in units of single words (bytes, integers, doubles, ...). In addition, the PRAM makes the strong assumption that processors operate synchronously in lock-step.

Real, shared memory systems are quite far from all these PRAM model assumptions. Memory access times are not on the order of a single or a few clock cycles, but take much longer than instructions carried out by the processor-cores. More importantly, access times are not uniform. Access to registers in the small register bank memory of the processor can indeed be fast, accesses to data stored in cache memories already slower (see Sect. 2.1.1), accesses to data in "main memory" again slower and so on. Memories, especially "main memory" is often divided into "banks" with some banks being "closer" to some processor-cores than to other processor-cores, and accesses to data in different banks can take different times for different processors. Memory can even be local to processors in the sense that some form of explicit communication is required for one processor to access memory that is controlled by another processor. These characteristics are loosely called *Non-Uniform Memory Access* (NUMA). More realistic computational models that capture aspects of these realities are much harder to formalize and use.

A third of these lectures is devoted to models, aspects, and concrete programming of so-called shared memory (multi-core) systems. The processor-cores in such systems exchange informations and solve computational problems by reading and writing from and to a quite large, but finite shared memory, somewhat like the PRAM. But the processors are not really synchronized and memory access times are both NUMA and higher than operations done by the processor. Memory is managed at different granularities. More about this will follow in Chapter 2.

Another third of these lectures is devoted to models, aspects, and concrete programming of so-called distributed memory (multi-node, multi-core) systems. Each multi-core processor-node has memory that is local to that node, and explicit communication between processors on different multi-core nodes is needed for exchanging information and solving computational problems. Communication is facilitated by a dedicated communication network. All this in Chapter 3.

1.1.8 Flynn's Taxonomy

A different, frequently used, less architecture-oriented and rather crude characterization of parallel machines, systems and even programs is the so-called *Flynn's taxonomy* [40]. This taxonomy looks at the instruction and data stream(s) of the computing system. A *Single-Instruction, Single-Data* (*SISD*) system is a sequential computer: one program is executed and the instructions operate on a single stream of data. This is, of course, a naive and simplified notion of the workings of a modern processor. A *Single-Instruction, Multiple-Data* (*SIMD*) system is one in which a single instruction can operate on a larger batch of data, like, for instance, a whole vector (array) of some size. Thus, classical *vector computers* that operate on long vectors, or modern processors with capabilities to operate on short vectors of a few words (with AVX or SSE instruction sets) are typical SIMD systems. A PRAM machine would be classified as *Multiple-Instruction, Multiple-Data* (*MIMD*), since each processor can execute its own instruction stream, each operating on its own stream of data. Finally, but not obviously, a

*Multiple-Instruction, Single-Data (*MISD*)* system could be a deeply pipelined system where a single stream of data passes through several processing stages. Many say that such systems do not exist, i.e., that this taxon in the taxonomy does not make sense.

Flynn's taxonomy is sometimes also used to characterize *programming models* by which we mean the abstractions under which a program can be described (threads, processes, data access patterns, synchronization and communication mechanisms, etc.). A SIMD model, for instance, is one in which there is a single "logical" instruction stream (that might, as in a PRAM, be executed by many processors) that operates on some abstract "vectors" [19].

The characterization *Single-Program, Multiple-Data (SPMD)* is sometimes used to describe the situation where all processors in a parallel system execute the same program, but each processor may, at any time instant, be in a different part of the program and thus operate on a different "data stream" than the other processors. Our PRAM pseudo-code is SPMD as is typical for most real parallel code, as we will see with OpenMP and MPI later in the lecture notes. There are relevant counter examples, though, where the processor-cores in a system actually do run different programs, but nevertheless cooperate to solve a given, computational problem. Complex simulations working at many levels at the same time with different program packages and code could be such an example. GPU programming models sometimes use the term *Single-Instruction Multiple-Threads (SIMT)* to emphasize that a single instruction can be executed simultaneously, concurrently by multiple threads, where batches of threads execute in lock-step as in the PRAM.

1.2 Second Block (1–2 Lectures)

The bar for Parallel Computing is high. We judge parallel algorithms and implementations by comparing them against the *best possible* sequential algorithm or implementation for solving the given computational problem, and in cases where the best possible (lower bound) is not known, against the *best known* sequential algorithm or implementation. The reasoning is that we, by using the dedicated parallel resources at hand, want to improve over what we can already do with a sequential algorithm on our system. With our parallel machine, we want to solve problems faster and/or better on some account.

For now, our parallel model and system will be left unspecified. Some number p of processor-cores interact to solve the problem at hand.

1.2.1 Sequential and Parallel Time

Parallel Computing is both a theoretical discipline and a practical/experimental endeavor. As a theoretical discipline, Parallel Computing is interested in the performance of algorithms in some models (RAM, PRAM, and more realistic settings), and typically looks at the performance in the worst possible case (worst possible inputs) when the input size is sufficiently large. Let Seq and Par denote sequential and parallel algorithms for a problem we are interested in solving. The parallel algorithm, in contrast to the sequential algorithm, additionally specifies how processors are to be employed in the solution, how they interact and coordinate, and how they exchange information. The sequential and parallel algorithms may be "similar" in idea and structure; they may also, as we have already seen (Theorem 1), be completely different. This is fine as long as we can argue or even prove that they both correctly solve the given problem.

By $T_{\text{seq}}(n)$ and $T_{\text{par}}^p(n)$ we denote the running times (depending on how our model accounts for time, for instance, number of steps taken) of Seq and Par on worst-case inputs of size n with one processor for the sequential algorithm Seq and with p processor-cores for the parallel algorithm Par. The best possible and best known algorithms for solving a given problem are those with the best worst-case asymptotic complexities. For a given problem, the best possible sequential running time is often denoted as $T^*(n)$, a function of the input size n [64, 90], which then defines the *sequential complexity* of the given problem. In the same way, we can define the *parallel time complexity* $T\infty(n)$ for a given parallel algorithm Par as the smallest running time that this algorithm can achieve using sufficiently many processors. The number of processors to use to achieve this best running time can then be turned into a function of the input size n. If the parallel algorithm is the *fastest possible* algorithm for our given problem, $T\infty(n)$ is the parallel time complexity of the problem.

As always, constants do matter (!), but they will often be ignored here and hidden behind $O, \Omega, \Theta, o, \omega$. Recall the definitions and rules for manipulating such expressions, see for instance [33] or any other algorithms text, and note that, for parallel algorithms, the worst-case time is a function of two variables, problem size n and number of processor-cores p. Saying that some $T_{\text{par}}^p(n)$ is in $O(f(p, n))$ then means that

$$\exists C > 0, \exists N, P > 0 : \forall n \geq N, p \geq P : 0 \leq T_{\text{par}}^p(n) \leq Cf(p,n)$$

and that some $T_{\text{par}}^p(n)$ is in $\Theta(f(p, n))$ that

$$\exists C_0, C_1 > 0, \exists N, P > 0 : \forall n \geq N, p \geq P : 0 \leq C_0 f(p,n) \leq T_{\text{par}}^p(n) \leq C_1 f(p,n) \ .$$

We may sometimes let the number of processors p change as a function of the problem size, $p = f(n)$ ("What is the best number of processors for this problem size?" as in the definition of parallel time complexity), or the problem size change as a function of the number of processors, $n = g(p)$ ("What is a good problem size for this number of processors?"), in which case the asymptotics are of one variable.

Typical sequential, best known/best possible worst-case complexities for some of our computational problems are [33]:

1.2 Second Block (1–2 Lectures)

- $\Theta(\log n)$: Searching for an element in an ordered array of size n.
- $\Theta(n)$: Maximum finding in an unordered n element sequence, computing the sum of the elements in an array (reduction), computing all prefix sums over an array.
- $\Theta(n \log n)$: Comparison-based sorting of an n element array.
- $\Theta(n^2)$: Matrix–vector multiplication with dense, square matrices of order n (inputs of size $\Theta(n^2)$).
- $O(n^3)$: Dense matrix–matrix multiplication, which we will take as the best bound known to us in this lecture (but far from best known, see, e.g., [107]).
- $O(n+m)$: Breadth-First Search (BFS) and Depth-First Search (DFS) in graphs with n vertices and m edges.
- $\Theta(m+n)$: Merging two ordered sequences of length n and m with a constant time comparison function, identifying the connected components of undirected graphs with n vertices and m edges.
- $O(n \log n + m)$: Dijkstra's Single-Source Shortest Problem algorithm on real, non-negative weight, directed graphs with n vertices and m arcs using a best known priority queue.

Regardless of how time per processor-core is accounted for, the time of the parallel algorithm Par when executed on p processor-cores is the time for the last processor-core to finish, assuming that all cores started at the same time. Note that we here make a lot of implicit assumptions, "same time" etc., that will not be discussed further but are worth thinking much more about. The rationale for this convention is twofold: Our problem is solved when the last processor has finished (and we know that this is the case), and since our parallel system is dedicated, it has to be paid for until all processor-cores are again free for something else.

In Parallel Computing as a practical, experimental endeavor, Seq and Par denote concrete implementations of the algorithms, and $T_{\text{seq}}(n)$ and $T_{\text{par}}^p(n)$ are measured running times for concrete, precisely specified inputs of size $O(n)$ on concrete and precisely specified systems. Designing measuring procedures and selecting inputs belong to experimental Computer Science and are highly non-trivial tasks; they will not be treated in great detail in these lectures. Suffice it to say that time is measured by starting the processor-cores at the same time as far as this is possible, and accounting for the time $T_{\text{par}}^p(n)$ by the last processor-core to finish. Inputs may be either single, concrete inputs or a whole larger set of inputs. Worst-case inputs may be difficult (impossible) to construct and are often also not interesting, so inputs are rather "typical" instances, "average-case" instances, randomly generated instances, inputs with particular structure, etc. (for recent criticism of and alternatives to worst-case analysis of algorithms, see [93]). The important point for now is that inputs and generally the whole experimental set-up be clearly described, so that claims and observations can be objectively verified (reproducibility).

1.2.2 Speed-up

We measure the gain of the parallel algorithm Par over the best known or possible sequential algorithm Seq for inputs of size $O(n)$ by relating the two running times. Parallel Computing aims to improve on the best that we can already do with a single processor-core. This is the fundamental notion of absolute *speed-up* over a given baseline:

Definition 4 (Absolute Speed-up). The *absolute speed-up* of parallel algorithm Par over best known or best possible sequential algorithm Seq (solving the same problem) for input of size $O(n)$ on a p processor-core parallel system is the ratio of sequential to parallel running time, i.e.,

$$\mathrm{SU}_p(n) = \frac{T_{\mathsf{seq}}(n)}{T^p_{\mathsf{par}}(n)} \quad .$$

The notion of speed-up is meaningful in both theoretical (analyzed, in some model) and practical (measured running times for specific inputs) settings. Often, speed-up is analyzed by keeping the problem size n fixed and varying the number of processor-cores p (strong scaling, see later). Sometimes (scaled speed-up, see later) both input size n and number of processor-cores p are varied. For the definition, it is assumed that $T^p_{\mathsf{par}}(n)$ is meaningful for any number of processors p (and any problem size n), which for concrete algorithms and implementations is not always the case: Some algorithms assume $p = 2^d$ for some d, a power-of-two number of processors, or $p = d^2, p = d^3$, a square or cubic number of processors, etc. The speed-up is well-defined only for the cases for which the algorithms actually work. For any input size n, there is obviously also a maximum number of processors beyond which the parallel algorithm does not become faster (or even work), namely when there is not enough computational work in the input of size n to keep any more processors busy with anything useful. Beyond this number, speed-up will decrease: Any additional processors are useless and wasted.

As an example, a parallel algorithm Par with $T^p_{\mathsf{par}}(n) = O(n/p)$ would have an absolute speed-up of $O(p)$ for a best known sequential algorithm with $T_{\mathsf{seq}}(n) = O(n)$, assuming that $n \geq p$ (p in $O(n)$ or, equivalently, n in $\Omega(p)$). If $T^p_{\mathsf{par}}(n) = O(n/\sqrt{p})$ the speed-up would be only $O(\sqrt{p})$.

A speed-up of $\Theta(p)$, with upper bounding constant of at most one and n allowed to increase with p, is said to be *linear*, and linear speed-up of p where both bounding constants are indeed close to one is said to be *perfect* (by measurement, or by analysis of constants). Perfect speed-up is rare and hardly achievable (sometimes provably not, an important example is given later in these lecture notes, see Theorem 10).

According to the definitions of linear and perfect speed-up, a parallel algorithm Par with running time of at most $T^p_{\mathsf{par}}(n) = c(\frac{n}{p} + \log n)$ for some constant c would have perfect speed-up relative to a best possible sequential algorithm with running time of at most $T_{\mathsf{seq}}(n) = cn$ steps. We have

$$\mathrm{SU}_p(n) = \frac{cn}{c(n/p + \log n)}$$
$$= \frac{p}{1 + (p\log n)/n}$$

which is as close to p as desired for $n/\log n > p$: For any $\varepsilon, \varepsilon > 0$, it holds that $(p \log n/n) < \varepsilon \Leftrightarrow n/\log n > p/\varepsilon$. If the sequential and parallel algorithms have different leading constants c_0 and c_1, respectively (with $c_0 < c_1$), the speed-up is linear with upper bounding constant $\frac{c_0}{c_1} < 1$. In other words, linear speed-up means that for any number of processors p, the parallel running time multiplied by p differs by a constant factor from the best (possible or known) sequential running time (the sequential time being lower) for sufficiently large n; perfect speed-up means that this constant is practically one.

1.2.3 "Linear Speed-up Is Best Possible"

Linear speed-up is the best that is possible. The argument for this is that a parallel algorithm running on p dedicated cores can be *simulated* on a single core in time no worse than $pT_{\mathsf{par}}^p(n)$ time steps by simulating the steps of the p processors one after the other in a round-robin fashion. If the speed-up would be more than linear, then $T_{\mathsf{seq}}(n) > pT_{\mathsf{par}}^p(n)$, and the simulated execution would run faster than the best known sequential algorithm for our problem, which cannot be. Or: in that case, an even better algorithm would have been constructed! Sometimes, indeed, a new parallel algorithm can by a clever simulation lead to a better than previously known sequential algorithm.

For the PRAM model, the simulation argument can be worked out in detail, for instance, by writing a sequential simulator for programs in our PRAM pseudo-code: Within each `par`-construct, execute the instructions of the assigned processors one after the other in a round-robin fashion, with some care taken to resolve concurrent writing correctly.

Despite this argument, *super-linear speed-up* larger than the number of processor-cores p is sometimes reported (mostly in practical settings) [38,57]. If the reasons for this are algorithmic, it can only be that the sequential and parallel algorithms are, on specific inputs, not doing the same amount of work (see below). Randomized algorithms, where more and different coin tosses are possibly done by the parallel algorithm than by the sequential algorithm, can likewise sometimes exhibit super-linear speed-up. But also deterministic algorithms, like search algorithms, can exhibit this behavior if the way the search space is divided over the parallel processors depends on the number of processor-cores causing the parallel algorithm to complete the search more than proportionally faster than the sequential algorithm. Finally, on "real" parallel computing systems, the memory system and in particular the average memory access times can differ between algorithms running on a single processor-core and on many processor-cores where memory is accessed in a distributed fashion and faster memory "closer to the core" can be used to a larger extent (see Sect. 2.1.1).

The argument that linear speed-up is best possible also tells us that for any parallel algorithm it holds that $T_{\text{par}}^p(n) \geq \frac{T_{\text{seq}}(n)}{p}$. In other words, the best possible parallel algorithm Par for the problem solved by Seq cannot run faster than $T_{\text{seq}}(n)/p$. This observation provides us with a first, useful *lower bound* on parallel running time.

For any parallel algorithm Par on concrete input of size $O(n)$, there is, of course a limit on the number of processor-cores that can be sensibly employed. For instance, putting in more processor-cores than there is actual work (operations) to be done makes no sense, and some processors would sit idle for parts of the computation. Specific speed-up claims are therefore (or should be) qualified with the range of processor-cores for which they apply.

1.2.4 Cost and Work

Our dedicated parallel system with p processor-cores running Par is kept occupied for $T_{\text{par}}^p(n)$ units of time, and this is what we have to "pay" for. The *cost* of a parallel algorithm is, accordingly, defined as the product $p \times T_{\text{par}}^p(n)$. If we picture a parallel computation as a rectangle with the processor-cores i on one axis, listed densely from 0 to $p-1$ and the time spent by the processor-cores on the other axis, the parallel time $T_{\text{par}}^p(n)$ is the largest time for some processor-core i, and the cost is the area of the rectangle $p \times T_{\text{par}}^p(n)$. The parallel algorithm Par exploits the parallel system well if the parallel cost invested for a given input is proportional to the cost of solving the given problem sequentially by Seq. This motivates the notion of *cost-optimality*.

Definition 5 (Cost-Optimal Parallel Algorithm). A parallel algorithm Par for a given problem is *cost-optimal* if its cost $pT_{\text{par}}^p(n)$ is in $O(T_{\text{seq}}(n))$ for a best known sequential algorithm Seq for any number of processors p up to some bound that is an increasing function of n.

Cost-optimality requires that, for any given input size n, there is a certain number of processors p for which the cost $p'T_{\text{par}}^{p'}(n)$ for any $p' \leq p$ is in $O(T_{\text{seq}}(n))$ and the bounding constant in $O(T_{\text{seq}}(n))$ does not depend on p' or p. The bound on the number of processors must be an increasing function of the problem size n. The intention is that the cost of Par is in the ballpark of the sequential running time of Seq. Almost per definition, cost-optimal algorithms have linear speed-up, since $pT_{\text{par}}^p(n) \leq cT_{\text{seq}}(n))$ implies $\frac{T_{\text{seq}}(n)}{T_{\text{par}}^p(n)} \geq \frac{p}{c}$ which is the speed-up. The requirement that the upper bound on the number of processors p increases with n makes it possible to find an increasing function of p for which the speed-up is in $\Theta(p)$. Cost-optimality is a strong property.

A different way of looking at cost-optimality is via the parallel time complexity and the number of processors needed to reach this fastest time. The product of

1.2 Second Block (1–2 Lectures)

this number of processors and this fastest possible time should still be in the order of the effort required by a best (known or possible) sequential algorithm. This is captured in the following definition.

Definition 6 (Asymptotically Cost-Optimal Parallel Algorithm). Let for some given problem Par be a parallel algorithm with parallel time complexity $T\infty(n)$. Let $P(n)$ be the smallest number of processors needed to reach $T\infty(n)$. The cost of Par with this number of processors is $P(n)T\infty(n)$ and Par is cost-optimal if $P(n)T\infty(n)$ is in $O(T_{\mathsf{seq}}(n))$ for a best known sequential algorithm Seq for the given problem.

We often use the term *work* to quantify the real "effort" that an algorithm puts into solving one of our computational problems. The work of a sequential algorithm Seq on input of size $O(n)$ is the number of operations (of some kind) carried out by the algorithm. Sequentially speaking, "work is time". The work of a parallel algorithm Par on a system with p processor-cores is the total work carried out by all of the p cores, excluding time and operations spent idling by some processors or by processors that are not assigned to do anything (useful). That is, anything that the cores might be doing that is not strictly related to the algorithm does not count as work. With a formal model like the PRAM, this can be given a precise definition ("work is the operations carried out by assigned processors"). In more realistic settings, we have to be careful which idle times should count and which not. The work of parallel algorithm Par on input n is denoted $W_{\mathsf{par}}^p(n)$. Ideally, work is independent of the number of processors p and we might write just $W_{\mathsf{par}}(n)$. This means that the work to be done by the algorithm Par has been separated from how the p processors that will eventually perform this work share the work. This is a very useful point of view which leads to a productive separation of concerns between what has to be done ("the work") and who does it ("which processors"). This point of view motivates the next definition.

Definition 7 (Work-Optimal Parallel Algorithm). A parallel algorithm Par with work $W_{\mathsf{par}}(n)$ is *work-optimal* if $W_{\mathsf{par}}(n)$ is $O(T_{\mathsf{seq}}(n))$ for a best known sequential algorithm Seq.

If an algorithm is work-optimal algorithms but not cost-optimal this indicates either that the way the processors are used in the parallel algorithms is not efficient (some processors sit idle for too long) or that most of the work must necessarily be done sequentially, one piece after the other (because of sequential dependencies). From a work-optimal algorithm that is not cost-optimal for the first reason, a better, cost-optimal algorithm with the same amount of work that runs on fewer processor-cores can sometimes be constructed, but this may not be easy.

A cost-optimal parallel algorithm is per definition work-optimal but not the other way around: A parallel algorithm that is not work-optimal cannot be cost-optimal. Thus, a first step towards designing a good parallel algorithm is to look for a solution that is (at least) work-optimal.

Another useful observation following from the notion of parallel work is that the best possible parallel running time of an algorithm with work $W_{\mathsf{par}}(n)$ is at least

$$T_{\mathsf{par}}^p(n) \geq \frac{W_{\mathsf{par}}(n)}{p} \quad .$$

This is another useful lower bound which is sometimes called the *Work Law* (See Sect. 1.3.1). The lower bound is met if the work $W_{\mathsf{par}}(n)$ that has to be done has been perfectly distributed over the p processors and no extra costs have been incurred.

As an extreme example, consider a "parallel" algorithm that is just a (best) sequential algorithm executed on one out of the p processors. This is a work-optimal parallel algorithm, but it is clearly not cost-optimal since all but one processor are idle. Its cost $O(pT_{\mathsf{seq}}(n))$ is optimal when running it on one or a small, constant number of processors p; but as long as the number of processors that can be efficiently exploited cannot be increased with increasing problem size, such an algorithm is not cost-optimal according to our definition, and speed-up beyond a limited, constant number of processors cannot be achieved. This is not what is desired of a good parallel algorithm. Cost- and work-optimality are asymptotic notions of properties that hold for large problems and large numbers of processors.

Algorithms that are not cost-optimal do not have linear speed-up. The PRAM maximum finding algorithm of Theorem 1 takes $O(1)$ time with $O(n^2)$ processors and therefore has cost $O(n^2)$, which is far from $T_{\mathsf{seq}}(n) = O(n)$. To determine the speed-up of this algorithm, we first have to observe that the algorithm can be simulated with $p \leq n^2$ processors in $O(n^2/p)$ parallel time steps. The speed-up is $\mathrm{SU}_p(n) = O(n/(n^2/p)) = p/n$. The speed-up is *not* independent of n, and actually decreases with n: The larger the input, the lower the speed-up.

The point of distinguishing work and cost is to separate the discovery of parallelism from an all too specific assignment of the work to the actually available processors. A good, parallel algorithm is work-optimal and can become fast when enough processors are given. A next design step is then to carefully assign the work to only as many processors as allowed to keep the algorithm cost-optimal. The PRAM abstraction supports this strategy well: Processors can be assigned freely (with the **par**-construct), and the analysis can focus on the number of operations actually done by the assigned processors (the work).

More precisely, let us assume that a work-optimal PRAM algorithm with work $W_{\mathsf{par}}(n)$ and parallel time complexity of $T\infty(n)$ has been found. Such an algorithm can (in principle) be implemented to run on a p-processor PRAM (same variant) in at most $\lfloor \frac{W_{\mathsf{par}}(n)}{p} \rfloor + T\infty(n)$ parallel time steps. This follows easily. In each of the $T\infty(n)$ parallel steps some amount of work $W_{\mathsf{par}}^i(n)$ has to be done. This work can be done in parallel on the p processors in $\lceil \frac{W_{\mathsf{par}}^i(n)}{p} \rceil$ time steps by a straightforward round-robin execution of the work units over the p processors. Summing over the steps gives

1.2 Second Block (1–2 Lectures)

$$\sum_{i=0}^{T\infty(n)-1} \lceil \frac{W_{\mathsf{par}}^i(n)}{p} \rceil \leq \sum_{i=0}^{T\infty(n)-1} (\lfloor \frac{W_{\mathsf{par}}^i(n)}{p} \rfloor + 1)$$

$$\leq \lfloor \frac{W_{\mathsf{par}}(n)}{p} \rfloor + T\infty(n)$$

This observation is also known as *Brent's Theorem* [23]. The observation only tells us that an efficient execution of the algorithm is possible on a p-processor PRAM, but not how the work units for each step can be identified. Sometimes this is obvious and sometimes not.

1.2.5 Relative Speed-up and Scalability

While the absolute speed-up measures how well a parallel algorithm can improve over its best known sequential counterpart, it does not measure whether the parallel algorithm by itself is able to exploit the p processors well. This notion of *scalability* is the *relative speed-up*.

Definition 8 (Relative Speed-up). The *relative speed-up* of a parallel algorithm Par is the ratio of the parallel running time with one processor-core to the parallel running time with p processor-cores, i.e.,

$$\mathrm{SUR}_p(n) = \frac{T_{\mathsf{par}}^1(n)}{T_{\mathsf{par}}^p(n)} \quad .$$

Assume that an arbitrary number of processors is available. Any parallel algorithm has, for any (fixed) input of size $O(n)$, a fastest running time that it can achieve, denoted by $T\infty(n)$ which is the time $T_{\mathsf{par}}^{p'}(n)$ for some number of processors p'; this was defined as the parallel time complexity (see Sect. 1.2.1). Per definition, $T_{\mathsf{par}}^p(n) \geq T\infty(n)$ for any number of processors p. It thus holds that $\mathrm{SUR}_p(n) = \frac{T_{\mathsf{par}}^1(n)}{T_{\mathsf{par}}^p(n)} \leq \frac{T_{\mathsf{par}}^1(n)}{T\infty(n)}$.

The ratio $\frac{T_{\mathsf{par}}^1(n)}{T\infty(n)}$ which is a function of the input size n only is called the *parallelism* of the parallel algorithm. It is clearly both the largest, relative speed-up that can be achieved, as well as an upper bound on the number of processors up to which linear, relative speed-up can be achieved. If some number of processors p' larger than the parallelism is chosen, the definition says that $\mathrm{SUR}_{p'}(n) < p'$, that is, less than linear speed-up. The parallelism is also the asymptotically smallest number of processor needed to achieve the best possible running time $T\infty(n)$.

It is important to clearly distinguish between absolute and relative speed-up. Relative speed-up compares a parallel algorithm or implementation against itself, and expresses to what extent the processors are exploited well (linear, relative speed-up). Absolute speed-up compares the parallel algorithm against a (best known or possible) baseline, and expresses how well it improves over the baseline. A parallel algorithm may have excellent relative speed-up, but poor absolute speed-up. Is such a good algorithm? In any case, reporting only the relative speed-up for a parallel algorithm or implementation can be grossly misleading and should never be done in serious Parallel Computing. An absolute baseline always must be defined (that which we want to improve over) and absolute running times also stated. There are plenty of examples of basing claims on relative speed-ups only also in the scientific literature. For more on such pitfalls and misrepresentations, see the now well-known and often paraphrased "...Ways to fool the masses..." [12], see also https://blogs.fau.de/hager/archives/5299.

The absolute speed-up compares the running time of the parallel algorithm against the running time of a best known or possible sequential algorithm. For such an algorithm it holds that $T_{\mathsf{seq}}(n) \leq T_{\mathsf{par}}^1(n)$ and therefore

$$\mathrm{SU}_p(n) \leq \mathrm{SUR}_p(n) \quad .$$

The absolute speed-up is at most as large as the relative speed-up and also in that sense a tougher measure.

1.2.6 Overhead and Load Balance

A parallel algorithm for a computational problem usually performs more work than a corresponding best known sequential algorithm. In summary, such work is termed *overhead*; thus, overhead is work incurred by the parallel algorithm that does not have to be done by the sequential algorithm. Beware that this definition tacitly assumes that sequential and parallel algorithms are somehow similar and can be compared ("extra work"). This is not always the case. Sometimes, a parallel algorithm is totally different from the best known sequential algorithm. Overheads can be caused by several factors, e.g.,

- preparation of data for other processor-cores,
- communication between and coordination of processor-cores,
- synchronization, and
- algorithmic overheads: extra or redundant work

when compared to a corresponding, somehow similar sequential algorithm. When a parallel algorithm Par is derived from a sequential algorithm Seq, we can loosely speak of *parallelization* and say that Seq has been *parallelized* into Par. Parallel algorithms implemented with OpenMP (see Sect. 2.3) are, for instance, often very concrete parallelizations of corresponding sequential algorithms. Again, it is important to stress that many parallel algorithms are specifically not parallelizations of some sequential algorithm.

Overheads are more or less inevitable, but if they are on the order of (within the bounds of) the sequential work, $O(T_{\mathsf{seq}}(n))$ the parallel algorithm can still be work- and cost-optimal, and thus have linear, although not perhaps perfect speed-up. Often, overheads increase with the number of processors p, giving, for fixed problem size n, a limit on the number of processors that can be used while

1.2 Second Block (1–2 Lectures)

still giving linear speed-up. If the overheads are asymptotically larger than the sequential work, the parallel algorithm will never have linear speed-up.

The overheads caused by communication and synchronization between processor-cores are often significant. Later in these lecture notes, we will introduce a simple model for accounting for communication operations. Suffice it here to say that a simple synchronization between p processors, which means ascertaining that a processor cannot continue beyond a certain point in its computation before all other processors have reached a certain point in their computations (see Sect. 1.3.14), may (and must) take $\Omega(\log p)$ operations. An exchange of data will typically take time proportional to the amount of the data (per processor) and an additive term dependent on the number of processors p.

Between communication operations, the processor-cores operate independently on parts of the problem although they could interfere indirectly through the memory and cache system (this will be discussed in later parts of these lecture notes, see Sect. 2.1.1). The length of the intervals between communication and synchronization operations is sometimes referred to as the *granularity* of the parallel algorithm. A parallel computation in which communication and synchronization occur rarely is called *coarse grained*. If communication and synchronization occur frequently, the computation is called *fine grained*. These are relative (and vague) terms. Machine models that can support fine grained algorithms, are also called fine grained. The PRAM is an extreme example: The processors can (and often do) communicate via the shared memory in every step, and they are lock-step synchronized with no overhead for synchronization.

In some parallel algorithms, the processors may not perform the same amount of work, and/or have different amounts of overhead. If we, for the moment, let $T_{\text{par}}^i(n)$ denote the time taken by some processor-core $i, 0 \leq i < p$ from the time this processor-core starts until it terminates, the (absolute) *load imbalance* is defined as

$$\max_{0 \leq i,j < p} |T_{\text{par}}^i(n) - T_{\text{par}}^j(n)| = \max_{0 \leq i < p} T_{\text{par}}^i(n) - \min_{0 \leq i < p} T_{\text{par}}^i(n) \quad .$$

The relative load imbalance is the ratio of absolute load balance to parallel time (completion time of slowest processor). Too large load imbalance is another reason that a parallel algorithm may have a too small (or non-linear) speed-up. Too large load imbalance may likewise be a reason why an otherwise work-optimal parallel algorithm is not cost-optimal: Too many processors take too small a share of the total work.

Good load balance means that $T_{\text{par}}^i(n) \approx T_{\text{par}}^j(n)$ for all pairs of processors (i, j). Achieving good, even load balance over the processors is called *load balancing* and is always an issue in designing a parallel algorithm, explicitly by the construction of the algorithm or implicitly by taking steps later to ensure a good load balance. We distinguish between *static load-balancing*, where the amount of work to be done can be divided upfront among the processors, and *dynamic load balancing*, where the processors have to communicate and exchange work during the execution of the parallel algorithm. Static load balancing can be further subdivided into *oblivious, static load-balancing*, where the problem can be divided over the processors based on the input size and structure alone but regardless of the actual input, and *adaptive, problem-dependent, static load-balancing*, where the input itself is needed in order to divide the work and preprocessing may be required. Some aspects of the load balancing problem (work-stealing, loop scheduling) will be discussed later in this part of the lecture notes. However, load balancing *per se* is too large a subfield of Parallel Computing to be treated in much detail here.

Problems and algorithms where the input and work can be statically distributed to the processors and where no further explicit interaction is required are called either *embarrassingly parallel*, *trivially parallel*, or *pleasantly parallel*. These are the best (but uninteresting, in the sense of being unchallenging) cases of easily parallelizable problems with linear or even perfect speed-up. The realization that the problem is trivially or embarrassingly parallel can, of course, be highly non-trivial and the way to see this unpleasant.

1.2.7 Amdahl's Law

Gene Amdahl made a simple observation on how to speed up programs [6], which when applied to Parallel Computing yields severe bounds on the speed-up that certain parallel algorithms can achieve. The observation assumes that the parallel algorithm is somehow derived by parallelization of the sequential algorithms.

Theorem 4 (Amdahl's Law). *Assume that the work performed by sequential algorithm* Seq *can be divided into a strictly sequential fraction* $s, 0 < s \leq 1$, *independent of* n, *that cannot be parallelized at all, and a fraction* $r = (1-s)$ *that can be perfectly parallelized. The parallelized algorithm is* Par. *Then, the maximum speed-up that can be achieved by* Par *over* Seq *is bounded by* $1/s$.

The proof is straightforward. With the assumption that

$$T^p_{\text{par}}(n) = sT_{\text{seq}}(n) + \frac{(1-s)T_{\text{seq}}(n)}{p}$$

we get

$$\begin{aligned}
\text{SU}_p(n) &= \frac{T_{\text{seq}}(n)}{sT_{\text{seq}}(n) + \frac{(1-s)T_{\text{seq}}(n)}{p}} \\
&= \frac{1}{s + \frac{1-s}{p}} \\
&\rightarrow \frac{1}{s} \text{ for } p \rightarrow \infty \ .
\end{aligned}$$

Amdahl's Law is devastating. Even the smallest, constant sequential fraction of the algorithm to be parallelized will limit and eventually kill speed-up. A sequential fraction of 10%, or 1%, sounds reasonable and harmless but limits the speed-up to 10, or 100, no matter what else is done, no matter how large the problem, and no matter how many processors are invested. Note that the parallelization considered is work-optimal; but it is surely not cost-optimal. The running time of the parallel algorithm is at least $sT_{\text{seq}}(n)$ and since $s, s < 1$ is constant, the cost is therefore $O(pT_{\text{seq}}(n))$ which is not in $O(T_{\text{seq}}(n))$.

A sequential algorithm which falls under Amdahl's Law cannot be used as the basis of a good, parallel algorithm: Its speed-up will be severely limited and bounded by a constant. Amdahl's Law is therefore rather an analysis tool: If it turns out that a (large) fraction of the algorithm at hand cannot be parallelized, we have to look for a different, better algorithm. This is what makes Parallel Computing a creative activity: Simple parallelization of a sequential algorithm will often not lead to a good, parallel counterpart. New ideas for old problems are sometimes needed.

Typical victims of Amdahl's Law are:

- Input/output: For linear work algorithms, reading the input and possibly also writing the output will take $\Omega(n)$ time steps, and thus be a constant fraction of $O(n)$.
- Sequential preprocessing: As above.
- Maintaining sequential data structures, in particular sequential initialization, can easily turn out to be a constant fraction of the total work.
- Hard-to-parallelize parts that are done sequentially (which might look innocent enough for just small parts): If such parts take a constant fraction of the total work, Amdahl's Law applies.
- Long chains of dependent operations (operations that have to be performed one after the other and cannot be done in parallel), not necessarily on the same processor-core.

When analyzing and benchmarking parallel algorithms, input/output is often disregarded when accounting for sequential and parallel time. The defensible reason for this is that we are interested in how the core parallel algorithm performs (speeds up), under the assumption that the input has already been read and properly distributed to the processor-cores according to the specification. In these lecture notes, our algorithms are small parts (building blocks) of larger applications and in this larger context would not need input/output: The data are already where they should be. Also results do not have to be output but should just stay and be available for the next building block to use. We, therefore, analyze the building blocks in isolation without the input/output parts that might fall victim to Amdahl's Law.

In a good parallel algorithm, not falling victim to Amdahl's Law, the sequential part $s(n)$ will not be a constant fraction of the total work but depend on and decrease with n. If such is the case, Amdahl's Law does not apply. Instead, a good speed-up can be achieved with large enough inputs. Parallel Computing is about solving large, work-intensive problems, and in good parallel algorithms the parts doing the parallel work dominate the total work as the input gets large enough.

1.2.8 Efficiency and Weak Scaling

As observed, there is, for any parallel algorithm on input of size $O(n)$, always a fastest possible time, $T\infty(n)$, that the algorithm can achieve (the parallel time complexity). Thus, the parallel running time of an algorithm with good, linear speed-up (up to the number of processor-cores determined by the parallelism), can be written as $T^p_{\text{par}}(n) = O(T(n)/p + t(n))$, that is, as a parallelizable term $T(n)$ and a non-parallelizable term $t(n) = T\infty(n)$. If speed-up is not linear, the parallel running time is instead something like $T^p_{\text{par}}(n) = O(T(n)/f(p) + t(n))$ strictly with $f(p) < p$ and $f(p)$ in $o(p)$, or $T(n)$ is not in $O(T_{\text{seq}}(n))$.

If we compare against a sequential algorithm with $T_{\text{seq}}(n) = O(T(n)) = O(T(n) + t(n))$, a parallel algorithm where $t(n)/T(n) \to 0$ as $n \to \infty$ is also good and can have linear speed-up for large enough n. The speed-up is namely

$$\text{SU}_p(n) = \frac{T_{\text{seq}}(n)}{T^p_{\text{par}}(n)} = O(\frac{T(n)}{T(n)/p + t(n)}) = O(\frac{1}{1/p + t(n)/T(n)}) \to O(p)$$

as n increases. This is called *scaled speed-up*, and the faster $t(n)/T(n)$ converges, the faster the speed-up becomes linear. Against Amdahl's Law, the sequential part $t(n)$ should be as small as possible and increase more slowly with n than the parallelizable part $T(n)$. Algorithms with this property are cost-optimal according to Definition 5.

It is a good way which we use throughout these lecture notes to state the performance of a (work-optimal) parallel algorithm as $T^p_{\text{par}}(n) = O(T(n)/p + t(n,p))$ with the assumption that $t(n,p)$ is in $O(T(n))$ for fixed p, and $T_{\text{seq}}(n) = O(T(n))$. That is, we allow the non-parallelizable part to depend on both n and p. Often, however, $t(n,p)$ is just $t(n)$ independent of p or $t(p)$ depending on p only (synchronization costs). An iterative parallel algorithm with a convergence check involving synchronization could, for instance, run in $O(n/p + \log n \log p)$ parallel time with $t(n,p) = O(\log n \log p)$. Such an algorithm would perform total linear $O(n)$ work which has been well distributed over the p processors; the algorithm performs $O(\log n)$ iterations each of which incurs a synchronization overhead of $O(\log p)$ operations.

The *parallel efficiency* of a parallel algorithm Par is measured by comparing Par against a best possible parallelization of Seq as given by the Work Law (see Sect. 1.3.1).

Definition 9 (Parallel Efficiency). The efficiency $E_p(n)$ for input of size $O(n)$ and p processors of parallel algorithm Par compared to sequential algorithm Seq is defined as

$$E_p(n) = \frac{T_{\text{seq}}(n)}{p}/T^p_{\text{par}}(n) = \frac{T_{\text{seq}}(n)}{pT^p_{\text{par}}(n)} = \frac{\text{SU}_p(n)}{p} .$$

As worked out in the definition, the efficiency is also the achieved speed-up divided by p as well as the sequential time divided by the cost of the parallel algorithm. It therefore holds that

- $E_p(n) \leq 1$.
- If $E_p(n) = e$ for some constant $e, 0 < e \leq 1$, the speed-up is linear.
- Cost-optimal algorithms have constant efficiency.

Should it happen that the efficiency $E_p(n)$, contrary to the statement above, for some n and number of processors p is larger than 1, equivalently that the absolute speed-up is larger than p, this tells us that the sequential baseline is not the best (known) possible. It can be replaced by some variation of the parallel algorithm. In such a case, Parallel Computing has helped to discover a better sequential algorithm for the given problem.

We note that this is a definition of *algorithmic efficiency*: How close is the time of the parallel algorithm with p processors to that of a best possible parallelization of a best (known) sequential algorithm? This definition does not say anything about how well the parallel or sequential algorithm exploits the hardware capabilities and how close the performance can come to the nominal performance of the parallel processor system at hand. This notion of *hardware efficiency* plays a role in High-Performance Computing (HPC), understood here as the discipline of getting the best out of the given system.

If an algorithm does not have constant efficiency and linear speed-up for fixed, constant input sizes n, we can try to maintain a desired, constant e efficiency by instead increasing the problem size n with the number of processors p. This is the notion of *iso-efficiency* [48, 49] and can be achieved for cost-optimal algorithms.

Definition 10 (Weak Scalability (Constant Efficiency)). A parallel algorithm Par is said to be *weakly scaling* relative to sequential algorithm Seq if, for a desired, constant efficiency e, there is a slowly growing function $f(p)$ such that the efficiency is $E_p(n) = e$ for n in $\Omega(f(p))$. The function $f(p)$ is called the *iso-efficiency function*.

How slowly should $f(p)$ grow? A possible answer is found in another definition of weak scaling.

Definition 11 (Weak Scalability (Constant Work)). A parallel algorithm Par with work $W_{\mathsf{par}}(n)$ is said to be *weakly scaling* relative to sequential algorithm Seq if, by keeping the average work per processor $W_{\mathsf{par}}(n)/p$ constant at w, the running time of the parallel algorithm $T_{\mathsf{par}}^p(n)$ remains constant. The input size scaling function is $g(p) = T_{\mathsf{seq}}^{-1}(pw)$.

Ideally, the iso-efficiency function $f(p)$, which tells how n should grow as a function of p to maintain constant efficiency, should not grow faster than the input size scaling function $g(p)$, which tells how much n can at most grow if the average work is to be kept constant: $f(p)$ should be $O(g(p))$. The two notions may contradict. Constant efficiency could require larger n than permitted for maintaining constant average work. This happens if the sequential running time is more than linear. Keeping constant efficiency requires n to increase faster than allowed by constant work weak scaling. For such algorithms, constant work is maintained with decreasing efficiency.

1.2.9 Scalability Analysis

How well does a parallel algorithm or implementation now perform against a sequential counterpart for the problem that we are interested in, in particular how well can it exploit the available processor resources? *Scalability analysis* examines this, theoretically and practically by analyzing (measuring) the parallel time that can be reached for different number of processors p and possibly different problem sizes n.

- Strong scaling analysis: Keep the input (size) n constant. The algorithm is *strongly scalable* up to some maximum number of processors, as expressed by the parallelism of the algorithm if the parallel time decreases proportionally to p (linear speed-up).
- Weak scaling analysis: Keep the average work per processor constant by increasing n with the number of processors p. The algorithm is *weakly scalable* if the parallel running time remains constant with increasing number of processors.

A strongly scaling algorithm, a strong property, is able to speed up the solution of the given problem for some fixed size n (large enough for parallel execution to make sense) proportionally to the number of employed processor-cores: our primary Parallel Computing goal. A weakly scaling algorithm in the sense of constant work per processor is able to solve larger and larger instances of the problem within an allotted time frame. Ideally, the time spent when the processor-cores are performing the same amount of work remains constant regardless of the number of processors employed. If this is not the case, and the parallel time is increasing with the number of processors, this indicates that the parallelization overhead (due to communication, synchronization, unfavorable load balancing, or redundant computation) is increasing with p.

1.2.10 Relativized Speed-up and Efficiency

For very large parallel systems with tens or hundred thousands of processor-cores, measuring speed-up relative to a sequential baseline running on one processor may not make sense or even be possible. The problem size needed to keep the extreme number of processors busy may simply be too large (and time consuming) to run on a single processor. Scalability analysis may in such cases use as baseline the parallel algorithm running on some number p' of processors (say, $p' = 2, p' = 100, p' = 1000$ processor-cores). What happens if the number of processors is doubled? What happens when going from p' to $2p'$ to $10p'$ or to some $p > p'$ processors? Does the problem size need to increase to maintain a certain efficiency? The definitions of relative speed-up and (relative) efficiency can easily be modified to use a different processor baseline p'.

1.2.11 Measuring Parallel Time and Speed-up Empirically

Running parallel programs on a parallel multi-core processor or a large parallel computing system requires quite considerable support from the system's runtime system: Processor-cores must be allocated to the program, the program's active entities (processes, threads, ...) must be started and so on, the execution monitored, the program execution terminated, and the resources be given free for the next program to use. The measured time for running a full, parallel application is taken as the *wall-clock time* from starting the application until the system is free again, in accordance with our definition of parallel time and assuming that accurate timers are available, and therefore includes all these surrounding "overheads". Benchmarking and assessing the performance ("is this good enough?") of an application in this context is done by varying the inputs, the number of processors used, the system, and other relevant factors in a systematic and well-documented way.

Parallel Computing is most often concerned with the algorithmic building blocks of such larger applications and these building blocks are the computational problems we are studying. Benchmarking and performance assessment is therefore rather done by conducting dedicated experiments, possibly using specific benchmarking tools, with our developed kernels and building blocks. A benchmarking program or tool will invoke the kernel to be benchmarked in a controlled manner. For Parallel Computing with our definition of parallel time, it is thus common to assume (and therefore ensure) that the available processor-cores to be used in the assessment will start at the same time (as far as this makes sense). This will entail some form of *temporal synchronization* between the processor-cores, which is in itself a non-trivial problem in Parallel Computing. Also some means of detecting which processor-core was the last to finish is needed, possibly by again synchronizing the processor-cores. As always in experimental science, measurement and synchronization should be non-intrusive and not affect or distort the experimental outcome, which is another highly non-trivial issue. Since computer systems are effectively not deterministic objects (with respect to timing) and measured run-times may fluctuate from run to run, kernel benchmarks are repeated a certain number of times, say 10, 30, 100 times, or until results are considered stable enough under some statistical measure, or until the experimenter runs out of time. The time reported by the experiment as the parallel running time of the algorithm in question may be based on a statistical measure like average time of the slowest processor-core over the repetitions or the median time of the measured times. Sometimes, the fastest time over the repetitions of the slowest processor-core in each repetition is taken as the parallel running time. The argument for this is that this best time that the system could produce can be reproducible and stable over repeated experiments. A good experiment will clearly describe the experimental setup and the statistics used in computing and reporting the run-times. For others to reproduce an experiment and verify claims on performance, a precise description of the parallel systems is likewise required: Processor architecture, instruction set, number of processor-cores, organization and grouping of the cores, clock frequency, memory, cache sizes, etc.

In these lecture notes, asymptotic worst-case analysis is used to judge and compare algorithms, but most often worst-case inputs are not known and may also not be interesting, common use-cases at all. Experimental analysis aims at showing performance under many different inputs, in particular those that are realistic and typical for the uses of the algorithmic kernel under examination. Experiment design deals with the construction of good experimental inputs. For non-oblivious algorithms that are sensitive to the actual input (and not only the size of the input) it is good practice to always consider extreme and otherwise special case inputs, such as are expected to lead to either extremely good or extremely bad performance. Average case and otherwise "typical" inputs are likewise probably of interest and should be considered.

In Parallel Computing we are most often interested in aspects of scalability in problem size and in particular in number of processor-cores. On both accounts, it can be considered bad practice to focus only on input sizes n and especially number of processors p that are powers of two. The reason for this is that in many algorithms, powers-of-two are special, and performance in these cases might be either extremely good or extremely bad. In particular, parallel algorithms are sometimes designed around communication structures or patterns where the number of processors is first considered to be some $p = 2^q$. Likewise, some algorithms, for instance, dealing with two-dimensional matrices, may be special for inputs and number of processor-cores that are square numbers. Benchmarking for only inputs n and p that are squares can likewise be highly misleading. Excluded from these considerations are of course algorithms and kernels that only work for such special numbers.

1.2.12 Examples

It is illustrative (!) to strengthen intuition to visualize parallel running time, (absolute) speed-up, efficiency, and iso-efficiency as functions of the number of processors put into solving a problem of size n (for different n). Let some such problems be given with best known sequential running times $O(n) \leq cn$, $O(n \log n) \leq c(n \log n)$, and $O(n^2) \leq cn^2$ as seen many times now in these lecture notes, for some bounding constant $c, c > 0$ (the notation is sloppy: We mean that the constant of the dominating term hidden within the O is c).

We first assume that the linear $O(n)$ algorithm has been parallelized by algorithms running work-optimally in $O(n/p + 1) \leq C(n/p + 1)$, $O(n/p + \log p) \leq C(n/p + \log p)$, $O(n/p + \log n) \leq C(n/p + \log n)$, and $O(n/p + p) \leq C(n/p + p)$, respectively, for some bounding constant $C, C > 0$: Also many examples of such algorithms have been (and will be) seen in the lecture notes.

We first assume that the bounding constants in our sequential and parallel algorithms are "in the same ballpark", and normalize both constants to $c = C = 1$. We plot the parallel running time as functions of the number of processors p for $1 \leq p \leq 128$, and take $n = 128, 128^2$, respectively; these are really small

1.2 Second Block (1–2 Lectures)

problems for a linear time algorithm, $128^2 = 16K$ (and even $128^3 = 2M$). The running times are shown in Figure 1.1 and Figure 1.2.

The running time (number of steps) plots do not very well differentiate the four different parallel algorithms. For the larger problem size, $n = 128^2$, there is virtually no difference to be seen. The shape of the curves for these linearly (perfect) scaling algorithms is hyperbolic (like $1/p$). The parallel algorithm with running time $O(n/p+p)$ is interesting: For the small input with $n = 128$, running time decreases until about $p = 10$ processors, and then increases. Indeed the best

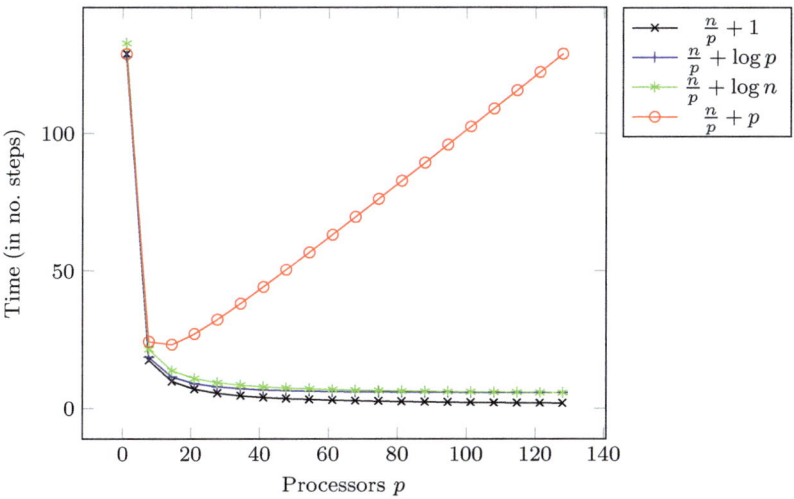

Fig. 1.1 Parallel time for $n = 128$ and $C = 1$

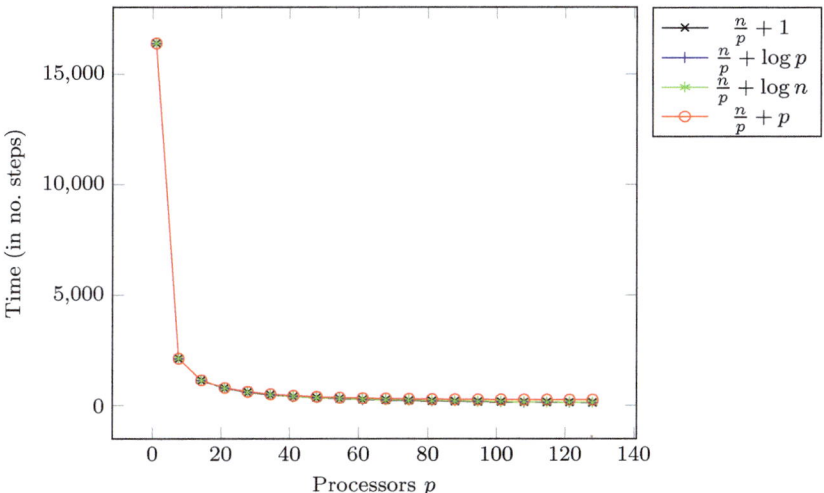

Fig. 1.2 Parallel time for $n = 128^2$ and $C = 1$

possible running time of this algorithm is $T\infty(n) = \sqrt{n}$, and the parallelism is also $n/\sqrt{n} = \sqrt{n}$. This can be seen by minimizing $C(n/p + p)$ for p, which can be done by solving $Cn/p = Cp$ for p, giving $p = \sqrt{n}$ (or more tediously, by calculus).

Plotting instead the absolute (unit-less) speed-up against the linear (best known) $O(n)$ algorithm (with $c = C = 1$) can highlight the actually different behavior of the four parallel algorithms. We plot for three problem sizes $n = 128, 128^2, 128^3$. The speed-ups are shown in Figures 1.3, 1.4, and 1.5.

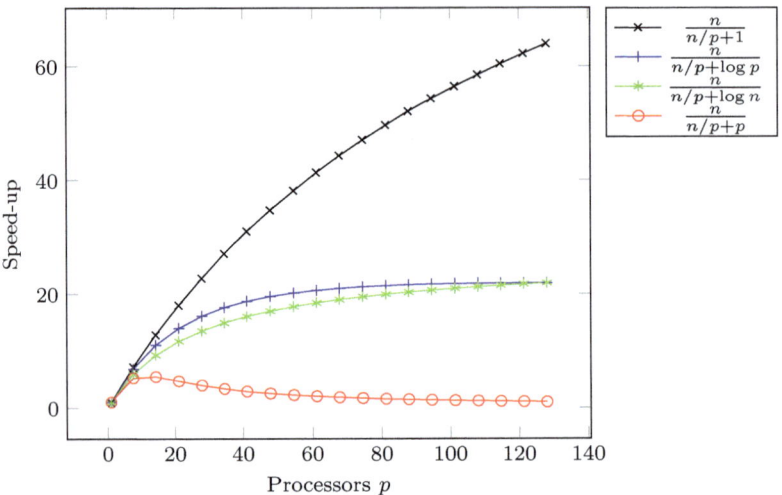

Fig. 1.3 Speed-up for $n = 128$ and $c = C = 1$

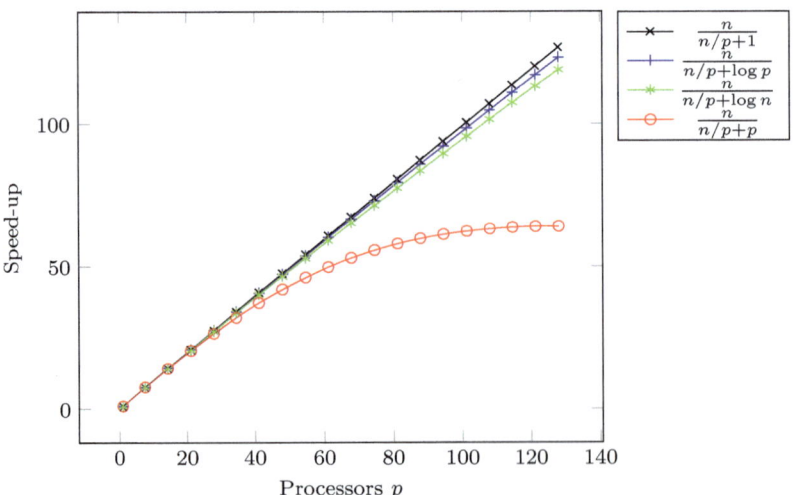

Fig. 1.4 Speed-up for $n = 128^2$ and $c = C = 1$

1.2 Second Block (1–2 Lectures)

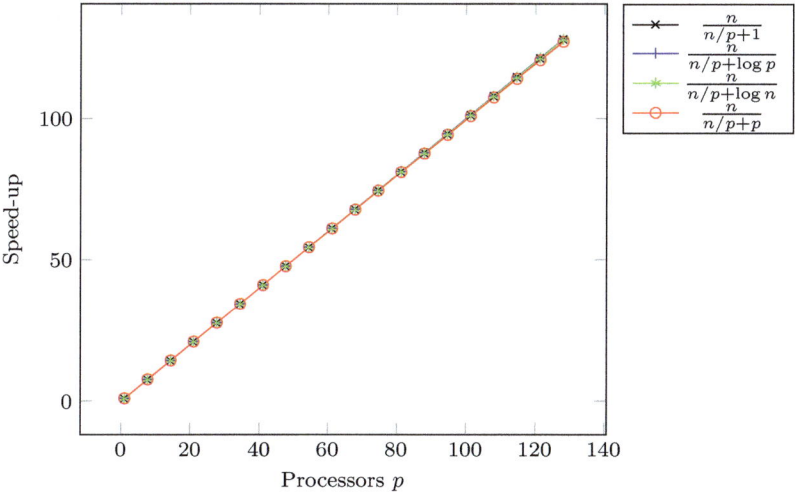

Fig. 1.5 Speed-up for $n = 128^3$ and $c = C = 1$

Speed-up for the small problem size $n = 128$ is not impressive and as we would like, except for the first parallel algorithm. This changes drastically and impressively as n grows. Indeed, for the "large" $n = 128^3$ problem, all four parallel algorithms show perfect speed-up of almost 128 for $p = 128$.

If there is a difference in the bounding constants between sequential and parallel algorithms, say $c = 1$ and $C = 10$, which means that the parallel algorithm is a constant factor of 10 slower than the sequential one when executed with only one processor, speed-ups change proportionally. This is shown in Figure 1.6.

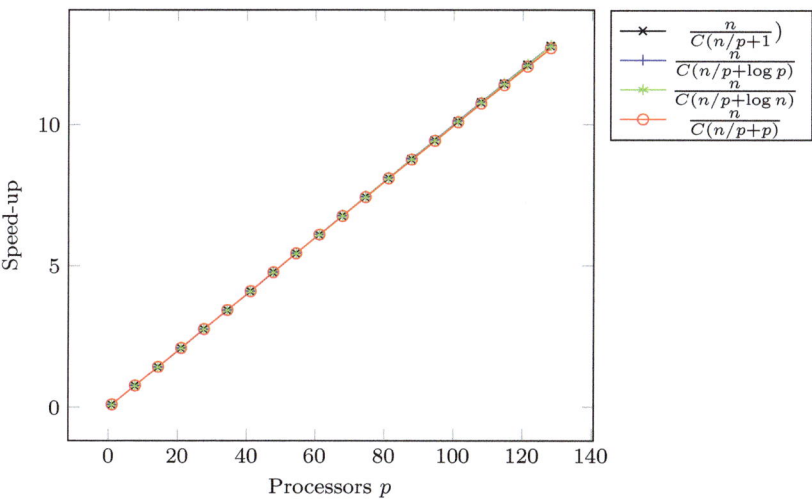

Fig. 1.6 Speed-up for $n = 128^3$ and $c = 1, C = 10$

Here, only $1/C$th of the processors are doing productive work in comparison to the sequential algorithm. Constants *do* matter, and it is obviously important that sequential and parallel algorithms have leading constants in the same ballpark. Otherwise, a proportional part of the processors is somehow wasted.

The parallel efficiency indicates how well the parallel algorithms behave in comparison to a best possible parallelization with running time cn/p. The (unitless) parallel efficiencies for the four parallel algorithms are plotted for $n = 128, 128^2, 128^3$ and shown in Figures 1.7, 1.8, and 1.9.

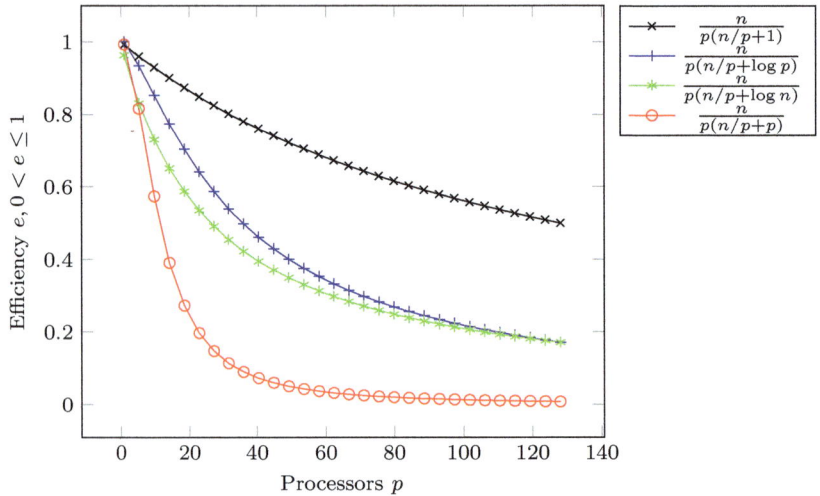

Fig. 1.7 Parallel efficiency for $n = 128$ and $c = C = 1$

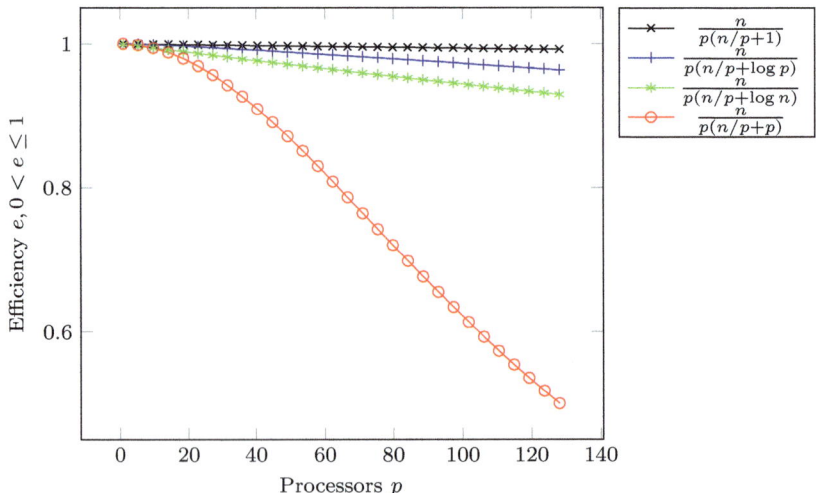

Fig. 1.8 Parallel efficiency for $n = 128^2$ and $c = C = 1$

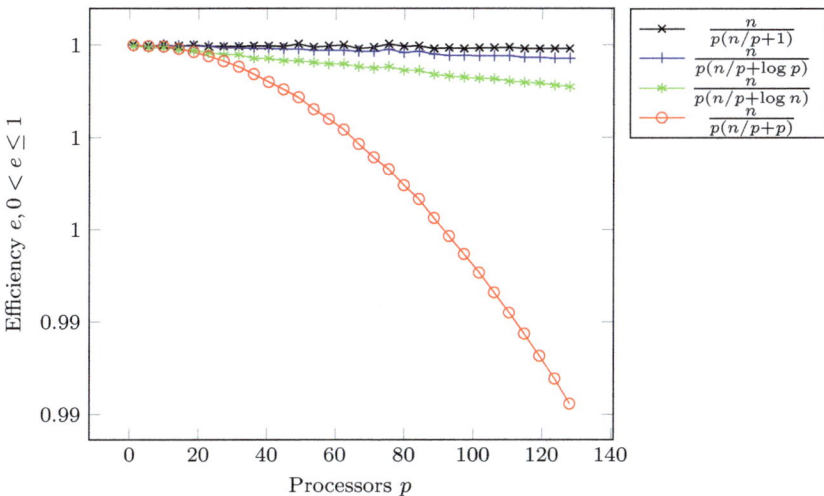

Fig. 1.9 Parallel efficiency for $n = 128^3$ and $c = C = 1$

Indeed, for work-optimal parallelizations, the efficiency improves greatly with growing problem size n and is already for $n = 128^3$ very close to 1 for all of the four parallelizations. The iso-efficiency functions more precisely tell how problem size must increase with p in order to maintain a given constant efficiency e. We calculate the iso-efficiency functions for the parallel algorithms as follows.

- For parallel running time $n/p + 1$ and desired efficiency e, we have $e = \frac{n}{p(n/p+1)} = n/(n+p) \Leftrightarrow e(n+p) = n \Leftrightarrow n = ep/(1-e)$.
- For parallel running time $n/p + \log p$ and desired efficiency e, we have $e = \frac{n}{p(n/p+\log p)} = n/(n+p\log p) \Leftrightarrow e(n+p\log p) = n \Leftrightarrow n = ep\log p/(1-e)$
- For parallel running time $n/p + p$ and desired efficiency e, we have $e = \frac{n}{p(n/p+p)} = n/(n+p^2) \Leftrightarrow e(n+p^2) = n \Leftrightarrow n = ep^2/(1-e)$

The case with parallel running time $n/p + \log n$ is more difficult. The efficiency calculation gives $e = \frac{n}{p(n/p+\log n)} = n/(n + p\log n)$ and therefore $n/\log n = ep/(1-e)$, for which we do not know an analytic, closed-form solution.

We plot the three analytic iso-efficiency functions for $p, 1 \leq p \leq 512$ and $e = 90\%$. The iso-efficiencies are shown in Figure 1.10.

For the first two parallel algorithms, the iso-efficiency function is indeed "slowly growing", and according to the first definition of weak scalability, these algorithms are both strongly and weakly scaling. With the last function, where the iso-efficiency function is in $O(p^2)$, it is a matter of taste whether to still consider it slowly growing. In the speed-up plots, we indeed let n grow exponentially $n = 128, 128^2, 128^3$, and the speed-up for the latter algorithms was excellent.

We now look at non-linear time sequential algorithms. The $O(n \log n)$ algorithm could be a sorting algorithm (mergesort, say) which could have been parallelized with running time $O(\frac{n \log n}{p} + \log^2 n)$. The second algorithm is perhaps matrix–vector multiplication, which can easily be done work-optimally in parallel time $O(\frac{n^2}{p}+n)$ (but also faster).

The corresponding speed-ups for $n = 100, 1\,000, 10\,000, 100\,000$ and $p, 1 \leq p \leq 1000$ are shown in Figures 1.11, 1.12, 1.13, and 1.14.

The parallelization of the low complexity algorithm with sequential running time $O(n \log n)$ does not scale as well as the other algorithm. For an $O(n^2)$ algorithm, an input of size $n = 100\,000$ is already large, and we did not plot for this large n here. However, both algorithms clearly approach a perfect speed-up with growing n.

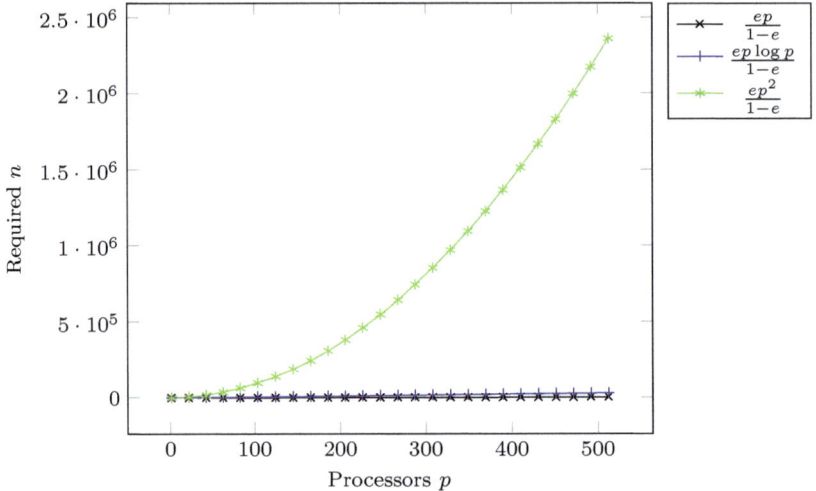

Fig. 1.10 Iso-efficiency functions for desired efficiency $e = 90\%$

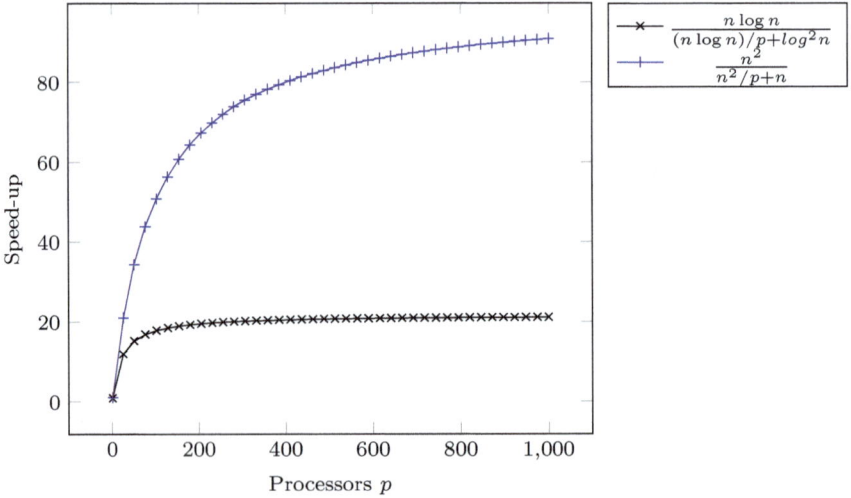

Fig. 1.11 Speed-up for $n = 100$ and $c = C = 1$

1.2 Second Block (1–2 Lectures)

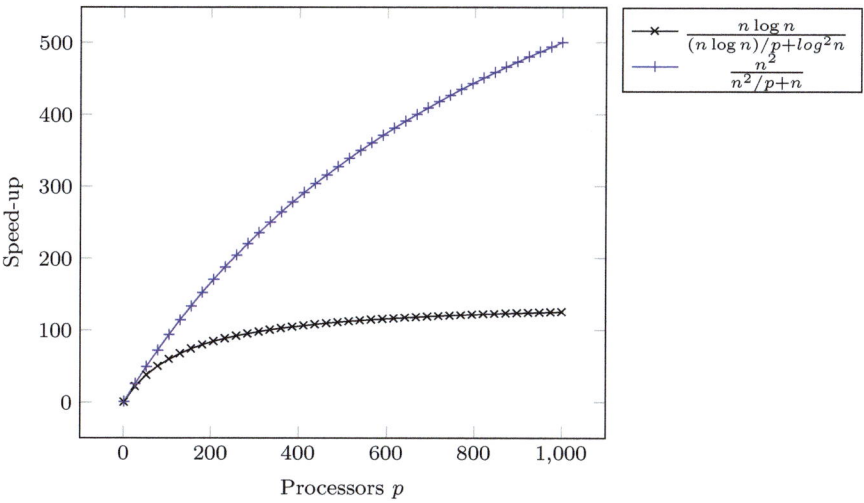

Fig. 1.12 Speed-up for $n = 1\,000$ and $c = C = 1$

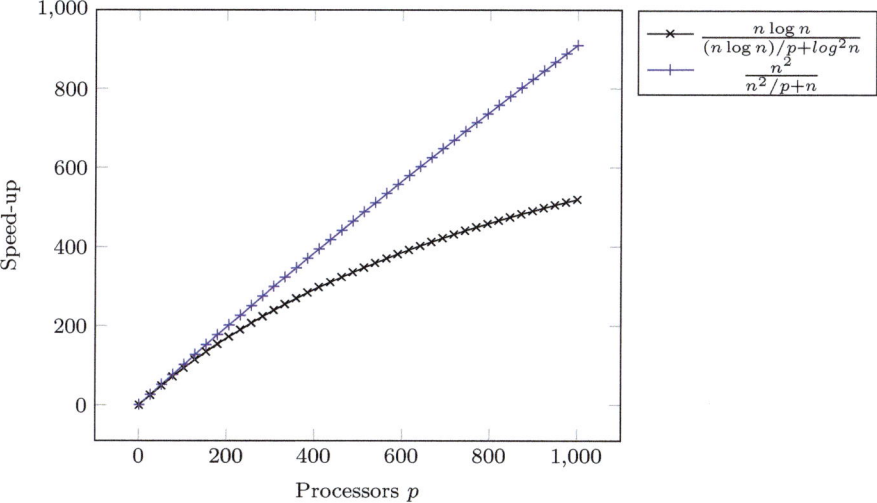

Fig. 1.13 Speed-up for $n = 10\,000$ and $c = C = 1$

Finally, we illustrate what happens with non work-optimal parallel algorithms. Assume we have a parallel algorithm with running times $O(\frac{n \log n}{p} + 1)$ relative to a linear time sequential algorithm, an $O(\frac{n^2}{p} + n)$ parallel algorithm relative to an $O(n \log n)$ best possible sequential algorithm, and an Amdahl case where the parallel algorithm has a sequential fraction $s, 0 < s < 1$ and parallel running time $O(sn + \frac{(1-s)n}{p})$. Lastly, a parallel algorithm with a running time of $O(\frac{n}{\sqrt{p}} + \sqrt{p}) = O(\frac{n\sqrt{p}}{p} + \sqrt{p})$ relative to an algorithm that solves an $O(n)$ problem. Speed-

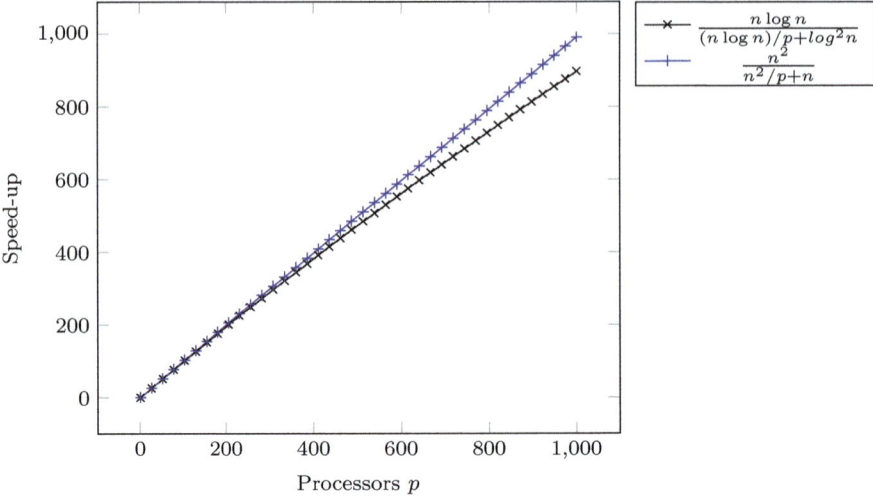

Fig. 1.14 Speed-up for $n = 100\,000$ and $c = C = 1$

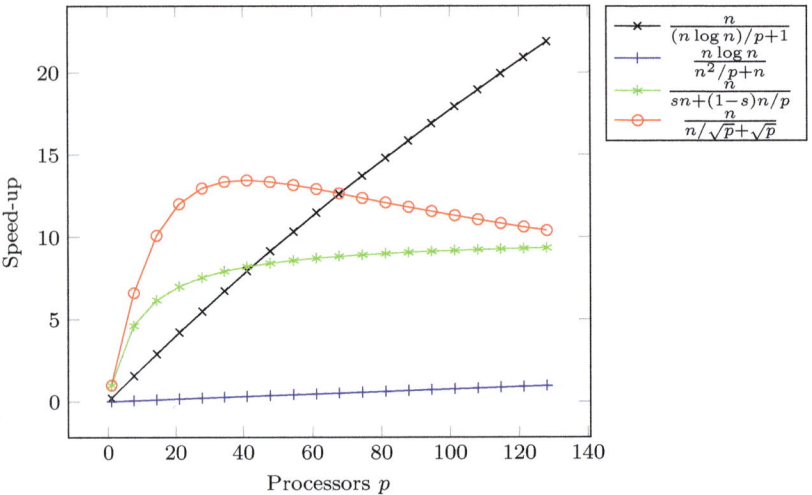

Fig. 1.15 Speed-up for $n = 128$ and $c = C = 1$ and sequential fraction $s = 0.1$

ups with $n = 128$ and $n = 128^2$ are shown in Figure 1.15 and Figure 1.16, respectively.

The two plots illustrate the Amdahl case well: Speed-up is bounded by $1/s$ (here 10 for $s = 10\%$) independently of n. The first two algorithms have a diminishing speed-up with increasing n. These two algorithms have parallel work determined by the problem size which is asymptotically larger than the sequential work. For the last algorithm, the parallel work increases "slowly" by a factor of $\sqrt{p}$ with p, and therefore the speed-up of this algorithm does indeed improve with increasing problem size n, but is $o(p)$ and not linear.

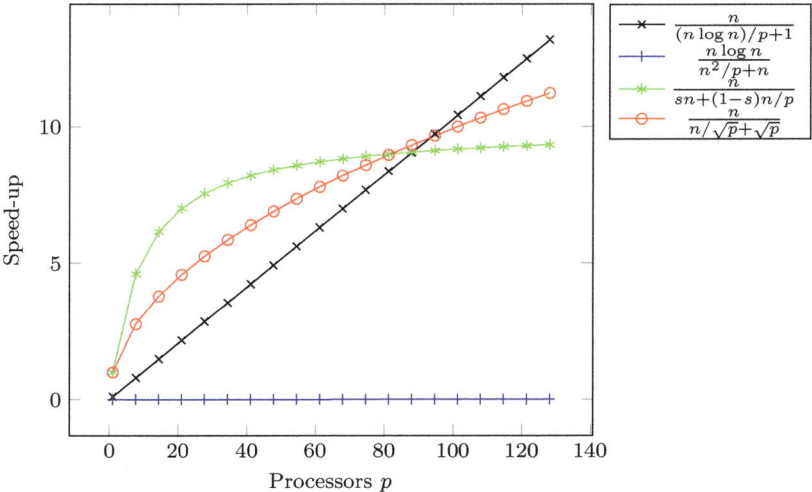

Fig. 1.16 Speed-up for $n = 128^2$ and $c = C = 1$ and sequential fraction $s = 0.1$

1.3 Third Block (1–2 Lectures)

In this part of the lecture notes, we take a closer look at the way (parallel) work may be structured. The most important structures discussed are work expressed as dependent tasks and work expressed as loops of independent iterations. The latter can be seen as an expression of recurring, similar computations in algorithms, pseudo-code and actual programs: A *parallel programming pattern* or *parallel design pattern*. The later part of this lecture block gives further examples of parallel algorithmic design patterns for which (good) parallelizations are known, including pipeline, stencil, master-slave/master-worker, reductions, data redistribution, and barrier synchronization. Parallel design patterns can, explicitly and implicitly, provide useful guidance for building parallel applications, sometimes even as concrete building blocks [76, 77]. We illustrate many of the patterns by sequential code snippets using C [68] to specify the intended outcome and semantics and use these descriptions to argue for lower bounds on the parallel performance with given numbers of processors p.

1.3.1 Directed Acyclic Task Graphs*

A *Directed Acyclic (task) Graph (DAG)*, $G = (V, E)$, consists of a set of *tasks*, $t_i \in V$, which are sequential computations that will not be analyzed further (sometimes also called *strands*). Tasks are connected by directed *dependency edges*, $(t_i, t_j) \in E$. An edge (t_i, t_j) means that task t_j is *directly dependent* on task t_i and cannot be executed before task t_i has completed, for instance, because the input data for task t_j are produced as output data by task t_i. In general, a task t_j is *dependent* on a task t_i if there is a directed path from t_i to t_j in G. If there is neither a directed path from t_i to t_j nor a directed path from t_j to t_i in G, the two tasks t_i and t_j are said to be *independent*. Independent tasks could possibly be executed in parallel, if enough processor-cores are available, since neither task needs input from nor produces output to the other. A task t_i may produce data for more than one other task, so there may be several outgoing edges from t_i. Likewise, a task t_j may need immediate input from more than one task, so there may be several incoming edges to t_j. Since G is acyclic, there is at least one task t_r in G with no incoming edges; such tasks are called *root* or *start* tasks. Likewise, there is at least one task t_f with no outgoing edges. Such tasks are called *final*.

Many computations can be pictured as task graphs. Consider as a first example the execution of the recursive Quicksort algorithm. The tasks may be the computations done in pivot selection and partitioning with a dependency from a pivoting task to the ensuing partitioning task. The root task will be the initial pivot selection in the input array, followed by the first partitioning task of the whole array. Dependent tasks will now be the pivot selection and partitioning of the two independent parts of the partitioned array and so on and so on. A final task will depend on all partitioning tasks to have completed and will indicate that the array has been Quicksorted. We will see later in these lecture notes how such task graphs suitable for parallel execution can be generated dynamically as OpenMP tasks or with Cilk. Another often encountered type of task DAG is the *fork-join* DAG: A dependent sequence of fork-join tasks, where each task has a number of dependent, forked tasks that are all connected to the next join task. A fork-join DAG is the standard structure of OpenMP programs corresponding to a sequence of loops of independent iterations, each of which can be executed in parallel as a set of forked, independent tasks.

For computations structured as task graphs, there is normally a single start task taking input of size $O(n)$ and a single, final task producing the results of the computation. In a dynamic setting, the task graph typically depends on the input, which can be emphasized by writing $G(n)$. This n is not to be confused with the number of tasks in G.

Each task t_i has an associated amount of work and takes sequential time $T(t_i)$, typically also depending on n. The total amount of work of a given task graph $G = (V, E)$ with k tasks $t_0, t_1, \ldots, t_{k-1}$ is given by the total time of all tasks and is denoted by $T_1(n) = \sum_{i=0}^{k-1} T(t_i)$. We will again compare against a best known sequential algorithm for the problem we are solving, so it holds that $T_1(n) \geq T_{\mathsf{seq}}(n)$.

Doing a computation as specified by a task graph G sequentially on a single processor-core amounts to the following: Pick a task t with no incoming edges and execute it. Remove all outgoing edges (t, t') from G. Continue this process until there are no more tasks in G. Since G is acyclic, there is at least one root task from which the execution can be started. After execution of this t, if t is not the last task, there will be at least one task with no incoming edges, etc. (if not, G would not be acyclic). Sequential execution of a task graph, therefore, amounts to executing the tasks (nodes) in some *topological order*. Any DAG has a topological order (as can be determined sequentially in $O(k)$ time steps [33]). A task that has become eligible for execution by having no incoming edges is said to be *ready*. Since all tasks of G are executed, each task exactly once, and since there is at least one ready task after completion of a task, the time taken for the sequential execution is $O(T_1(n))$.

Imagine that several processor-cores are available. A parallel execution of a computation specified by a task graph G could proceed as follows: Pick a ready task. If there is a processor-core that is not busy executing, assign the task to this core. When a task is completed, remove all outgoing edges, possibly giving rise to further, ready tasks (but also possibly not, tasks may have many incoming edges). Continue this process until there are no more ready tasks. The resulting order of tasks and assignment to processor-cores is called a *schedule*. The central property of a schedule is that both dependencies and processor availability are respected: A task is not executed before all incoming edges have been removed, which means that dependencies have been resolved and data been made available to the task; at no time, a processor-core is assigned more than one task; but at times, cores may be unassigned and idle.

We are interested in the time taken to execute the work $T_1(n)$ with some schedule with p processors. This is given by the time for the last task to finish. We denote the execution time of a (for now not further specified) p processor schedule by $T_p(n)$ and are, of course, interested in finding fast schedules.

No matter how scheduling is done, the total amount of work $T_1(n)$ can never be completed faster than $T_1(n)/p$, the best possible parallelization. Also, no matter how scheduling is done, tasks that are dependent on each other must be executed in order. Consider a heaviest path $(t_r, t_1, \ldots, t_f)$ from the start task t_r to a final task t_f with the largest amount of total work over the tasks t_i on the path and define $T\infty(n) = T(t_r) + T(t_1) + \ldots + T(T_f)$ as the work along such a heaviest path. With sufficiently many processor-cores available (this number is suggested by ∞), indeed a schedule exists that can achieve running time $T\infty(n)$ (think about this). Clearly, for any schedule, $T_p(n) \geq T\infty(n)$. These two observations are often summarized as follows:

- *Work Law*: $T_p(n) \geq T_1(n)/p \geq T_{\text{seq}}(n)/p$,
- *Depth Law*: $T_p(n) \geq T\infty(n)$.

The work on a heaviest path in a task graph G is often also called the *span* or the *depth* of the DAG. A heaviest path is commonly referred to as a *critical path* with *length* or *weight* $T\infty$. It is also the parallel time complexity of the DAG.

As an example, consider a fork-join DAG with start and final tasks t_r and t_f, with $T(t_r) = 1$ and $T(t_f) = 1$. The start task forks to a heavier task t_1 with $T(t_1) = 4$, and to, say, 27 light tasks with one unit of work. All forked tasks join at the final task. Thus, $T_1(n) = 1 + 4 + 27 + 1 = 33$ and $T\infty(n) = 1 + 4 + 1 = 6$. With $p = 3$, the Work Law says that $T_p(n) \geq 33/3 = 11$ with a (relative) speed-up of at most $T_p(n)/T_1(n) = 3$ and the Depth Law that $T_p(n) \geq 6$.

With more than, say, $p = 10$ processors, the Work law gives a running time of at least $T_1(n)/p \geq 33/11 = 3$ which is less than $T\infty(n) = 6$ and, therefore, not possible according to the Depth Law. The maximum speed-up achievable is obviously given by $T_1(n)/T\infty(n) = 33/6 = 5.5$.

For any schedule, the speed-up is bounded as follows:

$$\mathrm{SU}_p(n) = \frac{T_{\mathsf{seq}}(n)}{T^p_{\mathsf{par}}(n)} \leq \frac{T_1(n)}{T_p(n)} \leq \frac{T_1(n)}{T\infty(n)} \ .$$

The *parallelism* $\frac{T_1(n)}{T\infty(n)}$ is, therefore, an upper bound on the achievable relative speed-up and also gives the largest number of processor-cores for which linear speed-up could be possible.

Critical path analysis consisting in finding the longest chain of dependent, sequential work over all tasks, as used in the Depth Law, is an important tool to analyze the potential for parallelizing a computation when thinking of the computation as a task graph. If, for instance, the critical path $T\infty(n)$ is a constant fraction of $T_1(n)$, Amdahl's Law applies, which is a sign that a better algorithm and a better DAG must be found.

We now consider a specific scheduling strategy, so-called *greedy scheduling*. A greedy scheduler assigns a ready task to an available processor as soon as possible (task ready and processor available), meaning that a processor-core is idle only in the case when there is no ready task. Greedy schedules have a nice upper bound on the achieved running time, which is captured in the following theorem.

Theorem 5 (Two-Optimality of Greedy Scheduling). *Let $T_p(n)$ be the execution time of a DAG $G(n)$ with any greedy schedule on p processors, and let $T^*_p(n)$ be the execution time with a best possible p processor schedule. It holds that*

$$T_p(n) \leq \lfloor T_1(n)/p \rfloor + T\infty(n)$$
$$\leq 2T^*_p(n) \ .$$

The proof can be sketched as follows: Divide the work of the scheduler into discrete steps. A step is called *complete* if all processor-cores are busy on some tasks and *incomplete* if some cores are idle, which is the case when there are less ready tasks than processor-cores in that step. Then, the number of complete steps is bounded by $\lfloor T_1(n)/p \rfloor$; if there were more, more than the total work $T_1(n)$ would have been executed. The number of incomplete steps is bounded by $T\infty(n)$, since each incomplete step reduces the work on a critical path. The Work and the Depth Law hold for any p processor schedule, in particular for a best possible schedule, so $T_1(n)/p \leq T^*(n)$ and $T\infty(n) \leq T^*(n)$ and the last upper bound follows. The theorem, therefore, states that the execution time that can be achieved by a greedy schedule is bounded by two times what can be achieved by a best possible schedule, a guaranteed two-approximation!

Neither the definition of greedy schedules nor the theorem says how a greedy scheduler can or should be implemented. But if it can be shown by some means that a proposed scheduling algorithm is greedy, the greedy scheduling theorem says that the running time is within a factor two of best possible. Greedy scheduling is sometimes called *list scheduling* and the argument for Theorem 5 is also known as Brent's Theorem as discussed in Sect. 1.2.4. Later in these lecture notes, we will briefly touch on *work-stealing* which is a decentralized, randomized, greedy scheduling strategy for certain kinds of DAGs, like the one explained for Quicksort (called strict, spawn-join DAG's) [8].

Some parallel programming models and frameworks make it possible to dynamically construct what effectively amounts to directed acyclic task graphs, sometimes with additional structural properties, as the parallel execution progresses. The run-time system for such frameworks execute a (greedy) scheduling algorithm using the properties of the task graph. With the help of Theorem 5, it is sometimes possible to give provable time bounds for programs executed on such systems. Examples are OpenMP tasks, which will be covered in detail later (see Sect. 2.3.13), and Cilk, which we will briefly touch upon [20, 73, 95][2].

1.3.2 Loops of Independent Iterations

Computations are often expressed as loops, in algorithmic pseudo-code and in real programs. A computation is to be performed for the different values of the loop iteration variable in the range of this variable, typically in increasing order of the loop variable:

```
for (i=0; i<n; i++) {
  c[i] = F(a[i],b[i]);
}
```

In this loop, the iterations (different values of the iteration variable i) are *independent* of each other (provided the function F has no side effects): No computation for iteration i is affected by any computation for iteration i' before i, $i' < i$, and no computation for a later iteration i'', $i'' > i$, could possibly affect the computation for iteration i. In such a case, the loop could be trivially parallelized by dividing the iteration space into p roughly equal-sized blocks of about n/p iterations and letting each block be executed by a chosen processor-core.

The assignment of blocks, more generally individual iterations, to processor-cores is called *loop scheduling* and can be done either fully explicitly (as sometimes needed when parallelizing with pthreads, see Sect. 2.2.4, or with MPI, see lecture Block 3.2) or implicitly with the aid of a suitable compiler and runtime system by marking the loop (actually a bad name, since "loop" normally implies an order) as consisting of independent iterations (another misnomer in this context, "iteration" implies sequential dependency) and, therefore, parallelizable. An example, which we will see again in much detail later, is the following OpenMP style parallelization of a loop:

---
[2] See also www.opencilk.org

```
#pragma omp parallel for
for (i=0; i<n; i++) {
  c[i] = F(a[i],b[i]);
}
```

With the PRAM model, independent loop computations were handled by simply assigning a processor to each iteration with the **par**-construct:

```
par (0<=i<n) {
  c[i] = F(a[i],b[i]);
}
```

The parallel time of this "loop" on a PRAM would be $O(1)$ steps and the total number of operations $O(n)$ assuming that each evaluation of the function F also takes only a constant number of time steps. On a parallel computer with p processor-cores, optimistically, the parallel loop can be executed in $\Omega(n/p+1)$ time steps by splitting the n iterations roughly evenly between the p processors. The constant term is supposed to account for overheads in splitting and assigning the iterations to the processors. This assumes that also the number of iterations n is known in advance and that this n is not changed during the iterations. On parallel computers where the processors are not operating synchronously in lock-step like in the PRAM, a barrier synchronization (see Sect. 1.3.14) may be needed after the processor-cores have finished their iterations in order to ensure that the results in the c-array are all available to all processors. The parallel time of a "parallel loop" may, thus, have to include the time needed for the barrier synchronization and will be determined by the slowest processor-core to finish. Load imbalance could become an issue.

The loop of independent iterations pattern with the function **F** being a simple, arithmetic-logic expression with the same number of primitive instructions to be executed independently of the actual argument values is a standard way of expressing a SIMD parallel computation. One single stream of instructions controls the computations on multiple data, namely for all the n inputs of the iteration space. If the processor-architecture has actual SIMD instructions, the loop of independent instructions could be a way to instruct the compiler to use these instructions (see Sect. 2.3.16).

1.3.3 Independence of Program Fragments*

Independent loop iterations, in general, independent program fragments (which could be tasks as in Sect. 1.3.1) could possibly be executed concurrently, in parallel by different, available processor-cores. Independence of program fragments is a sufficient condition for allowing parallel execution.

Straightforward conditions for independence of program fragments are the three *Bernstein conditions* [15]. Let P_i and P_j be two program fragments, with P_j following after P_i in the program flow. Each of P_i and P_j has a set of (potential) input variables I_i and a set of (potential) output variables O_i. These sets can be determined statically by program analysis, but whether a potential output variable will actually be assigned is, in general, undecidable. The fragments P_i and P_j are *dependent* if either

1. $O_i \cap I_j \neq \emptyset$ (a *true dependency*, or *flow dependency*), or
2. $I_i \cap O_j \neq \emptyset$ (an *anti-dependency*), or
3. $O_i \cap O_j \neq \emptyset$ (an *output dependency*).

The conditions are obviously sufficient but not necessary: Either may hold, but input or output may not be read or written by the program fragment or read or written in some specific order such that the outcome of the parallel execution is still correct.

Dependencies between the iterations of a loop are called *loop carried dependencies*, and there are three types, corresponding to the three Bernstein conditions.

In a *loop carried flow dependency*, the outcome of an earlier iteration affects the computation of a later iteration:

```
for (i=k; i<n; i++) {
   a[i] = a[i]+a[i-k];
}
```

Here, the simple computation in iteration i is dependent on output in variable a[i-k] produced in iteration $i - k$ (assuming that $k > 0$; for $k = 0$ there would be no such dependency), an earlier iteration if the iterations were executed in increasing, sequential order. Such iterations can, therefore, not be done in parallel when expecting a correct outcome.

In a *loop carried anti-dependency*, the outcome of a later iteration affects an earlier iteration, if the two iterations were reversed or carried out simultaneously:

```
for (i=0; i<n-k; i++) {
   a[i] = a[i]+a[i+k];
}
```

The later iteration $i + k$ updates a variable that is used in iteration i, so if iteration $i + k$ would have been executed before or concurrently with iteration i, the output would be different than expected from a sequential execution in increasing iteration order and presumably not be correct.

Finally, in a *loop carried output dependency*, two or more iterations write to the same output variable(s). If executed simultaneously, the output would not be well-defined unless the same value is written for all iterations i (as allowed on the Common CRCW PRAM).

```
for (i=0; i<n-k; i++) {
   a[0] = a[i];
}
```

This is a first example of a *race condition*, about which we will learn more in later parts of the lecture notes.

Some loop carried dependencies can be removed by appropriate program transformations. For instance, the loop carried anti-dependency can be eliminated by introducing an auxiliary array b into which the results from the computations on array a are written:

```
for (i=0; i<n-k; i++) {
  a[i] = a[i]+a[i+k];
}
```

$\longrightarrow$

```
for (i=0; i<n-k; i++) {
  b[i] = a[i]+a[i+k];
}
```

The transformed (rewritten) loop now consists of independent iterations, and, therefore, the iterations can be executed in any order and concurrently, in parallel. Depending on the surrounding program logic (where is the result expected?), this may have to be followed by a loop (of independent iterations) to copy b back to a, taking $O(n)$ operations, or by a swapping of the two arrays, taking $O(1)$ operations. By similar tricks, sometimes other types of dependencies can be eliminated.

A *parallelizing compiler* would analyze loops and other constructs for dependencies and remove dependencies where possible by appropriate transformations in order to generate code that can exploit a large number of available processor-cores. Since the dependency problem is in general undecidable, there is a limit to what such compilers can do. Results may be modest [79].

1.3.4 Pipeline*

Consider the following nested loop computation:

```
for (i=0; i<m; i++) {
  for (j=1; j<n; j++) {
    a[i][j] = a[i][j-1]+a[i][j];
  }
}
```

The inner loop on j clearly contains a loop carried flow-dependency and, therefore, cannot be parallelized without sacrificing correctness as defined by the sequential loop order. The outer loop on i contains $O(n)$ work per iteration which could be performed in parallel with up to m processors. The parallel time would be $\Omega(\frac{m}{p}n)$ with up to at most m processors. We write this argument compactly as $\Omega(\frac{m}{p}n + n)$ parallel time.

A different way of assigning processors to the doubly nested loop work would be the following: Assume that up to n processor-cores are available. A processor is assigned for each index j in the inner loop. The jth such processor first sits idle for $j - 1$ rounds to wait for a[0][j-1] to have been computed by processors $0, 1, \ldots, j - 1$ before j. Now, processor j can compute the value a[0][j] followed by the values a[i][j] for $i = 1, \ldots, m - 1$. This latter viewpoint of the computation is a *linear pipeline*. The parallel running time for computing all the values in the two-dimensional array can be found by looking at the last processor $n-1$. This processor will have to wait for $n-1$ rounds before it can start computing values, after which it can compute a new value for the remaining $m-1$ elements. This gives a running time of $O(n+m-1) = O(n+m)$ with $p = n$ processor-cores. For $p \leq n$ processors the running time can be stated as $\Omega(\frac{n}{p}m + m + n)$.

The general, linear pipelining pattern assumes that a number of m work items are to be processed, each requiring work that can be structured into a sequence of n successive *stages* that have to be carried out one after the other and each take roughly the same (not necessarily constant amount of) time. The pipelining pattern allows parallelization by assigning up to n processors to the individual stages.

Pipelining is a surprisingly versatile technique, which can lead to highly efficient and fast parallel algorithms for some problems. Pipelining is, for instance, used in algorithms for data exchange problems (see Sect. 3.1.4). Pipelines can be more complex, directed, acyclic dependency graphs like, for instance, series-parallel graphs. The essence is that work items pass through the stages of the pipeline, perhaps being split or combined, and that the parallelism comes from stages that work in parallel on different work items. The number of processors that can be employed is thus determined by the number of pipeline stages and the parallel time by the number of work items.

1.3.5 Stencil

Here is another frequently occurring nested loop computation. An element of a two-dimensional $m \times n$ matrix `b[i][j]` is updated with the result of a constant time computation on a small set of elements of another matrix `a[i][j]`. In the example here, each update is a simple average function `avg` on eight elements neighboring `a[i][j]` and `a[i][j]` itself. The updates in the b matrix depend on the a matrix, but by using two matrices, the computation has no loop carried dependencies. Therefore, both loops could possibly be perfectly parallelized. Since each element update takes constant time, the total amount of computation for updating all elements is in $O(mn)$ which with p processor-cores could ideally be done in $\Theta(mn/p + 1)$ time steps. In a PRAM implementation, a processor is assigned to each matrix element to do the update; with less than mn processor-cores, the matrix is thought of as divided into p parts, typically block submatrices, and a processor-core is assigned to each block to do all the updates for the block (see Sect. 3.2.14 and 3.2.8). After the update step, the two matrices are swapped. The computation is repeated until some criteria is met and the **done**-flag is set to **true**. The pattern is an example of a two-dimensional, so-called 9-point *stencil computation*. The matrices are assumed to have also rows and columns indexed as `a[-1][j]`, `a[m][j]`, `a[i][-1]` and `a[i][n]`, respectively. This border of *ghost* rows and columns is sometimes called the *halo*, and in this example, the halo is of depth one.

As an aside, the code snippet illustrates a handy way of handling two-dimensional arrays in C (for the best introduction to C, see [68]). Matrices are stored in row-major order: one row of consecutive elements (here of type `double`) after the other. Each matrix is declared as a pointer to an array of rows of $n+2$ elements, and $(m+2)(n+2)$ elements are allocated for each matrix. By pointer arithmetic, adding one full row and one element, the matrix with halo can be conveniently addressed. The C compiler can, since $n+2$ is known (although not static), compute the starting address of each row in the allocated storage. This will also work for higher-dimensional matrices as long as the sizes of the lowest, faster changing dimensions are given in the declaration.

```
double (*a)[n+2];
double (*b)[n+2];
double (*c)[n+2];
double (*aa)[n+2];
double (*bb)[n+2];

a  = (double(*)[n+2])malloc((m+2)*(n+2)*sizeof(double));
aa = a; // save allocated address
// and shift address by one row and one column
a  = (double(*)[n+2])((char*)a+(n+2+1)*sizeof(double));

b  = (double(*)[n+2])malloc((m+2)*(n+2)*sizeof(double));
bb = b; // same for b
b  = (double(*)[n+2])((char*)b+(n+2+1)*sizeof(double));

int done = 0;
while (!done) {
  for (i=0; i<m; i++) {
    for (j=0; j<n; j++) {
      // 9-point stencil
      b[i][j] =
        avg(a[i][j-1],a[i+1][j-1],a[i+1][j],a[i+1][j+1],
            a[i][j+1],a[i-1][j+1],a[i-1][j],a[i-1][j-1],
            a[i][j]);
    }
  }
  c = a; a = b; b = c; // swap matrices a and b

  done = ... ; // set when done
}

free(aa); // free as allocated
free(bb);
```

A stencil computation on a d-dimensional matrix consists in updating all matrix elements according to a (most often) constant-time *stencil rule* that depends on and describes a small, bounded, constant-sized neighborhood of each matrix element. The total amount of computation per stencil iteration is then proportional to the size of the d-dimensional matrix. The 9-point stencil above has as neighbors of matrix element `a[i][j]` the elements whose distance is at most one in the maximum metric (Chebyshev distance), which is sometimes called a Moore-neighborhood. A two-dimensional, 5-point stencil would have as neighbors the elements whose Manhattan distance (taxi cab metric) is at most one. This is sometimes called a von Neumann neighborhood. Both are examples of first-order stencils; higher order stencils include neighbors that are farther away in the chosen metric. The stencil rule above is simply a computation of the average of the nine stencil elements but could be any other constant-time function, for instance, the rules of Conway's amazing *Game of Life* [14]. In Conway's game, life evolves in a two-dimensional, but potentially infinite universe. It is an example of a cellular automaton [28, 111] and, thus, not strictly a stencil computation; but a finite universe could easily be imagined and perhaps still be interesting. The standard use of a 5-point stencil computation is Jacobi's method for solving the Poisson differential equation, where the matrix updates are repeated until convergence [36, Chapter 16]. The values in the ghost rows and columns define the *boundary conditions*. Other, higher-dimensional stencils, e.g., 27-point (Chebyshev) or 7-point (Manhattan) in three dimensions are also frequently used, as are many other, sometimes also asymmetric stencils of higher order. Accordingly, there are much terminology and different notations for stencils in different application areas.

A single iteration of a one-dimensional, second-order stencil computation is expressed by the following loop.

```
for (i=0; i<n; i++) {
  b[i] = a[i-2]+a[i-1]+a[i]+a[i+1]+a[i+2];
}
```

It can be parallelized to run in $\Theta(n/p+1)$ parallel time with p processor-cores.

1.3.6 Work Pool

A *work pool* maintains items of work to be performed to solve our given computational problem in a data structure, the *pool*, which makes it possible to insert and remove items in no particular order. A work item being solved may give rise to new work items that are inserted into the pool. The process is repeated until all work items have been processed and the work pool is empty. Work items can be many things, like tasks ready for execution, parts of the input, partial results, depending on the situation. The work pool pattern is clearly attractive for parallelization. Since there are no dependencies among items in the pool, several/many processors can conceivably remove and work on work items from the pool independently. A good parallelization in the sense of good load balance might be possible if there is at any time during the computation a sufficient number of work items in the pool (compared to the number of processor-cores). A non-trivial issue is the *parallel data structure* needed to allow many processors to concurrently remove and insert work items into the pool. A *centralized* work pool is maintained by a single processor, and may be easier to design and reason about, but may also become a sequential bottleneck for parallelization when a large number of processors at the same time access the work pool. A centralized design can easily fall victim to Amdahl's Law. In a *distributed* work pool, the pool consists in a number of local pools maintained by the individual processors. As long as there are enough items in any of the pools, the processors can be kept busy and good load balance guaranteed. Problems arise when work pools run out of work. There are two strategies for alleviating the ensuing load balancing problem. With *work-dealing*, processors whose pools are too full relative to some (static or dynamic) threshold spontaneously deal out work item(s) to other processors whose pools may have few(er) items. With *work-stealing*, processors whose pools have become empty *steal* work from other pools, and continue stealing until they have either been successful or until it can be inferred that all pools are empty and the computation has come to the end. Work-stealing is currently a favored strategy for which appropriate parallel data structures have been developed, and where sometimes strong bounds on parallel running time can be proven. Regardless of what strategy is chosen, the work pool pattern eventually needs to solve the (distributed) *termination detection problem*: when is the pool definitely empty?

1.3.7 Master-Worker/Master-Slave

The *master-slave* or *master-worker* pattern is sometimes used to implement the work pool pattern. A dedicated master processor maintains a central data structure, from which the slaves or workers are given work (data to work on, tasks to execute) upon explicit request. The pattern is often simple to implement but fully centralized, highly asymmetric, and, thus, easily subject to Amdahl's Law and similar serialization issues.

1.3.8 Domain Decomposition*

The stencil computation employs a localized, constant time, mostly position-oblivious update operation to each element of a structured domain, typically a d-dimensional matrix, which is iterated a (large) number of times until a convergence criterion is met. It appears easy to parallelize efficiently and can utilize a considerable number of processor-cores. In the more general *domain decomposition* pattern, a term which we use here very loosely to characterize a computational pattern and not necessarily in accordance with terminology from other domains, the situation is like this: A more or less abstract domain in d-dimensional space over which computations are to be performed is subdivided into subdomains (not necessarily disjunct) which are assigned to the available p processors. The work to be done in the subdomains, say on moving particles, may not be uniformly distributed over the domain and may possibly move around in the domain. The computation per work item may or may not be uniform and constant. The computation over the domain is, like in the stencil pattern, typically to be repeated a large number of times until convergence.

This pattern generalizes the stencil computation in several respects. Thus, a static decomposition of the domain may not perform well, since the subdomains can contain different amounts of work items. Since the items may move, and since the amount of required computation may change from iteration to iteration, this pattern will typically need dynamic load balancing to keep the p processors equally active throughout the computation.

1.3.9 Iteration Until Convergence, Bulk Synchronous Parallel

In both the stencil and the domain decomposition pattern, a parallel computation is iterated a known or unknown number of times k until some convergence criterion is met. The parallel time that can be achieved regardless of the number of processor-cores employed is, therefore, bounded from below by $\Omega(k)$. If k is large compared to the total work to be carried out, the achievable speed-up will be limited by Amdahl's Law.

Another way of looking at the pattern is as follows. Let some number p of processors be given. In each iteration, each processor is assigned a part of the computation to be done, ideally in such a way that the work load is evenly balanced over the processors. The processors perform their work and in cooperation decide whether the termination condition has been met or not. If not, work for the next iteration is redistributed over the processors. When the work per iteration is large compared to the coordination at the end of the iteration consisting in communication (data exchange) and synchronization, this is a typical *coarse grained* parallel computation, often referred to as a *Bulk Synchronous Parallel* (BSP) computation. The term was probably coined by Les Valiant [17, 119].

An interesting example that can be cast in the bulk synchronous parallel pattern is level-wise Breadth-First Search (BFS) in a(n un)directed graph from some starting vertex. Let $G = (V, E)$ be the given graph with $n = |V|$ vertices and $m = |E|$ arcs, and $s \in V$ a given source vertex. The problem is to find the distance from s to all other vertices $u \in V$ defined as the number of arcs on a shortest path from s to u. A standard, sequential BFS algorithm (see any algorithms textbook [33]) maintains a queue of vertices being explored in the current iteration and a queue of new vertices to be explored in the next iteration. It maintains a distance label for each vertex which is the length of a shortest path from s in the part of the graph that has been explored so far. Initially, all vertices have distance label ∞, except s which has distance label 0, since no part of G has been explored. The invariant to be maintained for iteration $k, k = 0, 1, \ldots$ is that all vertices in the queue of vertices to be explored have correct distance label k. In iteration k, all vertices u in this queue are explored by examining the outgoing arcs $(u, v) \in E$. If v has a finite distance label already, there is nothing to be done. If the distance label of v is ∞, it is updated to $k+1$ and v is put into the queue of vertices for the next iteration. At the end of iteration k, the two queues are swapped.

It is clear that the algorithm terminates after K iterations where K is the largest finite distance from s of some vertex in G. It is also clear that all arcs are examined at most once. Thus, assuming that all vertices are reachable from s, the complexity of this algorithm is $\Theta(n+m)$ if the queue operations are in $O(1)$.

There is much potential for parallelization. Vertices in the queue of vertices to be explored in iteration k can be processed in parallel since order is not important and all arcs out of such vertices can also be examined in parallel; provided that vertices and arcs are available to the processor-cores and that conflicts, for instance, when inserting vertices into the queue of vertices for the next iteration can be handled. By the end of an iteration, arcs and vertices may have to be exchanged between processor-cores and queues consolidated for the next iteration. In the best possible case, a parallel running time in $\Omega(\frac{n+m}{p} + K)$ could be possible. If m is large compared to K and perhaps n, that is if G is not sparse and has low diameter, there might be enough "bulk" work for the processor-cores so that reasonable speed-up can be achieved in practice.

1.3.10 Data Distribution

Parallel algorithms working on structured data often seek to split the data into (disjoint) parts on which processor-cores can work independently in an embarrassingly parallel fashion. We have seen this approach with the stencil pattern and will see it again many times. The splitting of the data can be explicit by reorganizing the data into disjoint parts accessible to the available processors; or implicit by providing naming schemes and transformations to conveniently access the different parts. Structured data are here thought of as arrays, vectors, matrices, higher-dimensional matrices, etc., of objects that can themselves be structured. We refer to the splitting of such objects as *data distribution*, which may be an active operation to be performed (repeatedly) during the execution of a parallel algorithm or a matter of providing means to refer to the parts of the data in the required fashion.

Let some linear data structure of n elements be given, e.g., an array, and let $b, b \geq 1$ be a chosen *block size*. Let p be the number of processors, numbered from 0 to $p-1$. In a *block cyclic* data distribution the n-element array is split into *blocks* of consecutive elements of b elements each; one last piece may have fewer than b elements, depending on whether b divides n. Number these blocks consecutively starting from 0. Then blocks $0, p, 2p, \ldots$ can or will be accessed by processor 0, blocks $1, p+1, 2p+1, \ldots$ by processor 1, blocks $2, p+2, 2p+2, \ldots$ by processor 2, in general blocks $i, p+i, 2p+i, \ldots$ by processor $i, 0 \leq i < p$.

A *cyclic* data distribution is the special case where $b = 1$. A *blockwise* data distribution is the special case where roughly $b = n/p$ and where rounding is done such that, as far as possible, each processor has a block of at least one element.

A higher-dimensional matrix may likewise be divided into smaller blocks (many possibilities) and distributed in a block cyclic way. Special cases for two-dimensional matrices are the *row-wise* distribution where each processor is assigned to work on a consecutive number of full rows of the matrix, and the *column-wise* distribution where each processor is assigned to work on a consecutive number of full columns of the matrix. We will see examples of the use of such distribution in Sect. 3.2.30.

1.3.11 Compaction, Gather and Scatter

Consider the loop below. A marker array mark[i] is given, which for each element of the array a[i], tells whether the element is to be kept or not for some later computation. The marked elements are copied into the b[j] array in the loop order of appearance.

```
j = 0;
for (i=0; i<n; i++) {
   if (mark[i]) b[j++] = a[i];
}
```

As we will see in the rest of these lectures, this pattern, which is called *array compaction*, is important and surprisingly versatile. Unfortunately, the loop has obvious dependencies, e.g., the increment of j, and so far we have no means of parallelizing it. Sect. 1.4.5 will be devoted to this problem.

The (dense) *gather* and *scatter* patterns rearrange array elements and are illustrated below. Given an index array ix[i] with values $0 \leq$ ix[i] $< n$ which is not necessarily required to be a permutation (it may be that ix[i]==ix[j] for some, different i and j), the gather pattern copies elements from b in the order given by the index array into a. The scatter pattern is the opposite and copies into the a array in index order from the b array in sequential loop order.

```
// gather
for (i=0; i<n; i++) {
   a[i] = b[ix[i]];
}
// scatter
for (i=0; i<n; i++) {
   b[ix[i]] = a[i];
}
```

Ideally, with p processor cores, both of the patterns can be parallelized to run in $\Theta(n/p+1)$ parallel time steps. Dependent on the index array, there may be concurrent reading in the gather pattern. If implemented on a PRAM, this pattern requires either concurrent read capabilities or prior knowledge that the index array is indeed a permutation. Likewise, the scatter pattern may incur concurrent writing. If implemented on a PRAM, sufficiently strong concurrent write capabilities are required depending on which values may be written. More liberal gather and scatter patterns would allow the b-array to be an m-element array with $m \geq n$ and indices in this array.

1.3.12 Data Exchange, Collective Communication

Different parts of the data being processed at different stages of the execution of a parallel algorithm may be managed by or have special affinity to different processor-cores; indeed, this was the case for many of the parallel patterns discussed above. It can, therefore, be convenient or even necessary (as will be seen in Chapter 3) to explicitly exchange or reorganize data between processor-cores at different stages of the computation. Such *exchange patterns* and operations are frequent in Parallel Computing and are also often referred to as *collective communication* or just *collectives* because all affected processor-cores jointly take part in and jointly, by appropriate underlying algorithms, effect the exchange.

The following exchange and reduction patterns are traditionally considered.

- *Broadcast* and all-to-all broadcast, in which one, or all, processors have data to be distributed to all other processors.
- *Gather*, in which a specific processor collects individual data from all other processors.
- *Scatter*, in which a specific processor has individual data to be transmitted to each of the other processors.
- *All-to-all*, in which all processors have specific, individual data to each of the other processors.
- *Reduction*, reduction-broadcast (all-reduce) and reduction-scatter, in which data are combined together under an associative operator, with the results stored either at a specific processor, at all processors, or distributed in parts over all processors.

All these patterns are explicitly found in, for instance, MPI and will be discussed in great detail in Sect. 3.2.28; but they do appear explicitly and implicitly in many other Parallel Computing contexts as well.

1.3.13 Reduction, Map-Reduce and Scan

Surprisingly many problems can be viewed as reduction problems: A (large) number of input values which can be numbers, vectors, matrices, complex objects (texts, pictures, data bases) etc. are combined together using an associative, functional rule to arrive at the solution. Subsets of elements can be assigned to processor-cores, and by associativity, the reduction can be performed by repeated reduction of disjoint pairs of sets of values. The reduction pattern is a well parallelizable design pattern. The pattern can be made more powerful and flexible by allowing, for instance, a precomputation on the input values before reduction; this operation is often, for instance, in functional programming, called a *map operation*, and the combined pattern has been popularized as *map-reduce* [34, 35]. Many variations are possible and have been proposed.

A related pattern is the *scan* or prefix sums where the associative rule is applied on the input values in sequence: The result for the ith input is the associative reduction of all inputs before, up to and possibly including input i. A scan computes these prefix sums for all i [18]. We will see many applications of prefix sums and the scan operation throughout these lectures, see Sect. 1.4.5 also for efficient algorithms for computing the scan.

1.3.14 Barrier Synchronization

Some of the patterns above divide a computation into separate stages that are executed one after the other, for instance, until some convergence criterion is met. Other patterns and computations assume that computations done by other processors have been completed before a processor can continue to its next phase of computation. Ensuring such requires some form of *synchronization* between processor-cores and is the task of what we here call the *barrier* parallel pattern. Semantic barrier synchronization means that a processor that has reached a certain point in its computation, called the *barrier* (point), is not allowed to proceed before all other processors have reached their barrier point. After the barrier synchronization, updates and computations performed by the other processors shall be available to the processor to the extent that this is required.

Barrier synchronization can be implicit or explicit; for arguing for the correctness of a particular parallel algorithms it is often required, though, to know at which points the processors are synchronized in the sense of all having reached a certain point and having a consistent view of the computation.

In a lock-step, synchronized model like the PRAM, explicit barrier synchronization is not (or rarely) needed. In asynchronous, shared memory models and systems, various forms of barriers are needed to ensure correctness, and they are typically provided. The OpenMP thread model, see Sect. 2.3, provides implicit barrier points as well as explicit barrier synchronization constructs. Unlike for the PRAM, barriers are typically not for free. In asynchronous models, barrier synchronization of p processors takes $\Theta(\log p)$ parallel time steps. Many interesting standard algorithms for barrier synchronization on non-synchronous (non-PRAM) shared memory systems can be found in [78].

In distributed memory models, the required synchronization is sometimes guaranteed by the semantics of the provided communication operations, whether implicit or explicit. Explicit semantic barrier constructs may be provided as well, although they may be needed less often (to preview: in MPI, an explicit barrier is almost never needed!). Also here, $\Theta(\log p)$ dependent parallel operations go into enforcing a semantic barrier.

1.4 Fourth Block (1 Lecture)

As examples of not quite trivial parallel algorithms for computational problems that are not obviously parallelizable, we now look at two concrete problems, namely merging of two ordered sequences and computing the prefix sums of elements in an array. The aim is to derive good, parallel algorithms that can actually be implemented on real, parallel systems with both shared- and distributed memory. While the usefulness of the merging problem is obvious, this part of the lecture also motivates why computing prefix sums is such an important Parallel Computing problem. We also state the so-called "Master Theorem", a useful tool that will immediately solve (most of) the recurrence relations of these lectures.

1.4.1 Merging Ordered Sequences in Arrays

The *merging problem* is the following: Given two ordered sequences stored in arrays A and B with n and m elements, respectively, from some universe with a total order $\leq$, construct an ordered $n + m$ element array C containing exactly the elements from A and B.

The standard, straightforward sequential algorithm for merging steps through the arrays A and B hand in hand and in each iteration writes out the smaller element to the C array. This is captured by the following `seq_merge` function (for arrays of C doubles).

```
void seq_merge(double A[], int n, double B[], int m,
               double C[]) {
  int i, j, k;

  i = 0; j = 0; k = 0;
  while (i<n&&j<m) {
    C[k++] = (A[i] <= B[j]) ? A[i++] : B[j++];
  }
  while (i<n) C[k++] = A[i++];
  while (j<m) C[k++] = B[j++];
}
```

This algorithm (which is not the best possible in terms of constants [69]) unfortunately seems strictly sequential: The output at position i of C depends on the relative order of all the previous elements in A and B, and there is not much that can be done in parallel. Possibly the last two copy loops of independent iterations could be parallelized, but it is not known in advance how many elements of the input will be handled by these loops and, therefore, the observation does not help much. The complexity of the standard algorithm is $T_{\text{seq}}(n) = \Theta(n+m)$. A new, different idea is required for a good parallel solution.

Recall that merging and sorting algorithms are called *stable* if the relative order of equal elements in the input is preserved. For the merging problem, this means that the relative order of equal elements in the inputs arrays A and B is preserved in the output and also that elements in array A that are equal to an element of B occur before the B element in the output array C. Stability is often a useful or even desired property. Some merging and sorting algorithms are naturally stable, like the standard, sequential merging algorithm listed above, some are not.

For some of the merging algorithms in the following, it is convenient to assume that all elements are distinct. Distinctness can be assumed without loss of generality, because elements can always be made distinct: Instead of merging elements, we merge triples (x, F, i) where x is an element from either A or B, F marks whether the element comes from A or from B, and i is the index of the element in its array, whether in A or in B. We use a lexicographic order, defined by $(x, F, i) < (x', F', i')$ if either $x < x'$, or if $x = x'$ and $F \neq F'$ and $F = A$, or if $x = x'$, $F = F'$ and $i < i'$.

Using this order will ensure stability of any merging or sorting algorithm. The cost is extra space and a more expensive comparison operation (which should not be neglected. Try it!). It is, therefore, most often better if the merging or sorting algorithm is stable by design, without resorting to the "make-distinct trick".

1.4.2 Merging by Ranking

A different approach to merging is the following: For each element $A[i]$ in A, find the position j in B such that $B[j-1] < A[i] < B[j]$; here we assume element distinctness and for convenience that $B[-1] = -\infty$ and $B[m] = \infty$. The position j is called the *rank* of $A[i]$ in B, denoted by $\text{rank}(A[i], B)$. The rank of $A[i]$ in B, thus, counts the number of B elements that are strictly smaller than $A[i]$. By knowing the rank of element $A[i]$ in B, we also know the position of $A[i]$ in the output array C: It is $i + \text{rank}(A[i], B)$.

We can now merge the elements of A and B into C by computing the ranks for all elements in A and in B in the respective other array. The rank of any element of A in B can be computed by binary search in $O(\log m)$ time steps. The sequential complexity of *merging by ranking* is, therefore, $O(n \log m + m \log n) = O((n+m) \log \max(n, m))$, far worse than the standard, sequential merging algorithm.

However, merging by ranking can be performed in parallel: Assign a processor to each element of A and of B, let it compute the rank of the element in the other array and write the element to its position in the output array C. With $n + m$ processors, the algorithm takes $O(\log \max(n, m))$ time steps, so it is fast, but it is clearly not work-optimal: The work is the same as sequential merging by the ranking algorithm, namely $O((m+n)\log\max(n,m))$. We note also that when ranking is done concurrently by many processors, concurrent read capabilities (as in the CREW PRAM) are required of our system.

To reduce the work, a new idea is needed. We want to design an algorithm using p processors. The idea is to rank only some of the elements, more precisely $O(p)$ of them. The input array A is divided into disjoint, consecutive *blocks* of size roughly n/p, and the first element of each A block is ranked in B (it is helpful to graphically visualize this). Now, the A block can be merged with the corresponding part of B determined by the rank of the first element of the A block in B and the first element of the next A block in B, using our best known sequential merging algorithm. These pairs of blocks are disjoint and can all be merged in parallel. We now have a work-optimal, parallel merging algorithm. There are p processors, which together spend $O(p\log\max(n,m))$ work on ranking the p elements from A and $O(n+m)$ time for merging pairs of blocks. It should also be obvious that the algorithm is correct (given the distinctness assumption; use pictures to see this).

Unfortunately, we cannot give a good bound on the time. Desired is $O(\frac{n+m}{p} + \log\max(n,m))$. Since we do not know the inputs, and the arrays A and B can be arbitrarily interleaved in C, it can happen for one of the A blocks that the first element has a rank in B close to 0 and the first element of the next A block a rank close to $m-1$. Merging this pair which is done by one processor would, therefore, take $\Omega(n/p+m)$ sequential time steps, and there would be no speed-up over the sequential algorithm. This is a classical *load balancing problem*: One processor is doing almost all of the work.

There are at least two possible solutions to this problem. Assume that for some block in A the ranks in B of the first element and the rank of the first element of the next A block in B are far apart (close to m elements). Such a *bad segment* in B could be divided roughly evenly into p blocks of size about m/p elements and the rank in A for the first elements of each of these blocks computed (in parallel). It can easily be seen (use a picture) that all these ranks in A will lie within the A block which gave rise to the bad segment in the first place. Therefore, the pairs of the blocks of the bad B segment and the blocks now found in the A block will all have size at most $n/p+m/p$, and can, therefore, be merged sequentially within the desired bound of $O(\frac{n+m}{p})$ time steps. This would lead to a fast and work-optimal parallel algorithm. The only problem remaining is to be able to identify the bad B segments (there could be more than one) and to re-allocate the processors to work on these segments. This problem can be solved with use of prefix sums (see later) [64, 101].

The other solution is to divide from the outset both the A and B arrays into blocks of roughly equal size n/p and m/p elements and rank the first elements of these blocks in the other sequence. This gives rise to $2p$ pairs of blocks of size at most $n/p+m/p$ that can be merged in parallel with p (or $2p$) processor-cores. However, seeing that the blocks are indeed disjoint and cover the A and B arrays takes some care [55, 113]. Nevertheless, we claim the following theorem.

1.4 Fourth Block (1 Lecture)

Theorem 6. *On a p processor system where binary search can be performed, two ordered arrays A and B can be merged cost-optimally in $O(\frac{n+m}{p} + \log \max(n, m))$ time steps.*

Concurrent, simultaneous reading of the same location in either A or B array can happen during the binary searches, dependent on both the timing of the processors and the input, so if implemented on a PRAM, CREW capabilities are required.

1.4.3 Merging by Co-ranking

A completely different approach turns the parallel merging problem upside-down and focuses on what will be written into the C array. The idea is to find for each position i in the output array C the unique positions j and k in the input arrays A and B, such that by (stably!) merging $A[0,\ldots,j-1]$ and $B[0,\ldots,k-1]$, we get exactly the i-element prefix $C[0,\ldots,i-1]$ of C. The positions j and k are called the *co-ranks* for i and the approach itself *merging by co-ranking* [102]. Note that it holds that $j + k = i$ which will be an essential invariant in the algorithm for finding the co-ranks. If a processor can determine the co-ranks for the first element of a block of $(n+m)/p$ elements of C and the co-ranks for the first element of the next block of C, the $(n + m)/p$ element block of C can be constructed by (sequentially) merging the blocks of A and B determined by the respective co-ranks.

By this approach, we can ensure that all of the p processors have blocks of exactly the same size by diving the C array into blocks of size $(n + m)/p$ (plus/minus one element, if p does not divide $(n+m)$), and in that sense arrive at a perfectly load-balanced merging algorithm.

The observation of the following lemma tells how co-ranks can be computed.

Lemma 1. *For any index $i, 0 \leq i < n + m$, there are unique co-ranks j and k with $j + k = i$ such that*

1. *either $j = 0$ or $A[j-1] \leq B[k]$, and*
2. *either $k = 0$ or $B[k-1] < A[j]$.*

To see this, consider the element $C[i-1]$ of the output array that corresponds to the co-ranks j and k. Since each C-element comes from either A or B, either $C[i-1] = A[j-1]$ or $C[i-1] = B[k-1]$. Consider first the case where $C[i-1] = A[j-1]$ and $j > 0$. Then $B[k]$ is the first element of B that is not in $C[0,\ldots,i-1]$, and since the merge is stable, it follows that $A[j-1] \leq B[k]$. Also $B[k-1] < A[j-1]$, and therefore, since A is ordered, $B[k-1] < A[j-1] \leq A[j]$. For the other case, $C[i-1] = B[k-1]$ and $k > 0$, it similarly follows that $B[k-1] < A[j]$ (since the merge is stable, equal elements of A are before elements of B), and also that $A[j-1] \leq B[k-1] \leq B[k]$.

To find the co-ranks j and k for a given index i in C, a binary-search like procedure in both A and B can be applied, halving intervals in A and B until the conditions of Lemma 1 are both fulfilled while maintaining throughout the invariant that $i = j + k$. This will take $O(\log(n+m))$ iterations. The co-ranking code is shown below, and a full merge algorithm can (for parallel systems with shared memory) readily be implemented.

```
j = min(i,m); k = i-j;
jlow = max(0,i-n);
klow = 0;

done = 0;
do { // invariant: i = j+k
  if (j>0&&k<n&&A[j-1]>B[k]) {
    // condition 1 violated
    d = (1+j-jlow)/2;
    klow = k;
    j -= d; k += d;
  } else if (k>0&&j<m&&B[k-1]>=A[j]) {
    // condition 2 violated
    d = (1+k-klow)/2;
    jlow = j;
    k -= d; j += d;
  } else done = 1;
  assert(i==j+k);
} while (!done);
```

We summarize in the following theorem.

Theorem 7. *On a p processor system where co-ranking can be performed, the merging problem can be solved cost-optimally in $O(\frac{n+m}{p} + \log(n+m))$ time steps with p processor-cores. The algorithm is perfectly load balanced and stable.*

Like for binary-search based merging, concurrent, simultaneous reading of the same location in either A or B array can also happen with co-ranking, dependent both on the timing of the processors and on the input, so if implemented on a PRAM, CREW capabilities are required.

Ranking and co-ranking are examples of static, problem-dependent load balancing: Eventually, the blocks of the A and B arrays assigned to the processors to be merged sequentially have approximately the same total size, for the co-ranking approach exactly so (± 1), but how exactly the blocks are cut depends on the input. The preprocessing needed for the load balancing step, after which the sequential block merging is done, takes $O(\log \max(n,m))$, which is not a constant fraction of the total work $O(n+m)$, so Amdahl's Law does not apply.

1.4.4 Bitonic Merge*

Bitonic merging is an example of an *oblivious* merging algorithm: The indices that are compared against each other depend only on n and m, the size of the input, and not the input itself. Bitonic merging does not require concurrent read capabilities of the system and can be implemented on an EREW PRAM.

1.4 Fourth Block (1 Lecture)

Bitonic merging is an important example algorithm and can in some situations have practical advantages over the merging algorithms in the previous sections. Bitonic merging and Bitonic mergesort were invented by Kenneth Batcher [13].

Let $a_0, a_1, \ldots a_{n-1}$ be a sequence of $n, n > 1$ comparable elements, $a_i \leq a_j$ or $a_j \leq a_i$. The sequence is a *Bitonic sequence* if either

1. there is an $i, 0 \leq i < n$ such that $a_0 \leq a_1 \leq \ldots \leq a_i$ and $a_{i+1} \geq a_{i+2} \geq \ldots \geq a_{n-1}$, or
2. there is a cyclic shift of the sequence, such that the first condition holds.

For convenience, a sequence of $n = 1$ elements is also Bitonic.

It is not so difficult to see that the following lemma holds.

Lemma 2. *Let $a_0, a_1, \ldots a_{n-1}$ be a Bitonic sequence of even length. The two sequences*

- $\min(a_0, a_{n/2}), \min(a_1, a_{n/2+1}), \ldots, \min(a_{n/2-1}, a_{n-1})$ *and*
- $\max(a_0, a_{n/2}), \max(a_1, a_{n/2+1}), \ldots, \max(a_{n/2-1}, a_{n-1})$

of length $n/2$ are Bitonic and all elements of the first sequence are smaller than or equal to the elements of the second sequence.

A Bitonic sequence of length $n = 2^d$ can recursively be put into non-decreasing order as follows: By Lemma 2, split the sequence into two Bitonic halves with all elements of the first half smaller than the elements of the second half and recursively order the two Bitonic halves. In each recursive call, the number of elements to split is halved, so the number of calls in any successive sequence of calls needed to arrive at a single element is $d = \log_2 n$. The total number of comparisons performed and thus the total work measured as the number of operations as a function of n is given by the recurrence relation

$$W(1) = 0$$
$$W(n) = 2W(n/2) + n/2$$

which has the solution $W(n) = (n/2)\log_2 n$. This can be seen by induction or estimated by the Master Theorem 9 (Case 2). It is plausible that this can be turned into a parallel algorithm with $\log_2 n$ parallel time steps, in each of which $n/2$ comparisons are performed by recursive calls being carried out in parallel by the available processors.

Bitonic ordering can be used to merge two ordered sequences. From the two ordered sequences in arrays A and B of length n and m, a Bitonic sequence is constructed by listing the n elements from A in increasing order, followed by listing the m elements of B in reverse, that is in decreasing order. Bitonic merging can be extended to sequences of any length by padding with virtual $-\infty$ elements in front of the first sequence to get a virtual sequence of length some power of two. With some care, this can be made to work without doing any comparisons with the virtual $-\infty$ elements (outcome is always known). Compared to our sequential merge algorithm, this approach is not work-optimal.

Bitonic merge can elegantly be employed to sort a given n-element sequence in $O(\log^2 n)$ parallel time steps and $O(n \log^2 n)$ work (total number of operations).

Parallel Bitonic merging and sorting is commonly analyzed using another model of parallel computation: *comparator networks*. Bitonic ordering can be implemented on such a network of depth $\log_2 n$ and $(n/2) \log_2 n$ comparators. Bitonic mergesort, which can also be implemented on such a *sorting network*, is not work-optimal, and it was a long standing open question of theoretical importance whether sorting networks of depth $O(\log n)$ and size $O(n \log n)$ (number of comparators) exist [69, Sect. 5.3.4, Exercise 51]. The question was answered affirmatively in a famous paper by Ajtai, Komlós, and Szemerédi [5]. Another important result is "Cole's parallel mergesort", which shows that sorting by merging can be done in $O(\log n)$ parallel time steps with n processors on a(n EREW) PRAM [29, 30]. Both results have very large constants hidden in the Os and are in their original forms not practically relevant [11, 88].

1.4.5 The Prefix Sums Problem

We now turn our attention to another immensely important problem whose usefulness may not be obvious at first glance. Let an input array A of n elements from a set with a binary, associative operator $\oplus$ be given. The ith *inclusive prefix sum* for $0 \leq i < n$ is

$$B[i] = \bigoplus_{j=0}^{i} A[j]$$

and the ith *exclusive prefix sum* for $0 < i < n$ is

$$B[i] = \bigoplus_{j=0}^{i-1} A[j] \quad .$$

where the exclusive prefix sum $B[0]$ is left undefined.

The *prefix sums problem* is to compute the (exclusive or inclusive) prefix sums for all indices i. Computing all prefix sums over an array is sometimes also called *scan* which mostly, but not always, denotes the inclusive prefix sums, with *exscan* for the exclusive prefix sums [18]. Note that the ith inclusive prefix sum can be computed from the ith exclusive prefix sum by just adding the ith input element with $\oplus$ to the ith exclusive prefix sum. The converse does not hold, unless an inverse of the $\oplus$ operator is given, and that may not always be the case.

The prefix sums problem is a generalization of the *reduction problem* which is to compute only the last, inclusive prefix sum

$$B[n-1] = \bigoplus_{j=0}^{n-1} A[j] \quad .$$

1.4 Fourth Block (1 Lecture)

Since the operator $\oplus$ is associative, the sums are well-defined with $A[i] \oplus A[i+1] \oplus A[i+2] = (A[i] \oplus A[i+1]) \oplus A[i+2] = A[i] \oplus (A[i+1] \oplus A[i+2])$. If the $\oplus$ operator is in addition commutative, any two summands can be swapped and still yield the same result. Commutativity can give more freedom to algorithms to apply $\oplus$ in a convenient order but is normally not assumed.

Both problems are trivial to solve sequentially by a scan through the A array (thus the term), keeping a running sum in a register and writing it to $B[i]$.

```
sum = A[0]; // running sum in register
B[0] = sum;
for (i=1; i<n; i++) {
  sum = sum+A[i];
  B[i] = sum;
}
```

Improvements are possible by exploiting vector (SIMD) capabilities of the processor which is not quite trivial but can to some extent be accomplished by compiler loop unrolling. The sequential complexity is $\Theta(n)$ steps, since $n-1$ sum computations are necessary because the result can depend on any element.

Both reduction and prefix sums are examples of *parallel patterns* or *collective operations* (see Sect. 1.3.12): Each of the p processors contributes some of the $n, n \geq p$ elements, and the processors together perform a reduction or compute the prefix sums with results stored at the processors (prefix sums) or some selected processor (reduction).

1.4.6 Load Balancing with Prefix Sums

The reduction operation is clearly useful. A frequently occurring book-keeping task in parallel computations is for the processor-cores to agree on some common value (could be a flag, telling whether the computation is done). This common value is computed by a parallel reduction. A *broadcast operation* may be needed to provide the outcome to all processors or, even better, a combined reduce-broadcast, which is commonly called an *allreduce operation* (a somewhat unfortunate name choice).

Applications of the prefix sums problem are perhaps less obvious. Consider the following situation: Some expensive computation is to be done on some elements of a large array of n elements. It is not known a priori where these elements are. Instead, there is an associated marker array, also of size n, that for each index tells whether the associated element is to be worked on or not. All computations are independent of each other. Thus, there is potential for doing the work in parallel. We want to assign the element computations to p processors. The strategies for parallelizing loops that we have seen before (splitting the iteration range into p disjoint blocks, one for each of the p processor-cores) will not work well. Since it is not known which element indices are marked, it can easily happen that some blocks have many marked elements, while other blocks have no marked elements at all and, therefore, little to do apart from checking n/p indices and finding them unmarked. This is a typical load-balancing problem; the blocked merging by ranking algorithm had a similar problem. One processor may end up with all the work with no speed-up possible. Prefix sums solve this load balancing problem. This application is one of the most important uses of prefix sums in Parallel Computing and one reason why the problem is so important.

The solution is as follows: In a marker-index array M of size n, put a 1 for each marked element and a 0 for each non-marked element. This can be done in $O(n/p)$ parallel time steps by a parallelized loop of independent iterations. Perform an exclusive prefix sums computation on M. Now for each marked element, $M[i]$ is the number of marked elements up to (but not including) element i. It can, therefore, be used as index into another array which stores the marked elements consecutively. Assume that there are m marked elements; m can be computed by a reduction over the array of marks or directly from $M[n-1]$. Since these are now stored consecutively, the array of the marked elements can be partitioned into p blocks of about m/p elements, on all of which the expensive computation has to be performed. All p processors now have about the same amount of non-trivial work to do, and much better load balance is achieved, especially if the element computations all take about the same time.

This pattern, often called *parallel array compaction* (see Sect. 1.3.11), occurs in many guises. One is parallelizing the sequential, linear-time partitioning step of the Quicksort algorithm. We do three mark-and-compact steps. First, the elements strictly smaller than the pivot are marked and compacted into an array for the recursive call on the smaller elements. Second, the elements equal to the pivot (no recursive call needed) are compacted, and third, the elements strictly larger than the pivot are compacted into an array of the larger elements. The total work is $O(n)$, although the constants are larger than in standard, sequential partition implementations. How fast this is, depends on how fast the prefix sums problem can be solved. The two Quicksort calls (on smaller and larger elements) are independent of each other and can possibly be done in parallel as will be discussed in later parts of these lectures.

If the partitioning step is not parallelized, it will become a severe bottleneck for a parallel Quicksort implementation, consuming $O(n)$ time steps for the first Quicksort recursion level out of $O(n \log n)$ work in total, resulting in parallelism in the best case of only $O(\frac{n \log n}{n}) = O(\log n)$. The scan operation (parallel pattern) is similarly useful for sorting by counting and bucket sorting (see Sect. 3.2.31).

We now discuss three different solutions to the inclusive prefix sums.

1.4.7 Recursive Prefix Sums

The first algorithm is a recursive, divide-and-conquer approach. Let A be an array of n elements for which to compute the inclusive prefix sums. We reduce

1.4 Fourth Block (1 Lecture)

the problem to a prefix sums problem of only $\lfloor n/2 \rfloor$ elements by computing into an array B the sums of pairs of immediately consecutive elements of A: $B[i] = A[2i] \oplus A[2i+1]$, and recursively solve the prefix sums problem on B. The prefix sums of the A array can be constructed from B: $A[2i] = B[i-1] \oplus A[2i]$ and $A[2i+1] = B[i]$ with some care for the first and for the last element when n is odd. This can be implemented as shown below by a recursive function Scan that computes the prefix sums of the n-element A.

```
void Scan(int A[], int n) {
  if (n==1) return;

  int B[n/2]; // careful with stack allocation for large n
  int i;
  for (i=0; i<n/2; i++) B[i] = A[2*i]+A[2*i+1];

  Scan(B,n/2);

  A[1] = B[0]; // A[0] is per definition correct
  for (i=1; i<n/2; i++) {
    A[2*i]   = B[i-1]+A[2*i];
    A[2*i+1] = B[i];
  }
  if (n%2==1) A[n-1] = B[n/2-1]+A[n-1];
}
```

It is easy to see by an inductive argument that the recursive algorithm and program correctly compute the inclusive prefix sums of A. If there is only one element in A ($n = 1$), $A[0]$ is indeed the prefix sum. Now, assume that the function correctly computes the prefix sums of an array B of $\lfloor n/2 \rfloor$ elements. For $i > 0$, the ith prefix sum of A can be written as $\oplus_{j=0}^{i} A[i] = \oplus_{j=0}^{\lfloor i/2 \rfloor}(A[2j] \oplus A[2j+1])$ when i is odd and as $\oplus_{j=0}^{i} A[i] = \oplus_{j=0}^{\lfloor i/2 \rfloor}(A[2j] \oplus A[2j+1]) \oplus A[i]$ when i is even. By the initialization of B with $B[i] = A[2i] \oplus A[2i+1], 0 \le i < \lfloor n/2 \rfloor$, it will then hold by the induction hypothesis that $B[i] = \oplus_{j=0}^{i}(A[2j] \oplus A[2j+1])$ after the recursive call. Then, $\oplus_{j=0}^{i} A[i] = B[\lfloor i/2 \rfloor]$ when i is odd, and $\oplus_{j=0}^{i} A[i] = B[\lfloor i/2 \rfloor - 1] \oplus A[i]$ when i is even. This is what the program computes after the recursive call.

At each level of the recursion, there is $O(n)$ work to be done for computing the pairwise sums of the input array. Thus, the total work can be expressed by the following *recurrence relation*

$$W(n) = W(n/2) + O(n)$$
$$W(1) = 1$$

which can be solved by induction to give $W(n) = O(n)$. On each level of the recursion, the pairwise sums can be computed in parallel by a loop of independent iterations over the intermediate B array of size $\lfloor n/2 \rfloor$) in $O(n/p)$ time steps. Using, say, $\lfloor n/2 \rfloor$ processors, this is $O(1)$, and the parallel time over all recursion levels is, therefore, expressed by

$$T(n) = T(n/2) + O(1)$$
$$T(1) = 1$$

which by induction gives $T(n) = O(\log n)$. The parallel running time with p processors is, therefore, in the best case $O(n/p + \log n)$. The Master Theorem 9 applies to both recurrences.

To implement the algorithm with some fixed number p of processors, the pairwise summing (loop) must be parallelized. The recursive call is done by all processors, but before each call, the processors must wait for each other to have completed their part of the summing loop for which a *barrier synchronization* operation is needed. Likewise, after the recursive call, the processors must again wait for each other before they compute the results. Two barrier synchronizations are needed at each level for the recursion, for a total of $2\lfloor \log n \rfloor$.

Theorem 8. *The inclusive prefix sums problem can be solved in parallel time $\Omega(n/p + \log n)$.*

In the theorem, we have tacitly assumed that barrier synchronization is done in $O(1)$ parallel time, which would be the case on a PRAM. On other parallel computing systems, barrier synchronization takes $\Omega(\log p)$ parallel time steps. Therefore, a more realistic estimate of the algorithm is $O(n/p + \log n \log p)$ parallel time steps with p processor-cores.

The recursive prefix sums algorithm needs to allocate an intermediate array of size $\lfloor n/2 \rfloor$ elements at each recursive call (for a total of n elements). The pairwise summing has optimal spatial locality (see the next lecture) and can exploit the cache system well. It does about $2n$ summations with the $\oplus$ operator in the two parallel loops, about twice as many as the sequential algorithm.

1.4.8 Solving Recurrences with the Master Theorem

Recurrence relations, similar to the expression of work and time in the previous section, will often occur in the remainder of these lectures and many recursive algorithms give rise to this kind of very regular recurrence relations. Instead of doing an induction proof for each new recurrence, the solution to recurrences of this form can be summarized in a general theorem. This is often called the "Master Theorem" (for simple, regular divide-and-conquer recurrences), which exist in different versions. Here is one which covers most of the recurrences that will come up in these lectures:

Theorem 9. *Given a recurrence of the form*

$$T(n) = aT(n/b) + \Theta(n^d \log^e n)$$

for constants $a \geq 1$, $b > 1$, $d \geq 0$, $e \geq 0$, and $T(1)$ some constant. The recurrence has the following closed-form solution

1. $T(n) = \Theta(n^d \log^e n)$ if $a/b^d < 1$ (equivalently $b^d/a > 1$),
2. $T(n) = \Theta(n^d \log^{e+1} n)$ if $a/b^d = 1$ (equivalently $b^d/a = 1$), and
3. $T(n) = \Theta(n^{\log_b a})$ if $a/b^d > 1$ (equivalently $b^d/a < 1$).

When the recurrence relation models a recursive procedure, b is the shrinkage or reduction factor by which the subproblems get smaller, and a is the proliferation or expansion factor, roughly the "number" (not necessarily integer) of subproblems to be solved at each recursion level. It is clear that the number of levels of the recursion is $\lceil \log_b n \rceil$. A proof analyzes such recursion trees and can be found in any good algorithms' textbook, see for instance [3, 4, 33, 92, 99] and also a recent paper by Kuszmaul and Leiserson [70]. A proof can be found in the appendix to these lecture notes and is much recommended to study.

We can immediately apply the Master Theorem to the simple parallel prefix sums recurrences. For the $W(n)$ recurrence, $W(n) = W(n/2) + O(n)$, Case 1 applies (with $a = 1, b = 2, d = 1, e = 0$) which gives $W(n) = O(n)$. For the $T(n)$ recurrence, $T(n/2) + O(1)$, Case 2 applies (with $a = 1, b = 2, d = 0, e = 0$) and gives $T(n) = O(\log n)$.

1.4.9 Iterative Prefix Sums

Theorem 8 can be achieved by a different looking, iterative algorithm. In fact, the iterative algorithm can be found by unfolding the recursions of the recursive algorithm. An advantage of the iterative prefix sums algorithm is that no intermediate array has to be allocated.

The algorithm has two phases, an up-phase, corresponding to the pairwise sum computations before the recursive call and a down-phase, corresponding to the sum computations on return from the recursive call. Both up- and down-phases take $\lfloor \log_2 n \rfloor$ iterations.

In the first up-phase iteration, sums of even-odd pairs are computed. In the next iteration, sums of pairs of every second element are computed, in the third iteration, sums of pairs of every fourth element, and so on. The down-phase reverses this pattern. The following code illustrates the algorithm.

```
int k, kk;
int i;

// up-phase
for (k=1; k<n; k=kk) {
  kk = k<<1; // double the loop increment
  for (i=kk-1; i<n; i+=kk) A[i] = A[i-k]+A[i];
}
// down-phase
for (k=k>>1; k>1; k=kk) {
  kk = k>>1; // halve the loop increment
  for (i=k-1; i<n-kk; i+=k) A[i+kk] = A[i]+A[i+kk];
}
```

The correctness of the up-down-phase inclusive prefix sums algorithm (and implementation) can be proven by showing that certain invariant properties are maintained for each iteration and, in the end, imply the desired end result. To formulate the invariants, let $a_i, 0 \le i < n$ be the input sequence for which the inclusive prefix sums are to be computed in $A[i]$, that is, $A[i] = \oplus_{j=0}^{i} a_i$.

For the up-phase, the following invariant will hold before iteration $k, k = 0, 1, \ldots, \lfloor \log_2 n \rfloor$: For each $i, i < n$ of the form $i = j2^k - 1$ for some $j > 0$,

$$A[i] = \oplus_{j=i+1-2^k}^{i} a_j$$

That is, every 2^kth $A[i]$ will store the sum of the 2^k previous elements up to and including the ith element itself. This clearly holds before the first iteration ($k = 0$), since the input array is $A[i] = a_i = \sum_{j=i}^{i} a_j$. Assuming that the property holds before iteration $k, k > 0$, we have for that iteration which computes $A[i - 2^k] \oplus A[i]$ into $A[i]$ for elements $i = j2^{k+1}$ that

$$A[i] = (\oplus_{j=i-2^k+1-2^k}^{i-2^k} a_j) \oplus (\oplus_{j=i+1-2^k}^{i} a_j) = \oplus_{j=i+1-2^{k+1}}^{i} a_j$$

for all i of the form $i = j2^{k+1} + 1$. Thus, the invariant holds before the start of iteration $k + 1$. We can, by the way, observe that all $A[i]$ with $i = 2^k - 1$ for $k = 0, \ldots, \lfloor \log_2 n \rfloor$ are "good" in the sense of correctly containing the ith prefix sum. The task of the down-phase is to make all other elements in A "good" as well. Also note here that the variables k and kk in the program are 2^k and 2^{k+1}, respectively, for the iteration count k in the proof.

The down-phase starts with the results computed in the A array by the up-phase. The invariant for the kth iteration for $k = \lfloor \log_2 n \rfloor, \lfloor \log_2 n \rfloor - 1, \ldots, 0$ is that each 2^kth element is "good" in the sense that $A[i] = \oplus_{j=0}^{i} a_j$ for i of the form $i = j2^k - 1$. From the up-phase, this holds before the first iteration. In the iteration, the program computes $A[i] \oplus A[i+2^{k-1}]$ into $A[i+2^{k-1}]$. So, assuming the invariant to hold, we have that

$$A[i + 2^{k-1}] = (\oplus_{j=0}^{i} a_j) \oplus (\oplus_{j=i+2^{k-1}+1-2^{k-1}}^{i+2^{k-1}} a_j) = \oplus_{j=0}^{i+2^{k-1}} a_j$$

by the "goodness" of $A[i]$ and the invariant from the up-phase for $A[i + 2^{k-1}]$. The iteration, therefore, makes $A[i + 2^{k-1}]$ "good", and $i + 2^{k-1}$ is of the form $j2^{k-1} - 1$ for the next iteration. After the last iteration when $k = 1$, this implies that $A[i] = \oplus_{j=0}^{i} a_i$ for all i. Thus, the prefix sums for all indices are correctly computed in the A array.

The algorithm consists of the $\lfloor \log_2 n \rfloor$ iterations of loops of independent iterations with successively half the number of elements in the up-phase and double the number of elements in the down-phase. After each loop, the processor-cores employed have to be synchronized with some form of barrier synchronization, each of which may take $\Omega(\log p)$ parallel time steps.

The algorithm achieves the bound stated in Theorem 8. It also does about $2n$ summations with the $\oplus$ operator in the up- and down-phase parallel loops, about twice as many as the sequential algorithm. A drawback is that the pairs being summed are farther and farther apart $(1, 2, 4, \ldots)$. Thus, the iterative algorithm has worse *spatial locality* than the recursive algorithm (more on spatial locality in the next lecture).

It is an important theoretical result that any logarithmically fast parallel prefix sums algorithm has to do twice the number of sequentially required $\oplus$ operations. Paraphrasing, something like the following result has been proved (using yet another model of parallel computation: the *arithmetical circuit*).

Theorem 10. *For computing the inclusive prefix sums of an n-element input sequence, the following trade-off holds between* size s *(roughly number of $\oplus$ operations done by gates) and* depth t *(parallel time, longest path from an input to an output):* $s + t \geq 2n - 2$.

This was proved by Snir [104], a more intuitive proof can be found in [124].

The theorem says that for any fast (sublinear, logarithmic) parallel prefix sums algorithm, the speed-up (when counting the possibly expensive $\oplus$ operations) is at most $p/2$. This is bad news for highly parallel algorithms running on a large number of processors which may use prefix sums for array compaction and other important computations. The trade-off also tells us how many operations a best possible parallel prefix sums algorithm is allowed to perform.

1.4.10 Non-work-optimal, Faster Prefix Sums

The two previous algorithms executed the loops summing pairs of elements $2\lfloor \log n \rfloor$ times. The next algorithm will reduce this to about $\lceil \log n \rceil$ loops, but the price is that it is no longer work-optimal. The algorithm has been discovered many times, and in these lectures we use the name Hillis–Steele after some of the discoverers [60]. The algorithm computes the inclusive prefix sums in-place in an array A.

In the Hillis–Steele algorithm, a $\oplus$ operation is done for (almost) all of the n array elements in each iteration. In the first iteration, for each element i, except the first, $A[i]$ is updated by summing with its adjacent element, $A[i] = A[i-1] \oplus A[i]$. In the next iterations, the update is $A[i] = A[i-2] \oplus A[i]$, in the third iteration $A[i] = A[i-4] \oplus A[i]$, and so on, in iteration k, $A[i] = A[i-2^k] \oplus A[i]$ when $i - 2^k \geq 0$. Each iteration can be written as a loop of, unfortunately, flow (forward) dependent iterations. The dependencies can easily be eliminated by performing the updates into a result array B and swapping A and B after the iteration. The following code snippet shows how.

```
int *a, *b, *t;

a = A; b = B;

k = 1;
```

```
while (k<n) {
  // update into B
  for (i=0; i<k; i++) b[i] = a[i];
  for (i=k; i<n; i++) b[i] = a[i-k]+a[i];

  t = a; a = b; b = t;   // swap
  k <<= 1;               // and double
}
if (a!=A) {
  for (i=0; i<n; i++) A[i] = B[i];  // copy back as necessary
}
```

It is easy to prove by invariants that the Hillis–Steele algorithm correctly computes all inclusive prefix sums. Assuming that $a[i] = a_i$ for the input sequence a_i, one invariant is clearly that before iteration k it holds that $a[i] = \oplus_{\max(i-2^k+1,0)}^{i} a_i$ for each $i > 0$, which implies the claim when $2^k \geq n - 1$. As in the iterative prefix sums program, the variable k is 2^k for iteration count $k, k \geq 0$. The number of iterations is clearly $\lceil \log n \rceil$. The work of the algorithm is $O(n \log n)$ and it is therefore not work-optimal. This is summarized in the theorem below.

Theorem 11. *The inclusive prefix sums problem can be solved in parallel time* $O(\frac{n \log n}{p} + \log n)$.

Also, for this algorithm to be correct, the processor-cores must all have completed their part of the parallel loop before moving on to the next iteration, so a barrier synchronization is needed in each iteration of the while-loop. This would increase the parallel time by an $\Theta(\log p)$ factor.

1.4.11 Blocking

What is the use of a prefix sums algorithm that is not work-optimal? In itself, for solving the problem on input of size n, it is not useful, as the larger the n, the smaller the absolute speed-up.

Algorithms that are not work-optimal can, however, be useful in context, as building blocks, where some of their advantages (like being fast) may pay off without being hurt by the extra work they perform. The situation is like this: If p processors have already been allocated, we may as well use them to reduce the parallel time. There is no point in rescheduling the work to fewer processors. The processors are there anyway and have to be paid for.

The general idea is to reduce the problem at hand to a smaller (possibly different) problem that can be solved on p processors, and then use this solution to compute the solution to the original problem, both steps by the use of work-optimal algorithms. For the whole algorithm to be work-optimal, the problem reduction and computation of the final solution must be done by work-optimal algorithms, but for the step in the middle, where a smaller problem is solved on the available p processors, there may be "room enough" to employ a faster, but not work-optimal algorithm. Overall, this can pay off to arrive at an algorithm that is work-optimal and possibly faster.

When applied to the prefix sums problem, the idea is sometimes called *blocking*. An n-element input array A is given, as are p processors to solve the problem. The input array is divided into p *blocks* of about n/p elements each, one for each of the processors. Each processor performs a sequential reduction on its block of elements and puts the results into an array B of p elements, one for each processor. Now, all prefix sums of B are computed (by either of the parallel prefix sums algorithm). After this, each processor i adds $B[i]$ to the first element of its A block and computes the prefix sums over its A block. This completes the computation of the prefix sums of A.

The time complexity of this blocked prefix sums algorithm, using Hillis–Steele as building block is $O(n/p + (p \log p)/p + n/p) = O(n/p + \log p)$, since Hillis–Steele is applied to an array of p elements only. In contrast to Theorem 8, the non-parallelizable term is $\log p$, not $\log n$, and with Hillis–Steele, the constant is 1, and not 2, as would have been the case with the recursive or iterative prefix sums algorithm.

Theorem 12. *The inclusive prefix sums problem can be solved in parallel time $\Omega(n/p + \log p)$.*

The saving of a factor of 2 in the $\log p$ term does not sound like much. However, if pairwise summing involves expensive communication as is the case when the algorithm is used for distributed memory systems and implemented with MPI (see Sect. 3.2), such a factor can be worthwhile. There are more dramatic applications of the blocking technique in the literature. For instance, the fast, but not work-optimal Common CRCW PRAM maximum finding algorithm of Theorem 1 can be used to devise a work-optimal Common CRCW PRAM maximum finding algorithm running in $O(\log \log n)$ time steps (see Sect. 1.4.14).

1.4.12 Related Problems

In the prefix sums and reduction problems, the elements were given in an array, and the array order determined the order of the application of the associative function $\oplus$. In this sense, the prefix sums problems are *oblivious, data-independent* problems. It is a natural and, it has turned out, extremely useful generalization to consider also *data-dependent* prefix sums problems where the order in which to apply the associative function is determined by an additional list structure (array of pointers) given as part of the input. To solve the data-dependent prefix sums problem it would suffice to traverse the list structure and for each element count the distance to the last (tail) element of the list (or from the first, head, element of the list). Based on this, the input elements can easily be put into an array in list-order on which the prefix sums can be computed by either of our efficient algorithms. Doing the traversal efficiently in parallel is the notorious *list ranking problem*.

Although similar to the prefix sums problem, the list ranking problem turns out to be much more intricate and much more difficult to solve. For instance, although there are list ranking algorithms similar to the Hillis–Steele algorithm, the simple blocking technique does not work here. It was a long standing problem to devise a fast, work-optimal, deterministic list ranking algorithm. The best deterministic result on an EREW PRAM is the $O(n/p + \log n)$ time algorithm of Anderson and Miller [7].

1.4.13 A Careful Application of Blocking★

By a more careful application of blocking as described in Sect. 1.4.11, we can arrive at a parallel inclusive prefix sums algorithm that achieves the best possible combination of time and work, t and s both measured as the number of $\oplus$ operations, captured in Theorem 10, namely $s + t = 2n - 2$ (equality). The algorithm is reasonably fast when n is large compared to p. The trick is to divide the input sequence of n elements into $p+1$ blocks (for p processors) of $n/(p+1)$ elements, instead of just p blocks as was done above. Assume now that $(p+1)$ divides n; this assumption can with a little care easily be lifted by dealing with some blocks of $\lceil n/(p+1) \rceil$ elements and some blocks with $\lfloor n/(p+1) \rfloor$ elements. The blocks are ordered, the first block contains the first $n/(p+1)$ input elements (block 0), the second block the next $n/(p+1)$ elements (block 1), and so on; the last block (block p) contains the last $n/(p+1)$ elements.

We measure the time t in the number of $\oplus$ computations that have to be carried out in sequence (the depth) and work (or size) s as the total number of $\oplus$ operations carried out by the p processors. The prefix sums algorithm consists of three steps.

1. Compute, for each of the first p blocks, the inclusive prefix sums for the $n/(p+1)$ elements in the block. This takes $t_1 = \frac{n}{p+1} - 1$ operations (time) and requires a total work of $s_1 = p\left(\frac{n}{p+1} - 1\right)$ operations.
2. Compute the inclusive prefix sums for a sequence consisting of the p sums of the elements in each of the first p blocks: This is for each block the prefix sum for the last element computed in Step 1. This takes time $t_2 = p - 1$ and work $s_2 = p - 1$ operations.
3. For the $p-1$ blocks $1, 2, \ldots, p-1$, excluding the first block 0, which is done (all prefix sums computed by the first step), and the last block p, which is special, add the prefix sum for the last block to the first $\frac{n}{p+1} - 1$ elements of the block. This results in the correct prefix sums for all elements, since the prefix sum for the last element of each block is the prefix sum for the block that was computed in Step 2. This takes time $t_3 = \frac{n}{p+1} - 1$ and work of $(p-1)\left(\frac{n}{p+1} - 1\right)$ operations. For the last block (block p), instead the prefix

sum for block $p-1$ is added to the first element of the block, and the inclusive prefix sums for the $n/(p+1)$ elements of the block are computed. This takes time of $n/(p+1) = t_3 + 1$ operations and another $n/(p+1)$ operations of work. The total work (number of operations) for the last step is therefore $s_3 = 1 + p\left(\frac{n}{p+1} - 1\right)$.

The total time for this algorithm is

$$t = t_1 + t_2 + t_3$$
$$= \left(\frac{n}{p+1} - 1\right) + p - 1 + \left(\frac{n}{p+1} - 1\right) + 1$$
$$= 2\left(\frac{n}{p+1} - 1\right) + p$$

The total work for this algorithm is

$$s = s_1 + s_2 + s_3$$
$$= p\left(\frac{n}{p+1} - 1\right) + (p-1) + p\left(\frac{n}{p+1} - 1\right) + 1$$
$$= 2p\left(\frac{n}{p+1} - 1\right) + p$$

The sum of work and time is

$$s + t = 2p\left(\frac{n}{p+1} - 1\right) + p + 2\left(\frac{n}{p+1} - 1\right) + p$$
$$= 2(p+1)\frac{n}{p+1} - 2(p+1) + 2p$$
$$= 2n - 2$$

which meets the trade-off of Theorem 10. When carefully implemented, the algorithm could run in $O(n/p+p)$ parallel time steps not counting the two barrier synchronization operations that may be needed.

The same trick of dividing an input sequence into $p+1$ blocks was used by Snir [103] to speed up binary search from $\log_2 n$ to $\log_{p+1} n$ comparison steps. It was also shown that this is the best possible. Note that the number of steps is constant if n is in $O(p^k)$ for some constant $k \geq 1$.

1.4.14 A Very Fast, Work-Optimal Maximum Algorithm

Can the maximally fast, $O(1)$ time step Common CRCW PRAM algorithm of Theorem 1 be made work-optimal or more efficient? By itself not, but with the blocking technique, it can be put to use to achieve a very fast and work-optimal algorithm for finding the maximum of a sequence of n numbers. We prove the following theorem constructively by outlining the corresponding algorithm.

Theorem 13. *The maximum of n numbers stored in an array can be found in $O(\log \log n)$ parallel time steps, using $n/\log \log n$ processors and performing $O(n)$ operations on a Common CRCW PRAM.*

Divide the array into blocks of roughly $\sqrt{n}$ numbers. Assume (recursively) that the maximum has been found for each of these $\sqrt{n}$ blocks. Now, we can employ the optimally fast maximum finding algorithm to find the maximum among these $\sqrt{n}$ block maxima in $O(1)$ time steps and $O((\sqrt{n})^2) = O((n^{\frac{1}{2}})^2) = O(n)$ work. The time and work, including the recursive solution to the $\sqrt{n}$ subproblems of roughly $\sqrt{n}$ numbers each, is given by the following recurrence relations.

For the time, we have

$$T(n) = T(\sqrt{n}) + 1$$
$$T(1) = 1$$

and for the work

$$W(n) = \sqrt{n} W(\sqrt{n}) + n$$
$$W(1) = 1$$

Neither of these recursions are covered by the Master Theorem 9. Its is, however, easy to guess a closed form and verify the guess by induction. For the time recurrence $T(n)$, we see that we have to repeat taking the square root of n until we get down to some constant. We conjecture that $T(n) = \log \log n$. With this as induction hypothesis, we get $T(n) = T(\sqrt{n}) + 1 = \log \log \sqrt{n} + 1 = \log(\frac{1}{2} \log n) + 1 = \log \frac{1}{2} + \log \log n + 1 = -1 + \log \log n + 1 = \log \log n$. Similarly, we can find that $W(n) = n \log \log n$.

This recursive algorithm gives the time claimed in Theorem 13, but the work of $O(n \log \log n)$ operations is still too much. Precomputation, in parallel, by blocking, with the right number of processors, decreases the work to the desired $O(n)$ operations. Let n be the size of the given input array. The work-optimal algorithm does the following.

1. Divide the input into $n/\log \log n$ blocks of roughly $\log \log n$ elements. Assign a processor to each of the blocks to find a maximum for each block. This preprocessing has reduced the problem size to $n/\log \log n$ block maxima and takes $O(\log \log n)$ parallel time steps and $O(n)$ work using the $n/\log \log n$ processors.
2. Apply the fast, recursive algorithm with $n/\log \log n$ processors to the reduced problem to find the maximum (of the original input) in

$$O(\log \log(n/\log \log n)) = O(\log \log n)$$

parallel time steps. The parallel work is

$$O((n/\log\log n)\log\log(n/\log\log n)) = O(n)$$

as desired.

This very fast maximum finding algorithm dates back to early work on fast and efficient PRAM algorithms [31, 101].

1.4.15 Do Fast Parallel Algorithms Always Exist?*

Cost-optimal algorithms have linear speed-up, and are especially attractive if they are efficient with a slowly growing iso-efficiency function. For a fixed input size, they can possibly achieve a solid, linear speed-up up to some large number of processors. Many examples of work- and cost-optimal algorithms were discussed (for merging and prefix sums, for instance) that can even achieve a logarithmic running time in the size of the input given that enough processors are available. It is a central question, not only for complexity theory and theoretical computer science, but also for the practitioner, whether such fast parallel algorithms exist for all problems.

A qualified, "most likely no" answer is given by standard complexity theory [86]. In outline, the answer is as follows.

The complexity class $\mathcal{P}$ is the class of *tractable problems*, computational problems that can be solved in polynomial time in the worst case in the size of the input on a Random Access Machine (RAM) or other reasonable model of sequential computation. The question is now whether all tractable problems can be solved fast in parallel. The complexity class $\mathcal{NC}$ ("Nick's class", after Nicholas Pippenger) is the class of problems than can be solved in poly-logarithmic parallel time $O(\log^c n)$ in the size of the input n for some constant c with a polynomial number of processors in the size of the input n on a Parallel Random Access Machine (PRAM) or other reasonable model of parallel computation. Put differently, $\mathcal{NC}$ is the class of parallel algorithms with tractable (polynomial) costs and poly-logarithmic parallel time complexity. By a simulation argument, it is clear that $\mathcal{NC} \subseteq \mathcal{P}$. So, the question is whether $\mathcal{NC} = \mathcal{P}$ or $\mathcal{NC} \neq \mathcal{P}$. That is, whether all tractable problems can be solved by an algorithm that runs in poly-logarithmic time with a polynomial number of processors on a suitable PRAM variant. A complete problem in $\mathcal{P}$ with respect to $\mathcal{NC}$-reduction is a problem to which all other problems in $\mathcal{P}$ can be reduced by a reduction (an algorithm) that is also in $\mathcal{NC}$. If such a complete problem is in $\mathcal{NC}$, then all problems in $\mathcal{P}$ would be in $\mathcal{NC}$ and admit fast parallelization.

Interestingly, it can be shown that there are indeed such complete problems in $\mathcal{P}$ and that many important, classic, and practically relevant problems are $\mathcal{P}$-complete. Furthermore, for none of these problems, an $\mathcal{NC}$-algorithm has been found despite much effort. It may, therefore, be that $\mathcal{NC} \subset \mathcal{P}$ and that there are problems (namely the $\mathcal{P}$-complete problems) that do not admit fast parallel solution with "only" polynomial resources (number of processors). Some problems that are $\mathcal{P}$-complete under $\mathcal{NC}$-reduction are (ordered) depth-first search (ODFS), maximum flow, and linear programming [32, 50, 64, 65]. These problems may, thus, turn out to be, in a sense *inherently sequential*.

The emphasis in these lectures has been on work- and cost-optimal algorithms with provable, linear speed-up. The emphasis in the small part of parallel complexity theory outlined in this section is on tractable problems with polylogarithmically fast running time (parallel time complexity) and not on work- and cost-optimality. A problem in $\mathcal{NC}$ may or may not be cost- or work-optimal and indeed very far removed from that. Establishing membership in $\mathcal{NC}$ is, therefore, only one aspect of parallel algorithmics and parallel complexity theory.

1.5 Exercises

1. Is the PRAM a NUMA or a UMA model? Is the PRAM a SIMD or a MIMD model? Does the SPMD characterization apply to the PRAM? Anticipating the programming frameworks to come, what can the advantages of adhering to an SPMD style possibly be? Anticipating even further, is the PRAM a PGAS model?
2. Let A be a two-dimensional $m \times n$ element matrix stored as `A[i,j]` and x an n-element vector stored as `x[j]`. Consider the following PRAM algorithm:

```
par (0<=i<m) b[i] = 0;
par (0<=i<m) {
   for (j=0; j<n; j++) b[i] = b[i]+A[i,j]*x[j];
}
```

Explain what this PRAM algorithm accomplishes. What is the number of parallel steps of the algorithm? What is the parallel time taken for the algorithm to finish? What is the total number of operations performed by the processors of the algorithm (parallel work)? Which PRAM variant is needed for the algorithm to work correctly? Which PRAM variant is sufficient? Explain your answers.
3. Modify the parallel $O(\log n)$ time algorithm of Theorem 2 for finding a maximum among n elements stored in an array to perform a reduction, e.g., compute the sum, over the n elements for a given, associative operator $\oplus$. You may assume that the operator is commutative, so that the summands may be used in any order. What is the total number of operations performed by the assigned processors? Which PRAM variant is needed?
4. Give a different (or modify the above) algorithm for performing reductions over the elements in an n-element array `a` that works for not necessarily commutative but still associative operators $\odot$. This means that the sum must be computed in a fixed order as $a[0] + a[1] + \ldots + a[n-1]$ (summations can, of course, be grouped into smaller parts, but the order of the summands must not be changed). The algorithm must run in $O(\log n)$ parallel time steps.

1.5 Exercises 77

What is the total number of operations? Is the algorithm work-optimal? Is the algorithm cost-optimal?

5. Modify the parallel $O(\log n)$ time algorithm of Theorem 2 for finding a maximum among n elements stored in an array a to instead copy a specific element a[r] for some given index r between 0 and $n-1$ to all positions of a. The algorithm must work on an EREW PRAM. What is the total number of operations performed? Are any further assumptions needed to guarantee that the EREW PRAM capabilities suffice? What is the time and work of the copy operation on a CREW PRAM?

6. Show how to make $p-1$ additional copies of a large array of n elements on an EREW PRAM by actually writing a program in PRAM pseudo-code. More precisely, given an $n \times p$ matrix a[n][p], copy a specific column a[][r] to all other columns a[][i] for $0 \leq i < p, i \neq r$ for any given r as input. It may be assumed that n is (much) larger than p and that p divides n, $p|n$. Note that the total number of operations required is $\Omega(n(p-1))$ so that the best possible number of parallel time steps is $\Omega(n - n/p)$. What is the number of parallel time steps of your algorithm? What if p does not divide n, $p \nmid n$.

7. Give a PRAM algorithm for matrix–vector multiplication running in $O(\log n)$ time steps for vectors of n elements. Hint: Use and modify the $O(\log n)$ time algorithm for finding the maximum of n numbers. Which PRAM variant is needed? Can the algorithm be made to work on an EREW PRAM?

8. Give an EREW PRAM algorithm for adding two $m \times n$ matrices A and B, that is, for computing $C = A + B$. What is the number of parallel time steps of your algorithm? What is the total number of operations?

9. Give a work-optimal $n \times n$ matrix–matrix multiplication algorithm running in $O(\log n)$ time steps, first on a CREW PRAM, then on an EREW PRAM. You may assume that the optimal work of sequential matrix–matrix multiplication is in $O(n^3)$.

10. A collection of n list elements is stored in an array with an additional n-element array next that for each list element gives the index of a next following element. The indices in this array must fulfill that $0 \leq \text{next}[i] < n$ and that, for each $i, 0 \leq i < n$, there is at most one $j, 0 \leq j < n$ such that $\text{next}[j] = i$. A *tail* (final, last element of the list) is an element i with $\text{next}[i] = i$, that is, an element indexing itself. A *head* (initial, first, start element of the list) is an element i to which no other element points; that is, with no j such that $\text{next}[j] = i$. The next-array must have at least one tail and at least one head element. If there is one head and one tail element in the next-array, the collection is a single list, if there are more the collection consists of several shorter sublists.

Let now an n-element index array next be given. Devise a fast $O(\log n)$ time step PRAM algorithm to verify that the next array fulfils the conditions described above. Attach flags, stored in n-element arrays, which for each list element tell whether the element is a head or a tail element. How many operations does your algorithm perform? How does it compare in terms of number of operations to a sequential algorithm that analyzes and traverses the next-array? Are there interesting trade-offs between different PRAM variants? The Arbitrary CRCW PRAM may be relevant to consider. Is is possible to easily decide whether the next array represents exactly one list?

11. Given a collection of n list elements represented as described in the previous exercise: Devise a super fast and efficient (in number of operations performed) EREW PRAM algorithm to make this singly linked list a doubly linked list. The algorithm should compute an additional index array prev that for each list element i gives the previous (preceding) element. That is, it must hold for all $i, 0 \leq i < n$ that next[prev[i]] = i (the preceding element of a head element shall be the element itself).

12. Consider the following PRAM program. It is intended to work on a list defined by an n-element array of next indices, as described in the two previous exercises. The n-element arrays tail, dist, and sum store new information for the list elements and can be assumed to already have been allocated. The sum array stores results that have to be computed, and this array has been initialized with an input value for each list element.

```
par (0<=i<n) {
  tail[i] = next[i];
  if (tail[i]!=i) dist[i] = 1; else dist[i] = 0;
}
for (k=1; k<n; k<<=1) {
  par (0<=i<n) {
    if (tail[tail[i]]!=tail[i]) {
      dist[i] = dist[i]+dist[tail[i]];
      sum[i]  = sum[i]+sum[tail[i]];
      tail[i] = tail[tail[i]];
    }
  }
}
```

What does the algorithm accomplish, in particular, what will be the contents of the dist and sum arrays after it has finished? What is the number of parallel steps taken by the algorithm? What is the number of operations performed? Which PRAM variant does it need? Can you make the algorithm work on an EREW PRAM? Devise a sequential algorithm achieving the same result. What is the complexity of your sequential algorithm? Is the PRAM algorithm work-optimal when compared to your best possible sequential algorithm? Note: This algorithm is Wyllie's list ranking algorithm and illustrates an important technique (for achieving logarithmic parallel time complexity) called *pointer jumping*.

13. A directed graph $G = (V, E)$ with $n = |V|$ vertices numbered consecutively $0, \ldots, n-1$ is represented by an $n \times n$ adjacency (incidence) matrix A$[n, n]$. In the adjacency matrix, A$[i, j] = 1$ iff there is a directed edge in G from vertex i to vertex j and A$[i, j] = 0$ if there is no such edge. This is the input to the program you have to devise. It is not known from the input how many edges G has and neither are the out-degree and the in-degree of the vertices.
Write a (slow, i.e., not necessarily $O(\log n)$ steps) EREW PRAM program for computing the in-degree and the out-degree of all vertices V in G. The

out-degree and the in-degree of vertex i shall be stored as outdeg[i] and indeg[i], respectively. What is the running time (number of time steps) and the work (total number of operations) of your program? Write a fast $O(\log n)$ PRAM algorithm for computing m, the number of edges in G (hint: see the previous exercises). Which PRAM model is needed? What is the number of operations performed by your algorithm? How does it compare to a best sequential algorithm operating on the same representation of G?

14. A directed graph $G = (V, E)$ with $n = |V|$ vertices numbered consecutively $0, \ldots, n-1$ is represented as a set of n adjacency lists: For each vertex i, there is a list of adjacent vertices j, stored as a consecutive array with outdeg[i] elements. The outdeg-array is given as part of the input. It may be assumed that all adjacency lists are stored consecutively in a larger array with m elements in total, where m is the number of edges of G. In this array, the list of adjacent vertices for vertex i start at index adj[i]. Devise a sequential algorithm to compute the in-degree for each vertex i. What is the complexity of this best possible sequential algorithm? Devise a fast PRAM algorithm to accomplish the same task. Which PRAM variant does your solution require? Is the algorithm efficient in comparison to the sequential algorithm (in number of operations performed by the PRAM processors)?

    Now, extend the algorithm to compute for each vertex i an array storing the vertices that are adjacent to i (that is, the list of incoming edges of i). Hint: You will probably have to use extra space, n^2 instead of m, and perform asymptotically more operations than the sequential algorithm.

15. Consider the SPMD PRAM program execution of a conditional statement in which some processors execute one (the **true**) branch and some other processors execute the other (the **false**) branch. If the two branches consist of different numbers of instructions (as was disallowed for PRAM programs discussed in the text), the processors will not reach the end of the conditional statement in the same step (clock cycle) and in that sense they will not be synchronized anymore (at the algorithmic level) even though the individual instructions are executed in lock-step. Devise a PRAM barrier synchronization algorithm that will ensure that processors reach a specified synchronization point in the same instruction. Your algorithm should use $O(\log p)$ instructions on a p-processor PRAM. What is the smallest constant you can achieve? Your algorithm should preferably run on an EREW PRAM.

16. What is the parallel time complexity of finding the maximum of n numbers stored in an array?

17. Consider and give an example of a sequential algorithm running in $O(mn)$ operations: Which problem could be solved by such an algorithm? Is the algorithm of your choice in $\Theta(mn)$? Assume that different parallel algorithms for the problem have been developed running in parallel time $O(mn^2/p + n^2)$, $O((mn \log n)/p + n)$, $O(mn/p + n)$, $O(mn/p + \log n)$ and $O(mn/p + \log n \log p)$, respectively. Explain how these running times could possibly have been achieved, say, on a PRAM with different algorithmic approaches. Do these parallelizations have linear speed-up? Can they have perfect speed-up? Are thy work-optimal? Are they cost-optimal?

18. Repeat the previous exercise with a sequential algorithm running in $O(n+m)$ time steps and with parallel algorithms running in $O(((n+m)\log n)/p + n)$, $O((n+m)/p+n)$, $O((n+m)/p+n\log n)$, $O(n+m/p)$ and $O((n+(m\log n)/p+\log n)$, respectively.

19. Explain why standard BFS (Breadth-First Search) algorithms on graphs $G = (V, E)$ with n vertices and m edges starting from a given source vertex $s \in V$) are *not* in $\Theta(n+m)$. Give an example where the running time of the standard sequential algorithm is $o(n+m)$.

20. Some parallel algorithms with running times $O(n/p + \log n)$, $O((m+n)/p + \log n)$ and $O(n^2/p + \log n)$ are given. What are their parallel time complexities? Give expressions that characterize the number of processors needed to reach the claimed parallel time.

21. Different parallel algorithms for a computational problem that can be solved sequentially in $O(n)$ time have been given with running times $O(n/p + \log n)$, $O(\frac{n}{p/\log p} + \log n)$ and $O(\frac{n}{\sqrt{p}} + \log n)$, respectively. What is the parallel time complexity of the three parallel algorithms? Give expressions that characterize the number of processors needed to reach the minimum parallel running time. Which of the algorithms are cost-optimal? Which algorithm would be preferable for solving the problem and why?

22. Let $T_{\text{seq}}(n)$ for four computational problems be in $O(n)$, $O(n\log n)$, $O(n\sqrt{n})$, $O(n^2)$, respectively. For each, there is a parallel algorithm with fastest possible running time (parallel time complexity) in $O(\log^2 n)$. Give corresponding expressions for the parallel running times with p processors for four parallel algorithms that can give linear speed-up in the four cases.

23. What is the parallelism of an algorithm that is susceptible to Amdahl's Law? Can such an algorithm be cost-optimal? Is it work-optimal?

24. A program works on square matrices of order n and performs a large number of matrix–vector multiplications. The number of such multiplications is some constant k. Each iteration takes $O(n^2)$ time steps and has been perfectly parallelized (up to some number $p, p < n^2$ of processors). A sequential preprocessing of the input matrix is necessary and takes cn^2 operations for some (medium large) c.

    What is the maximum speed-up that this program can achieve as a function of c, k, n? Calculate the concrete speed-up with $c = 100, k = 10000, n = 1000$, and $p = 10$ and $p = 100$ processors, respectively. What is the maximum speed-up that can be achieved for $c = 1000, k = 100000, n = 10000$?

25. Two computational problems have best sequential algorithms with running times in $O(n)$ and $O(n^3)$, respectively. Assume they can be parallelized, however, with a fraction of the work being strictly sequential (non-parallelizable). With a multi-core processor with 64 cores, we want to achieve a speed-up of 60. How large can the sequential (non-parallelizable) fraction of the work be in the two cases? Are there differences between the two cases?

26. A simple, parallel, work-optimal matrix–vector multiplication algorithm is running in $O(n^2/p + n)$ time steps on input of $n^2 + n$ elements (matrix and vector), whereas sequential matrix–vector multiplication can be done (op-

1.5 Exercises   81

timally) in $O(n^2)$ time steps. How large must n be in order to achieve a speed-up of 64 on 128 processor-cores? Which assumptions on the constants in the parallel and sequential algorithms are needed for the calculation?

27. Consider two parallel matrix–matrix multiplication algorithms with running times $O(n^3/p+n)$ and $O(n^3/p+n^2)$, respectively. You are asked to perform a weak-scaling analysis of the algorithms. In this analysis, the average work over the p processors should stay fixed at some given number of operations w. Assuming that the sequential algorithm used as (best known) baseline has work $O(n^3)$, you have to determine up to how many processors our algorithms can work efficiently if the average work per processor is to be kept at $w = n^3/p$. How does n have to grow as a function of p to keep constant average work w? What are the asymptotic running times for the two algorithms as a function of w and p? Up to how many processors will the two algorithms be weakly scaling (that is, have constant running time independent of p)? When do the second terms in the parallel running times start to dominate?

28. We have a number of parallel algorithms for matrix–matrix multiplication at our disposal, running in time $O(n^3/p+\log n)$, $O(n^3/p+\sqrt{n})$, $O(n^3/p+n)$, $O(n^3/p+n\sqrt{n})$, $O(n^3/p+n^2)$, respectively. The best sequential algorithm known to us is, for now, running in $O(n^3)$ time.

    For each of the five cases, assuming the asymptotic constants have been normalized to 1, state the maximum number of processors that can sensibly be used, i.e., the maximum number of processors for which a linear speed-up can be achieved. For each of the five cases, state $T\infty(n)$ and state the parallelism. State the iso-efficiency functions for the five cases, that is, the smallest input size n as a function of p that is required to achieve a given, fixed parallel efficiency e. It may not be possible to give a closed-form formula in each case; if not state that: n must be at least... Compute a required (integer) input size n to maintain efficiency $e = 0.5$ and efficiency $e = 0.95$ for $p = 10$, $p = 100$ and $p = 1000$ for the parallel algorithm with running time $O(n^3/p+n)$.

29. Consider algorithms with $T_{\text{seq}}(n)$ in $O(n)$, in $O(n^2)$ and in $O(n^3)$. Assume we have found parallel algorithms with running times in $O(n/p+\log p)$, in $O(n^2/p+\log p)$ and in $O(n^3/p+\log p)$, respectively. Consider the two different definitions of weak scaling. Either we want to maintain a constant, given parallel efficiency e or we want to maintain constant average work w $(= n/p$, $= n^2/p$ or $= n^3/p$, respectively) by increasing n as a function of p. Compute the iso-efficiency function for the three algorithms. Compute the input size scaling function for constant average work for the three algorithms. Can the running times be kept constant under constant efficiency?

    Repeat the exercise with three different parallel algorithms now running $O(n/p+\sqrt{n})$, in $O(n^2/p+\sqrt{n})$ and in $O(n^3/p+\sqrt{n})$, respectively.

30. A (best known) sequential algorithm for some interesting problem runs in $T_{\text{seq}}(n) = O(n\log\log n)$ time steps for input of size n (for an example, see Sect. 2.3.18). A parallel algorithm for the same problem running in $T^p_{\text{par}}(n) = O((n\log\log n)/p+\sqrt{n})$ time steps has been found. Is this parallel algorithm work-optimal? Does the algorithm give linear speed-up and if so, up to which

number of processors p? Derive the iso-efficiency function for the parallel algorithm relative to the best known sequential algorithm. Is the parallel algorithm weakly scalable?

31. Implement the rank(x, A, n) operation for computing the number of elements in an ordered n-element array A that are smaller than the element x. Assume first that elements in A are distinct and different from x. What if this is not the case? Modify the definition of rank accordingly to either count elements $A[i] < x$ or elements $A[i] \leq x < A[i+1]$.

32. Implement the merging by binary search algorithm as a sequential program working on input arrays A and B of n and m elements, respectively. For each element in input array A, compute the rank of $A[i]$ in the other input array B. For each element in input array B, compute the rank of $B[j]$ in the other input array A. Use indices and ranks to put each element from A and B to its correct position in the output array C. What is the sequential complexity of this algorithm? Compare it (experimentally) against the standard seq_merge() function from Sect. 1.4.1. What are the assumptions on the inputs in A and B for your program to be correct? Find a way to make the algorithm *stable* (Hint: consider the previous exercise).

33. Implement a function

```
void corank(int A[], const int n, int B[], const int m,
            const int i, int *j, int *k);
```

for computing the co-ranks j, k for $i, 0 \leq i < n + m$ in arrays A and B.

34. Assume that p processors numbered from 0 to $p-1$ are available that can all access input arrays A and B from memory. Write out pseudo-code for a parallel merging by co-ranking algorithm describing what each processor $i, 0 \leq i < p$ has to do. Use the seq_merge() and corank() functions and indicate where barrier synchronization is required in order to guarantee correct output in the C array. You may look ahead and implement your parallel algorithm with OpenMP.

35. Devise a synchronization free merging by co-ranking algorithm, i.e., an implementation where no internal barrier synchronization is required. Hint: Use two corank() calls. Challenge: Can you do with only one call and still be synchronization free? Hint: By stability, this is possible, but requires a slight change in the sequential merge function.

36. Describe how to do mergesort (sorting by merging) in parallel by doing $\lceil \log n \rceil$ iterations of parallel merge operations. Here, n is the size of the input array A to be sorted. What is the running time with p processors of your algorithm? Is this cost-optimal? Hint: It is possible to achieve $O(\frac{n \log n}{p} + \log^2 n)$ by this approach.

37. Write a sequential (recursive) program using Bitonic merging to merge any two ordered sequences of n and m elements, respectively, for any $n, n > 0$ and $m, m > 0$ (not necessarily powers of two). Make sure that the implementation remains oblivious, meaning that the splitting of sequences depends only on length and position and never on actual values of elements.

1.5 Exercises

38. Consider the following two, semantically equivalent and correct implementations of sequential, inclusive prefix sums.

```
for (i=1; i<n; i++) {
  a[i] = a[i-1]+a[i];
}
```

and

```
register s = a[0]; // running sum
for (i=1; i<n; i++) {
  s += a[i];
  a[i] = s;
}
```

Create a benchmark to compare the performance of the two implementations (you can use the OpenMP framework and the `omp_get_wtime()` timing function; alternatively, implement with a C timing library like `time.h`) with large, preinitialized arrays. Experiment with various compiler optimization options, including no optimization. Are there notable differences? Use different element types for the array `a` (`int`, `double`, ...). Think of a model that can explain the expected and observed differences between the two implementations assuming the compiler does not transform one into the other. Study the assembly output (`gcc -s`).

39. What is the exact number of recursive calls performed by the Scan algorithm as a function of the input array size n? What is the exact number of applications of the + operator as a function of n? The function `popcount(n)` which counts the number of ones (set bits) in the binary representation of n will be helpful to express these numbers. Verify your solution by implementing the recursive `Scan()` function and instrumenting it with a count of the number of element + operations performed (not the i++ loop index increments). How much extra space for the intermediate B-arrays is allocated? Can you modify the program such that allocation is done only once and for all?

40. Devise an algorithm for recursively solving the exclusive prefix sums problem by modifying the Scan algorithm that motivated Theorem 8. What is the exact number of recursive calls as a function of the array size n? What is the exact number of applications of the + operator? Express as recurrence relations and solve by induction; be as general as possible (in the sense of exact solutions for as many n as possible). As above, the `popcount(n)` function will be helpful.

41. Implement the iterative inclusive prefix sums algorithm (up- and down-phases) as a sequential function `inclusive_prefix(int A[], int n)`. Modify the algorithm and your implementation to compute the exclusive prefix sums by a function `exclusive_prefix(int A[], int n)`.

42. Prove that $a[i] = \oplus_{\max(i-2^k+1,0)}^{i} a_i$ is an invariant for the non work-optimal inclusive prefix sums algorithm of Sect. 1.4.10.

43. Prove the claim that $W(n) = n \log \log n$ for the recurrence $W(n) = W(\sqrt{n}) + n$ for the very fast maximum finding algorithm in Sect. 1.4.14.

44. Implement the optimal trade-off inclusive prefix sums algorithm outlined in Sect. 1.4.13. The implementation should be entirely in-place, that is computation done on the input (and output) array with no extra arrays and only some constant number of additional variables (loop indices, running sums).
45. Give an algorithm for the exclusive prefix sums problem similar to the blocking algorithm of Sect. 1.4.13. Count the number of element + operations performed and the longest chain of dependent such operations. What does this imply for the trade-off between the total number of additions (size) and the longest chain of dependent additions (depth) of Theorem 10?
46. Give an algorithm for performing $p+1$-ary (instead of binary) search in ordered arrays of n elements with $p \geq 1$ processors. Show that the running time of your algorithm is $O(\log_{p+1} n)$ (as claimed in Sect. 1.4.13).
47. Explain why the following *Work Law* argument is incorrect and does not improve the work and depth lower bounds: With p processor-cores, assign one core permanently to the work on a critical path. This leaves $p-1$ processor-cores to work on the remaining work, which can in the best case be sped up by a factor of $p-1$. That is, for any p processor schedule it holds that $T_p(n) \geq \frac{T_1(n) - T_\infty(n)}{p-1}$.
48. You are tasked with inventing a parallel algorithm for Depth-First Search (DFS) that can provide provable speed-up for graphs that are not too sparse. Apparently and in contrast to Breadth-First Search (BFS) as discussed in Sect. 1.3.9, processing all the arcs directed from a found vertex in parallel will not help much. Another idea is needed. Describe your algorithm and state the parallel running time with p processors for graphs with n vertices and m arcs assuming a PRAM model of computation. Compare the complexity to a parallel BFS algorithm. Hint: See [114].
49. Which of the following parallel algorithms with running times $O(n/p+1)$, $O(n/\sqrt{p} + \log^3 p)$, $O(n/p + \sqrt{n})$, and $O(n^3/p + \log n \log p)$ would belong to the complexity class $\mathcal{NC}$? Defend your answers.
50. Write out an iterative Common CRCW PRAM implementation of the very fast $O(\log \log n)$ maximum finding algorithm of Theorem 13 in detail. Use additional arrays for bookkeeping to make it possible in each iteration to look up which part of the input array the processor is assigned and with how many processors it shares work (comparisons). Use n processors at first resulting in $O(n \log \log n)$ work and then improve by blocking to make the implementation work-optimal.

Chapter 2
Shared Memory Parallel Systems and OpenMP

Abstract. The middle third of the lectures on Parallel Computing is concerned with efficient, practical use of real, parallel computing systems with a shared memory through which processor(-core)s can exchange information. It deals first with structure and properties of and inevitable constraints on shared memory systems that must be understood and taken into account in the development of algorithms and implementations for such systems and distinguishes real shared memory systems from idealized constructs like the PRAM. As concrete programming frameworks for such systems that illustrate many fundamental ideas, the library `pthreads` and the programming language extension OpenMP (and, briefly touched upon, Cilk) are treated in detail.

2.1 Fifth Block (1 Lecture)

This block is an introduction to performance-relevant aspects of real, parallel, shared memory systems. The practical questions are how to deal with these aspects in order to get the best possible speed-up out of our parallel algorithms and whether there are architectural obstacles to achieving the linear-speed that our algorithm analysis might suggest. Concrete results are often modest and contradictory to first expectations. Practical Parallel Computing is challenging.

A naïve, parallel shared memory *system model* consists of a (fixed) number of processor-cores p connected to a large (but finite) shared memory. Every processor-core can read/write every location in memory, but memory access is significantly more expensive than performing operations in the processor-core. Furthermore, memory accesses are not uniform: From each processor-core's point of view, some locations can be accessed (much) faster than others. Processors are not synchronized. All these assumptions are in stark contrast to those made for the idealized PRAM model.

In a corresponding, shared memory *programming model*, processes or threads (being executed by the processor-cores) can likewise access objects in a shared memory space. Processes or threads also have their own, private memory spaces that cannot be directly accessed by other processes or threads. There may be more processes or threads than processor-cores. These are scheduled to run by the operating (runtime) system (OS). Processes or threads are not synchronized, but the programming model defines means for synchronization and exchange of information via shared objects. In the next lectures, concrete shared memory programming interfaces will be covered, namely the thread programming models `pthreads` and OpenMP. A programming model in which threads or processes can be executed by any of the processor-cores, as chosen by the OS, is called *Symmetric MultiProcessing* (SMP). We here define SMP as a property of the programming model; there are other uses of the term, as in *Symmetric Multi-Processor* where SMP is rather an architectural property. It can have advantages to leave it to the OS to exploit the processor-cores well, but it can also have drawbacks (for instance related to the cache system, see below). In Parallel Computing, where our system is dedicated (check again Definition 1), we often program with only as many threads or processes as there are processor-cores (dedicated to us for exclusive use) and make sure that each thread or process is executed by one specific core. Ensuring this binding is sometimes called *pinning* and will be discussed briefly in this lecture.

2.1.1 On Caches and Locality

The first difference between real shared memory systems and the naïve model is the existence of caches. A (hardware) *cache* is a small, fast memory close to the processor-core that is used to store frequently used values and, thus, to *amortize* the slow access times to the main memory. For instance, if a value that is read from memory can be reused 10 times, the effective main memory access time is one tenth of what it would have been if the value had to be read at every use. On the other hand, with no reuse, a cache might even introduce overhead in the memory access time. Note that *reuse* is an *algorithmic property*. Indeed, since many algorithms have locality of access properties (see next section), caches help immensely toward sustaining the illusion of fast, uniform memory access (the RAM model). However, some algorithms are truly "random access" and have no locality of accesses. For such algorithms, caches do not help. Instead, the speed of the main memory accesses determines the performance of such algorithms. Examples are graph search problems (DFS, BFS) on very large graphs, where the access pattern is determined by the input graph and the next graph vertex to be accessed would in most cases not be in the cache.

The ratio of the access times between data values fetched from main memory and from cache memory has increased over time. The ratio of accessing data in main memory and accessing data in the fastest cache (lowest level of the cache hierarchy) can easily be a factor of 10 or more. Also nominal processor performance has (up to the early 2000 years) increased dramatically. Effectively, improvements in memory performance have not kept up with improvements in nominal processor performance. As a consequence, caches have grown significantly and now typically take up a substantial amount of space and transistors of the multi-core processor-chip. Also, the cache system itself has become more and more elaborate. The behavior of the cache system of a standard processor can normally not be changed. The development in caches accounted for much of the "free lunch".

2.1.2 Cache System Recap

The cache system of a standard processor does not work on the granularity of single values or words in memory but on larger blocks of memory addresses. Also, caches map addresses (locations) of words in memory to addresses in the cache. The memory can be thought of as being segmented into small *blocks* (a typical block size could be 64 bytes). Each block can be mapped to some cache line. A *cache line*, thus, stores a memory block but also some additional meta information (bits and flags) needed by the cache system. The terms block and cache line are sometimes used interchangeably.

A cache in which each memory block is mapped to one, predetermined cache line is called *directly mapped*. The other extreme, a cache in which each memory block can be mapped to any cache line, is called *fully associative*. A cache where each memory can be mapped to some predetermined, small set of cache lines is called *set associative*. Modern processors have set associative caches with small k-set sizes with $k = 2, 4, 8, \ldots$ and are called k-*way set associative*. A directly mapped cache is a 1-way set associative cache. Direct cache mapping schemes can be easily implemented by means of a few integer division and modulo operations. Associative caches need additional search logic and are more involved.

When a processor reads a word, the memory block to which the word belongs is calculated, and it is checked whether this block is already in the cache. If so, the reference is a *cache hit*, and the word can be read fast from the cache. If not, the reference is a *cache miss*, and the block has to be read from slow memory into a corresponding cache line. In an application, the cache *miss/hit rate* is the ratio of cache hits/misses over a longer sequence of memory references.

On a cache miss, a new block has to be read into a corresponding cache line. Since the cache is finite and much smaller than the main memory, it can easily happen that the cache or cache line is full, in which case there is a conflict and some cache line has to be *evicted*.

There are three types of cache misses. A *compulsory (cold)* cache miss happens when there are no address blocks in the cache. In this case every first reference to some block address will lead to a cache miss. A *capacity miss* happens when the cache (every cache line) is full; it is inevitable that some line is evicted. Finally, a *conflict miss* happens when all cache lines in the set in which the block being read can fit are occupied. Thus, a conflict miss can happen even when the cache as a whole is not full. Conflict misses can be particularly frequent for directly mapped caches, where it is normally easy (if the mapping function is known) to construct cases where every memory access will be a conflict miss. Examples include strided memory accesses with a bad stride determined by the size of the cache. Conflict misses can happen only for directly mapped or set-associative caches. A fully associative cache would have only capacity misses; in general, a capacity miss is also a conflict miss. In a k-way set associative cache, either of the k cache lines can be evicted upon a conflict miss. The choice which cache line to evict is called the *eviction* or *replacement policy*. Typically used replacement policies are *least recently used (LRU)* and *least frequently used (LFU)*. Such concrete details of the processor and memory system may be difficult to find out.

On a write to a memory address, the workings of the cache system are a little more involved. If the block of the address written is already in the cache, it is (must be) overwritten; otherwise, a subsequent read from the cache would deliver an outdated value. If it is not in the cache, either a cache line for that block is *allocated* (thus, possibly resulting in a conflict miss), or the address is updated directly in memory. The former policy is called *write allocate*, the latter *write non-allocate*. On an update to a block already in a cache line, the value written may nevertheless be written to memory, which is called *write-through* cache. The other possibility, that the cache line is not written to memory but kept until it is eventually evicted, is called *write back*.

The *granularity* of the cache system is in units of memory blocks, which each hold several words (in todays processors, typically 64 bytes, i.e., 8 `double` floating point numbers). When an address is read into the cache, the whole memory block to which the address belongs is read. Thus, at the cost of one long read, a whole block of addresses will be in cache and some cache misses can be avoided. Such a cache system can benefit applications with two types of *locality of access*.

An application is said to have *temporal locality* if the content of a memory address is reused several time in brief succession with no or few other uses in between, so that eviction will not happen. An application is said to have *spatial locality* if addresses in the same block are also used (before the cache line is evicted). Again, we stress that access locality is a property of applications and algorithms, and only applications that have this property benefit from the cache system. It is a lucky incident that many applications have either or both temporal and spatial access locality, which is the reason why hardware caching is such a successful idea.

A good computer architecture textbook can provide additional detail on the cache system, some of which may be important for exploiting a given system efficiently, see for instance [25].

2.1.3 Cache System and Performance: Matrix–Matrix Multiplication

Access locality matters: A standard, and highly illustrative example application is matrix–matrix multiplication following the definition of the matrix–matrix product.

The matrix–matrix multiplication problem is to compute for an $m \times l$ input matrix A, and an $l \times n$ input matrix B, in an $m \times n$ output matrix C all product-sums $C[i,j] = \sum_{k=0}^{l-1} A[i,k]B[k,j]$. The straightforward sequential implementation takes three nested loops to do this, assuming that the C matrix has been initialized to all zeros (neutral element for addition). In C, the programming language [68], matrices are stored in row-major order, one row after the other, as in one-dimensional arrays. Thus, the three matrices are given by three one-dimensional arrays a, b and c, which we can cast into matrices (pointers to rows) and address in matrix-notation.

```
double (*A)[l] = (double(*)[l])a;  // indexing in mxl matrix a
double (*B)[n] = (double(*)[n])b;  // indexing in lxn matrix b
double (*C)[n] = (double(*)[n])c;  // indexing in mxn matrix c

...  // allocate, initialize

for (i=0; i<m; i++) {
  for (j=0; j<n; j++) {
    for (k=0; k<l; k++) {
      C[i][j] += A[i][k]*B[k][j];
    }
  }
}
```

The work (sequential time) of this algorithm is clearly $O(mnl)$, and $O(n^3)$ for square matrices of order n. How well does this implementation perform (and compared to what)? In Theorem 3, we observed that, in this implementation, two of the loops have independent iterations and can be parallelized. A further observation is that the three loops can be interchanged and essentially be done in any order.

There are six $3! = 3 \cdot 2 \cdot 1 = 6$ permutations of the three loops. We ran them all on a few standard (Intel, AMD) processors, on medium-large, square matrices of order $n = 1,000$, with and without compiler optimizations (gcc -O3) and for both C int and double matrix elements. The results are surprising, and illustrative (do try this at home)! Briefly, we observed a factor of about $20 - 40$ between the worst and the best loop orders. The worst are the versions where the i loop is the innermost; the best when the j loop is innermost.

The differences can be grossly explained by looking at the cache miss rate. Matrices in C are conventionally stored in row-major order with the elements of each row in consecutive memory addresses and the rows one after the other. We assume that the cache is large enough to hold a single row of each of the three matrices, but no more. In that case, for the worst variants (i-loop innermost), each load of A[i][k] and each write to C[i][j] would result in a cache miss. For the best variants (j-loop innermost), B[k][j] and C[i][j] are both accessed in row-order (best possible spatial locality): the miss rate is determined by the cache line size.

2.1.4 Recursive, Divide-and-Conquer Matrix–Matrix Multiplication

Other approaches to matrix–matrix multiplication solve the problem by doing the multiplications and additions not on individual elements, but instead on smaller submatrices that may fit better in the cache. A recursive formulation of such an approach splits the input matrices A and B roughly in half along both dimensions, recursively multiplies the submatrices, and computes the corresponding submatrices of C by adding the resulting submatrices.

Concretely, write the input matrices A and B as matrices of four submatrices.

$$A = \begin{pmatrix} A_{00} & A_{01} \\ A_{10} & A_{11} \end{pmatrix} \text{ and } B = \begin{pmatrix} B_{00} & B_{01} \\ B_{10} & B_{11} \end{pmatrix} \ .$$

Then

$$C = \begin{pmatrix} C_{00} & C_{01} \\ C_{10} & C_{11} \end{pmatrix} = \begin{pmatrix} A_{00}B_{00} + A_{01}B_{10} & A_{00}B_{01} + A_{01}B_{11} \\ A_{10}B_{00} + A_{11}B_{10} & A_{10}B_{01} + A_{11}B_{11} \end{pmatrix} \ .$$

where the submatrix products $A_{00}B_{00}$ etc. are all computed recursively. It is a good exercise to complete and implement this in C (and compare the performance to the loop-based implementations). Dealing with matrices in C is still cumbersome (see the code snippets for how to declare and allocate efficiently) and care is needed when allocating (and freeing) space for intermediate submatrices. Submatrices are given implicitly by the start and end row and column indices of the original input and output matrices. For performance reasons, we usually look for a good cutoff value; that is, the size of the matrix at which the recursive algorithm stops and the remaining subproblem (a submatrix–matrix multiplication) is solved iteratively. Our implementation performs similarly to the second best iterative implementation (see above), but can still be improved by careful attention to cutoff and memory allocation.

The recursive formulation does 8 (recursive) matrix–matrix multiplications and 4 matrix additions. The total amount of work performed by the algorithm can be estimated by the following recurrence relation:

$$W(n) = 8W(n/2) + O(n^2)$$
$$W(1) = O(1) \ .$$

The recursion depth can be estimated by the following recurrence relation. Here, we are assuming that matrix addition is also done recursively and has depth $O(\log n)$:

$$T(n) = T(n/2) + O(\log n),$$
$$T(1) = O(1) \ .$$

The recurrences are readily solved by the Master Theorem 9 which gives $W(n) = O(n^3)$ (Case 3 with $a = 8, b = 2, d = 2, e = 0$), and $T(n) = O(\log^2 n)$ (Case 2 with $a = 1, b = 1, d = 0, e = 1$). Thus, the work is of the same order as the straightforward implementation, and the length of the critical path(s) if the computation is viewed as a task graph is $O(\log^2 n)$. The parallel time complexity can be improved by doing the matrix additions with more processors in $O(1)$ time steps (depth).

Volker Strassen brilliantly discovered that it is possible to do with only 7 matrix–matrix multiplications and 18 matrix additions [107] which gives rise to an algorithm with $W(n) = O(n^{2.81})$ (Master Theorem again).

2.1.5 Blocked Matrix–Matrix Multiplication

Instead of splitting the matrices recursively, the matrices can be split into submatrices of size $k' \times k''$ up front for some k', k'' and the matrix–matrix multiplication performed as the three-loop iterative algorithm on these submatrices. This gives rise to an implementation with the same work but with 6 nested loops. If the submatrices are small enough to fit in cache, this implementation can perform better than the straightforward implementation. The choice of best k', k'' depends on the size of the cache. Such an algorithm which needs to know the sizes and other properties of the caches is called *cache-aware*, in contrast to a *cache-oblivious algorithm*. Cache-oblivious algorithms can have good or even optimal cache performance, regardless of the concrete size of the cache, which does not have to be known by the algorithm [42, 43].

2.1.6 Multi-core Caches

The cache system in modern multi-core processor systems is structured in several dimensions. First, there is a hierarchy of caches of increasing size, L1, L2, L3 (perhaps more), with L1 the lowest level, closest to the processor-core, smallest, but fastest cache (typically 16 KB), and L3 the *last level cache* (LLC), of typically several MB. The L1 cache is often divided into a data cache and an instruction cache. The memory management system has another cache, the virtual page cache or *translation look-aside buffer* (TLB). The L1 and sometimes also the

L2 caches are *private* to one processor-core (and, therefore, replicated for the number of cores of the multi-core processor), whereas from some level in the hierarchy, the caches are shared among more and more cores For example, the L2 cache might be shared among the cores on a single CPU "socket", the L3 among all cores in the parallel, multi-CPU "socket" system. Processors differ in the way the cache system is structured.

Caches in parallel multi-core systems pose new problems that do not manifest when a single processor-core works in isolation (doing, for instance, matrix–matrix multiplication), related to both semantics and performance.

The first is the *cache coherence problem* among private caches [84]. Assume that a memory block is in the private L1 caches of two different cores. What should happen if one core updates an address in the cache line where the block is kept? If the cache line will *eventually* be updated in the other core's cache to reflect the change, the cache system is said to be *coherent*. If the cache line is *never* updated as a response to the update of the other core, the cache system is *non-coherent*. Updated as a response can mean that either the cache line is indeed modified with the new value or that it is *invalidated* such that the next reference from the other core to the block in the cache line will result in a cache miss. Keeping caches coherent is a non-trivial task that requires a complex algorithm in the processor hardware, a *cache coherence protocol*. This protocol can affect performance by *cache coherence traffic*. The cache coherence protocol of a processor can normally not be influenced by the application programmer (or only with difficulty or to some extent). Cache coherence is a strong property, that guarantees that the processor-cores have a consistent view of individual memory addresses. Let a be an address (location) in memory. A cache coherent system fulfills:

1. If core c writes to a at time t_1 and reads a at a later time $t_2, t_2 > t_1$, and there are no other writes (by c or any other core) to a between t_1 and t_2, then c reads the value written at t_1 (local consistency).
2. If core c_1 writes to a at time t_1 and another core c_2 reads a at a later time $t_2, t_2 > t_1$ and no other core writes to a between t_1 and t_2, then c_2 reads the value written by c_1 at t_1 (update transfer).
3. If core c_1 and core c_2 write to a at the same time, then either the value written by c_1 or the value written by c_2 is stored at a (write consistency, order).

The terms *eventually, later, at the same time* are modalities: Something will happen. When something will happen is not said. Also, note that the term *later* assumes that the read and write *events* corresponding to the memory instructions performed by the core can be ordered relative to some (virtual) global time. It is possible to formulate the cache coherency axioms without any reference to such a virtual, global time.

Current, shared memory multi-core systems are cache coherent, but there have been exceptions (often in the HPC area) and it is frequently debated whether cache coherence is a reasonable expectation for many-core parallel systems with very large numbers of cores [74].

2.1 Fifth Block (1 Lecture)

The second problem is a phenomenon called *false sharing* which is caused by the granularity of the cache system. Recall that cache lines map blocks of consecutive addresses, say 8 double words for cache blocks of 64 bytes. If some block is in the private caches of two or more cores, any update that one core performs to an address of that block will affect the other core's cache, either by an update or by an invalidation of the cache line. In particular, updates to two different addresses &c0 and &c1 in the block mapped by the two cores, will create coherence traffic, even if the two variables c0 and c1 are not in any way related. This can degrade the expected performance significantly [112]. Here are two classical examples of false sharing with OpenMP (see Sect. 2.3). In the first example, the elements in a C structure of integers filling a memory block of 64 bytes (16 integers) are updated by individual threads. In the second example, the 16 integers are stored in an array, and again updated by the individual threads.

```
struct { // 16 int in consecutive block
    int c0, c1, c2, c3, c4, c5, c6, c7, c8, c9, c10,
        c11, c12, c13, c14, c15;
} cl;
int cs[16]; // 16 int in array

// false sharing in struct
#pragma omp parallel
{
  int t = omp_get_thread_num();
  for (i=0; i<r; i++) {
    switch (t) {
    case 0:
      if (i==0) cl.c0 = 0; else cl.c0 += t;
      break;
    case 1:
      if (i==0) cl.c1 = 0; else cl.c1 += t;
      break;
      ...
    }
  }
}
// false sharing in array
#pragma omp parallel
{
  int t = omp_get_thread_num();
  for (i=0; i<r; i++) {
    switch (t) {
    case 0:
      if (i==0) cs[t] = 0; else cs[t] += t;
      break;
    case 1:
      if (i==0) cs[t] = 0; else cs[t] += t;
      break;
      ...
    }
  }
}
```

```
// no false sharing, local variable
#pragma omp parallel
{
  int t = omp_get_thread_num();
  int c; // local variable, hopefully each on own cache line
  for (i=0; i<r; i++) {
    switch (t) {
    case 0:
      if (i==0) c = 0; else c += t;
      break;
    case 1:
      if (i==0) c = 0; else c += t;
      break;
      ...
    }
  }
}
```

In both cases, the updates to the different integer elements are in no way related, but by being on the same cache line, each update will lead to activity in the cache system. The different running times compared to when the updates are performed on a local variable for the threads that (presumably) will not share a cache line can be dramatic and indeed a large factor. It is extremely illustrative to try the example on a multi-core processor using at most 16 OpenMP threads with and without compiler optimizations with a sufficiently large number of iterations (the variable r).

Avoiding false sharing requires attention to allocation and use of variables, attempting to ensure that independent and frequently used and updated variables will always reside on different cache lines. The strategy called *padding* which ensures that there is only one critical variable per cache line by allocating variables at the granularity of the cache block size will work, but is obviously wasteful in memory; in the integer example above by a factor of 16. Using local variables for the threads and updating the global structure or array only once or rarely is often an effective solution as also illustrated in the example. The example shows that variables that are updated frequently by individual threads should not be put too spatially close to each other in arrays. Performance counters introduced in an application to may be an example of such variables.

2.1.7 The Memory System

The cache system is part of the *memory hierarchy* which, for our purposes, will mainly be the large *main memory*, beyond which there are disks and other types of *external memory*. The characteristic of the memory hierarchy is that as memory up (from L1 to L2 to L3 caches to main memory, etc.) in the hierarchy becomes larger and larger, the access times (and often also the granularity of access) also get larger and larger. Any textbook on computer architecture will give approximate ratios of access times and details on granularity [25, 58].

A final, important part of the memory system not mentioned so far, is the *write buffer* in which writes to the main memory are buffered and written to the memory in the pace that the memory system can process updates. The write buffer, as long as it has capacity, makes writes to memory appear fast. Write buffers may be simple FIFO buffers but can also be sorted and usually coalesce writes to the same address. The interaction with the cache system is highly non-trivial, but for single-core processors, write buffers (and caches) were part of the "free lunch" in that they transparently made (most) memory writes (and reads) appear much faster than the actual main memory access times. For multi-core processors, the existence of write buffers is no longer transparent, as will be explained below.

In a hierarchical memory system, memory access times are not uniform. The first time an address or block is accessed, access time depends on where in the hierarchy the address is located, and later accesses may be less expensive due to the cache system. Different addresses, residing in different parts of the hierarchy likewise have different access times. Modern memory systems are highly NUMA.

Memory systems for multi-core parallel systems have additional structure and additional restrictions. In a multi-core CPU, not every core has a direct connection to the main memory. Instead, the cores share a small(er than the number of cores) number of *memory controllers*. The memory is divided into separate banks over the memory controllers. The memory access times for a particular core depend on the "closeness" to the memory controller for the bank in which the accessed address is contained. Access times to different addresses are again non-uniform. The non-uniformity becomes even more prominent for parallel systems consisting of several multi-core CPUs. Access to memory that is controlled by a different CPU than the core issuing the access requires communication between the CPUs and can take significantly longer than access to memory controlled by the CPU of the core.

Not taking the NUMA architecture and behavior of the memory system into account can become a serious performance issue. To some extent, NUMA effects can be alleviated by paying attention to the placement of data used by an application. Partly, this is done automatically by the virtual memory system. An often used virtual memory page allocation policy is the so-called "first touch" policy, by which a virtual memory page will be put physically in the memory bank closest to the core that does the first access to the page. An application can attempt a good placement of virtual memory pages by first "touching" pages (addresses) by the cores that will later most heavily use the pages.

2.1.8 Super-linear Speed-up Caused by the Memory System

Although super-linear (absolute) speed-up was claimed to be impossible, it can nevertheless happen and be observed on real, parallel systems. What is wrong with the argument presented in Sect. 1.2.3?

The argument that linear (perfect) speed-up is best possible assumes that the sequential and parallel system behave identically, in particular that memory accesses behave identically and take the same time for the two systems. Due to the memory hierarchy with large caches, exactly this may not be the case. Assume for simplicity an algorithm that can be parallelized well in the sense that the working set with p processors is $1/p$ of the working set on just one processor. As p grows, the smaller and smaller working set will fit in faster and faster caches in the memory hierarchy, effectively leading the memory accesses of the parallel algorithm to be much faster than for the sequential algorithm. The speed-up can exceed p by a factor equal to the ratio between effective, average sequential memory access time and effective, average parallel memory access time. As a consequence, super-linear speed-up of the form kp with some constant $k > 1$ can indeed be possible and is indeed sometimes observed.

2.1.9 Application Performance and the Memory Hierarchy

The nominal performance of the CPU and processor-cores do not alone determine what the performance of some given application on a system will be. If the memory system is not able to supply data fast enough to the processor-cores, the performance of the memory system (access times) will eventually determine the performance. What "fast enough" is, is determined by the application.

We say that an application is

- *memory-bound*, if the operations to be performed per unit read from or written to the memory take less time than reading/writing a unit from/to memory, and
- *compute-bound*, if the operations to be performed per unit read from or written to the memory take more time than reading/writing a unit from/to memory.

In a memory-bound application, the memory system and memory access times will determine the application's performance including its speed-up, and in a compute bound application the nominal processor performance will determine the application performance. Thus, the application is the determining factor, whether a fast memory or on a fast processor would be the better investment. This trade-off is worked out quantitatively in the so-called *roofline performance model* [123].

2.1.10 Memory Consistency

While the memory hierarchy, cache system, and write buffer are all functionally and semantically transparent for a single core, this is no longer the case when multiple cores are doing Parallel Computing together.

2.1 Fifth Block (1 Lecture)

When a program is executing sequentially, reads and writes to memory addresses (appear to) take place in the execution order of the program's instructions (a read instruction of an address written by an already executed write to that address, will return the value that was written). This is called the *program order* which is assumed in order to prove properties of the program by state invariants. When two programs are being executed concurrently by our asynchronous, parallel, multi-core system, it is (probably) a natural expectation that the outcome will be as if some *interleaving* of the two program order executions has taken place; that is, that memory order follows program order. This is a particular kind of memory consistency which is called *sequential consistency* [71] which would allow us to prove properties of parallel programs much like we do for sequential programs. Only the possibility of different interleavings has to be considered.

Unfortunately, often due to the existence of per-core write buffers and the complex, banked structure of the memory system, modern multi-core systems are *not* sequentially consistent. The consequences of this can be seen by considering the simple example given below. Two cores execute the respective pieces of code. The idea is to protect the code which is in the body of the if-statement such that at most one of the cores will be executing this code body. The two flags f0 and f1 are in shared memory and can be read and written by both cores. The question is whether we can prove the property that "at most one of the two cores can execute the if-body"?

```
int f0, f1; // shared flags initialized to 0

// (thread) code for core 0
f0 = 0;

... // some code

f0 = 1; // core 0 now wants to enter
if (f1==0) {
    ... // protect: core 0 alone?
}
```

```
// (thread) code for core 1
f1 = 0;

... // some code

f1 = 1; // core 1 now wants to enter
if (f0==0) {
    ... // protect: core 1 alone?
}
```

We can try to argue by contradiction. Assume that one of the cores, say core 0, has entered the if-body. In that case, it has set its flag f0 to 1, and read the other flag f1 and found it to be 0. This means that core 1 cannot have reached the instruction where it sets its flag f1 to 1. Therefore, it is not in the if-body and will also not be able to enter, since f0 is still 1. If one of the cores is in the `if`-body, the other cannot be, and the desired property holds. There is no interleaving of the two pieces of code that will lead to both cores being in the `if`-body, and the parallel program has the desired effect under sequential consistency. As can easily be seen, though, it can of course happen that none of the cores enter, but that was not the claim.

The crucial observation is that the argument holds only under the assumption that reads and writes to memory happen in program order. If the memory system is not sequentially consistent, this might not be the case. For instance, with write buffers for the two cores, the following could happen. Both cores execute the initialization of the flags and the 0 values are written to memory. Now the cores proceed, execute their flag updates to 1, but these updates end up in the write buffers. Both cores execute the read of the flag in the `if`-expression, both return 0, and both enter the body, exactly what should not happen. What happened was that the outcome of the write and the read instructions in memory did not follow program order. This is a major problem: How can we reason about parallel programs running on such systems? How can we prove fundamental correctness properties?

Answering these questions is beyond these lectures. The programming interfaces that we will see in the next lectures (`pthreads`, OpenMP) will help us in that they give constructs to ensure guarantees that, at certain points in the execution, the memory is in a well-defined state. The guarantees are typically of the form that updates performed by one thread are, at this specific point in the execution after a particular construct visible to other threads. If used correctly, it will ideally not be necessary to pay attention to the exact behavior of the memory system. To do so, it is important that the hardware provides mechanisms to ensure that operations on memory (read and writes) have indeed been performed. Such mechanisms are operations to *flush* the write (and other) buffers. They are often called *memory fences*. Also so-called *atomic operations* can serve as memory fences. Another means of ensuring some (total) order between codes executed by different threads is to have special, privileged hardware instructions or mechanisms that always execute in order: If instruction `IA` happened before instruction `IB` as observed by some thread, the same order can be observed by (any other) thread(s). Often, *atomic operations* provide such ordering guarantees. In C, ordering between atomic (and other) instructions by different threads can be controlled and enforced, but this is beyond these lectures.

Memory and cache behavior for parallel multi-core systems is intriguingly and painfully intricate. Being aware of the issues is essential for writing correct programs and for getting the best possible performance out of the system at hand. We summarize the two kinds of issues we have discussed:

- The *cache coherence problem*: What happens when different cores read/write the same address?
- The *memory consistency problem*: What happens when different cores read/write different addresses?

2.2 Sixth Block (1–2 Lectures)

pthreads is our first example of a concrete programming interface in the form of a library that implements a shared memory programming model and is intended for running on parallel shared memory systems. pthreads is an early example of a thread programming interface for the C programming language [68], is still widely used, and has been taken as a blueprint for many subsequent thread interfaces(despite issues with correctly realizing a thread interface as a library [22]). Native threads in C are defined since C11 and essentially follow the pthreads interface but are often not supported and seem little used. pthreads is standardized in POSIX (Portable Operating Systems Interface for uniX) as an IEEE standard (IEEE POSIX 1003.1c).

From now on, the lectures will use C as programming language, and the practical projects given in the exercises are intended to be implemented in C. The standard reference text is the book by Kernighan and Ritchie [68]. For learning good programming style in C, the book by Kernighan and Pike [67] is likewise valuable.

2.2.1 pthreads Programming Model

A *thread* is the *smallest unit of execution* that can be scheduled and preempted by the operating system (OS). In C and Unix/Linux, threads live inside *processes* and different threads share information that is global to their process. Threads in C are functions, and shared information are, for instance, global variables, static variables, file pointers, and the heap used for dynamic memory allocation. Threads maintain their own stack. Also, the registers can be thought of as private to a thread. It is also possible to allocate thread-local storage: special memory that is bound to the allocating thread.

The main characteristics of the pthreads programming model are:

1. Fork-join parallelism. A thread can *spawn* any number of new threads (up to system limitations) and wait for completion of any other thread. Threads are referenced by *thread identifiers*. Initially, a single (master) thread is running.
2. Threads are symmetric *peers*. Any thread can wait for the completion of any other thread via the thread identifier.
3. Threads execute functions in the same program (SPMD model) but possibly different functions for different threads (MIMD model). Initially, only one main function thread is active.
4. Threads are scheduled by the operating system (OS) and may or may not run simultaneously on the different cores of the parallel system.
5. There is no implicit synchronization among threads. Threads progress independently of each other.
6. Threads share global objects and information.

7. Coordination constructs for synchronization and updates to shared objects are provided: mutexes, readers-writer locks, condition variables. All updates to shared information must be protected by coordination constructs. Otherwise, the program is illegal and the outcome undefined.

pthreads does not come with a performance model (for analyzing the performance of pthreads programs) and does not come with (much of) a memory model, either (for writing correct programs on hardware memory that is not sequentially consistent). It just requires that updates to shared information are done via the coordination constructs of pthreads.

pthreads allows any number of threads to be spawned (subject to system limitations). Spawning more threads than the number of available cores in the parallel system at hand is called *oversubscription*. It is up to the operating system (OS) how and when threads are scheduled to run (even when there are fewer threads than cores). Threads can also be preempted or suspend themselves, which can, to some extent, be influenced by (non-standard) pthreads functionality that we will not go into in these lectures.

Oversubscription can have advantages (hiding latencies, giving freedom to the OS), but the *pragmatics* of Parallel Computing is mostly to have only as many threads as there are processor-cores and to assume that these threads all run simultaneously.

2.2.2 pthreads in C

pthreads is a library and the thread functionality can be used by linking the code against the pthreads library. C code using pthreads must include the function prototype header with the #include <pthread.h> preprocessor directive. All functions and predefined objects relevant to pthreads are prefixed with pthread_ which identifies the pthreads "name space". With gcc, code can be compiled using the -pthread option which enables linking against the library.

Most pthreads functions return an error code, and it is good practice to check the error code (which is often not done). The error code 0 indicates success.

2.2.3 Creating Threads

When a C program with pthreads is started, the main() function is the only ("master") thread running. The master thread and any other thread can start new threads and wait for termination of any other thread. A thread is identified by an *opaque* object of type pthread_t which is set by the creation call and used to reference the now started thread. Such objects can be used and manipulated exclusively through defined operations; their implementation and structures is neither defined nor accessible. Thread identifiers can be compared for equality but otherwise not manipulated.

2.2 Sixth Block (1–2 Lectures)

```
int pthread_create(pthread_t *thread,
                   const pthread_attr_t *attr,
                   void *(*start_func)(void *), void *arg);
void pthread_exit(void *retval);
int pthread_join(pthread_t thread, void **retval);

pthread_t pthread_self(void); // return own thread identifier
int pthread_equal(pthread_t t1, pthread_t t2);
```

Code that is to run as a thread must be written as a C function with a single `void*` pointer argument. This pointer is used to point to a structure holding the actual arguments to the thread. The thread function will, therefore, often cast this void pointer to something more meaningful. The pointer to the function together with a pointer to the actual arguments are given as arguments to the thread creation call. Attributes will not be covered in these lectures; but they can be used to control the way the thread is to run. In most cases, NULL can just be given as the attribute argument. C programming is brittle: It is easy to make mistakes with function and argument pointers and such mistakes have grave consequences by leading to memory corruption and program crashes often much later than the call where the mistake was made.

Once a thread has been created, the corresponding function runs on its own, asynchronously and concurrently with other activities, possibly on its own processor-core. When a thread function comes to an end, it should terminate itself by making the exit call. This call also takes a pointer that can point to information to be given back to the thread that intercepts the terminating thread. If return information is used, it must be allocated on the heap and definitely not on the stack where it will sooner or later disappear. Waiting for a thread to exit is done by a join call, which will update its `void**` pointer argument to point to the structure returned by the exiting thread. Thread identifiers can be exchanged freely between threads, and any thread can wait for any other thread to finish. In that sense, threads are "peers".

The following simple, almost full-fledged `pthreads` program shows how to start p threads one after the other and assign each a "rank" (a unique identifier between 0 and p-1) by passing a corresponding argument.

```
#include <pthread.h>

typedef struct { // the real arguments to thread functions
  int rank;
} realargs;

void *hello(void *arguments) {
  realargs *args = (realargs*)arguments;

  // a classic race; try it, and see later
  printf("Thread %d starting\n",args->rank);
  pthread_exit(NULL);
}
```

```
int main(int argc, char *argv[])
{
  int p = ...; // (small) number of threads
  int i;

  pthread_t thread[p];
  realargs threadargs[p];

  // create and start the threads
  for (i=0; i<p; i++) {
    threadargs[i].rank = i;
    pthread_create(&thread[i],NULL,hello,&threadargs[i]);
  }

  // wait for termination and intercept return values (none)
  for (i=0; i<p; i++) {
    pthread_join(thread[i],NULL);
  }

  return 0;
}
```

The program snippet illustrates how threads are created and started, but is technically wrong. One problem is the call to the C `printf()` library function in the thread function `hello()` which will possibly lead to the threads competing for a resource: the printing device. This is a first example of a classic *race condition* (for more, see later). The general problem is that a function called from a thread may not work properly when other threads can also be calling the function; the calls to such functions are *unsafe*. Conversely, a function that can be called concurrently by any number of threads is called *thread safe*. Pure functions without side effects (for instance, not updating shared state in the form of global or static variables) are thread safe. System functions may or may not be thread safe, and one should always check; a notoriously thread unsafe function is the "old" random number generator `rand()`.

The binding of threads to processor cores can be controlled by the following (non-standard) `pthreads` functions. A `cpuset` is a set data structure (bit vector) representing a set of possible physical cores, numbered consecutively and corresponding to the numbering of the cores on the shared memory system. They should be manipulated through predefined macros.

```
int pthread_setaffinity_np(pthread_t thread,
                           size_t cpusetsize,
                           const cpu_set_t *cpuset);
int pthread_getaffinity_np(pthread_t thread,
                           size_t cpusetsize,
                           cpu_set_t *cpuset);
```

2.2.4 Loops of Independent Iterations in pthreads

The parallel patterns we have seen in the previous lectures (Sect. 1.3) can all be implemented with pthreads. A loop of independent iterations, for instance, can be parallelized by assigning each thread a consecutive range of iterations. The thread function performs the iterations, taking the arguments for the loop from a suitable argument data structure. This is shown in the code snippet below which reuses the linear thread creation and argument structure from the previous example.

```
#include <pthread.h>

typedef struct {
  int rank;
  int size;
  int n;
  int *a; // pointer to shared array
} realargs; // argument structure

void *loop(void *arguments)
{
  realargs *args = (realargs*)arguments;

  int i;
  int s, nn; // start index and number of iterations
  int *a = args->a;

  nn = (args->n)/args->size;
  s  = nn*args->rank;
  if (args->rank==args->size-1) nn = args->n-s;

  // part of loop for thread
  for (i=s; i<s+nn; i++) a[i] = i; // some loop body

  pthread_exit(NULL);
}
```

The thread function uses the total number of iterations to be parallelized to compute the iteration start and the local number of consecutive iterations that the thread will perform. This is taken as $\lfloor n/p \rfloor$ where n is the number of iterations and p the number of threads in the field size. The last thread will perform the remaining $n - (p-1)\lfloor n/p \rfloor$ iterations.

2.2.5 Race Conditions and Data Races

In a thread model with shared memory, executed on a shared memory multi-core system, it is possible for different threads to access and update shared variables. Since threads may execute concurrently, such updates may happen "at the same time". In such a situation the outcome is (for almost all systems, and we will assume this behavior) the update by one of the threads and not something in between (also not: no update). But which thread succeeds with its update is undetermined. We say that the outcome of a concurrent update to a shared variable is *non-deterministic*, and such non-determinism may affect the final result of the whole program, an often undesirable situation. Since threads execute asynchronously (our thread model makes this assumption: no synchronicity among threads, unlike the PRAM model), the order of updates to shared objects is not defined, and either thread can be the "last" thread to perform an update. Depending on the memory system behavior, updates may or may not become visible to the other threads in the order in which they were performed. Thread programs are inherently non-deterministic. In order to write correct programs that give a determinate, final output, we need to be able to deal with and restrict the non-determinism in updates and accesses to shared variables and objects.

Such non-deterministic updates to shared objects and variables in a program which can lead to different results of the program, some of which are not correct, are commonly called *race conditions*. It is important to keep in mind that asynchronous parallel programs are inherently non-deterministic. Non-determinism is the price for the potential performance benefits of asynchronous parallelism. Also, concurrent updates may not always lead to different, or wrong, final results.

Any thread programming model needs either means to reason about non-deterministic executions and updates to shared objects or means to restrict and control non-determinism wherever it is crucial that updates are done in a certain, specific order, or both.

A particular kind of race condition is the *data race*. Technically, a *data race* is a situation where two or more threads access a shared object, and at least one of the accesses is an update (write). It is undecidable to determine whether a program will have a data race, so automatically finding *all* race conditions (by a compiler) is algorithmically impossible.

Thread models like `pthreads` and OpenMP forbid uncontrolled, concurrent updates to shared variables and objects. In particular, they forbid data races. Instead, they have constructs for threads to access and update shared objects that make concurrent updates well-behaved and technically eliminate data races. Such programs are informally called *data race free* here. A way to look at such constructs is that they restrict the possible interleavings of asynchronous thread executions. We will see the main `pthreads` construct in the next section.

The following simple example shows why data races can be harmful and lead to race conditions. Let a be a variable that is shared among many threads. The threads all execute the following simple (but composite) update:

```
a = a+27;
```

With typical processors and instruction sets, this simple expression evaluation and assignment translates into at least three instructions, namely (1) a load of a into a register, (2) an addition with a constant (here 27), and (3) a write back to the location of a. The sequential semantics of the statement is that a is incremented by the constant 27. When several threads execute this code, the following can easily happen: The threads all read the old value of a, all perform the addition in their respective (private, non-shared) registers and then race on the update to a. Instead of each thread incrementing by 27, only one increment will effectively have happened. With many threads, many outputs of this sort are possible (any increment by some multiple k of 27 with $k < p$), most of which are probably not what was intended (intended was probably an increment of $27p$ when all threads have finished).

Not all updates that are technically data races may be harmful. For instance, it might be unproblematic if all threads write the same value to the shared variable, as allowed by the Common CRCW PRAM, for instance. In the above example, it was harmful and leading to very unintended outcome.

2.2.6 Critical Sections, Mutual Exclusion, Locks

pthreads and OpenMP programs (see later) with data races are technically not correct, and programs with updates to shared variables by several threads that could happen concurrently are illegal. pthreads provides constructs to control accesses and updates to shared variables and shared objects.

The problem in the example above is not so much the individual data races on the shared variable a but rather the whole sequence of read-modify-write instructions involved in the update. When two threads at the same time come to this little piece of code, what is required for the intended outcome is that either of the threads runs entirely before the other. In order to get the (presumably) desired behavior, we need to exclude the interleavings of read and modify and write sketched above from the possible interleavings of the two thread executions.

A piece of code that should not be executed concurrently by several threads is commonly called a *critical section*. A thread running code in a critical section must exclude other threads from doing so. Threads need to cooperate to ensure this, and guaranteeing that a critical section is indeed being executed by at most one thread is commonly called *mutual exclusion*. The *mutual exclusion problem* is to guarantee mutual exclusion and is not a trivial problem. It is not the purpose of this lecture to go into solutions or algorithms for the mutual exclusion problem which has a long and ongoing history [59, 91]. Note that the code in a thread's critical section must not necessarily be the same for all threads. Rather, a critical section is a piece of a thread's code that should not be executed concurrently, in parallel with certain other pieces of code of other threads. The mutual exclusion problem is to ensure that this is the case.

A programming model mechanism that guarantees mutual exclusion is commonly called a *lock*. Locks provide mutual exclusion as follows: A thread that wants to enter a critical section tries to *acquire* the corresponding lock. If it succeeds, the thread is on its own in the critical section and does what it needs to do, typically reading and writing shared variables. No other thread can acquire the same lock as long as it is being held. Therefore, there can be no data races on objects updated by the thread having the lock as long as they are not referenced by threads not being in the critical section. When finished, the thread exits the critical section by *releasing* the lock. Then, other threads can again enter the critical section by trying to acquire the lock. If a thread tries to but cannot acquire the lock, it cannot progress and is *blocked* waiting for the lock to become available. The lock acquire and release operations are often also called just *lock* and *unlock*. A lock is a brittle construct: If the unlock is forgotten or does not happen, either due to program logic or because the corresponding thread is not progressing (suspended by the operating system, sleeping or entirely gone), other threads wanting to acquire the lock will become blocked and eventually the whole program execution will grind to a stand-still (allusion intended). This is the dreaded *deadlock* situation. In general, a deadlock in Parallel Computing or Concurrent Computing is the following: A thread, a process or a processor-core needs a resource from another thread before it can proceed which in turn needs a resource from another thread and so on infinitely or cyclically, preventing the thread, process or core from ever getting the resource. All waiting entities are stuck forever or until the deadlock is dissolved from the outside by an arbiter that breaks the dependencies.

Apart from guaranteeing mutual exclusion (at most one thread at a time can hold a given lock), the fundamental property of a lock is that it must itself be *deadlock free*. This means that whenever a number of threads (one, some, many, all) is trying to acquire the lock, *eventually* one thread *must* succeed and get the lock. A perhaps desirable property is that whenever a specific thread is trying to acquire the lock it will *eventually* acquire the lock, no matter which other threads are also trying to acquire the lock. A lock is said to be *starvation free* if it has this property that no thread can be *starved* forever. Locks are said to be *fair* if they provide stronger starvation freedom guarantees, like that when a thread is trying to acquire a lock "before" some other thread it will also get the lock before that other thread (whatever "before" means).

In `pthreads` terminology, a lock is called a *mutex* (for mutual exclusion) and shared objects are only allowed to be updated by acquiring a mutex before doing so (concurrent reading alone is allowed). A mutex is identified by an opaque `pthread_mutex_t` type. Mutexes must be initialized either statically (by assigning `PTHREAD_MUTEX_INITIALIZER`) or dynamically before they can be used.

`pthreads` mutexes guarantee mutual exclusion and are deadlock free, but they are *not* starvation free. In addition, they guarantee that all memory updates performed by a thread in the critical section before the release of the mutex will be visible to any other thread acquiring the lock afterwards. This is the `pthreads` memory model.

```
int  pthread_mutex_init(pthread_mutex_t *restrict mutex,
                const pthread_mutexattr_t
                *restrict attr);
int  pthread_mutex_destroy(pthread_mutex_t *mutex);

int  pthread_mutex_lock(pthread_mutex_t *mutex);
int  pthread_mutex_trylock(pthread_mutex_t *mutex);
int  pthread_mutex_unlock(pthread_mutex_t *mutex);
```

2.2 Sixth Block (1–2 Lectures)

```
int pthread_rwlock_init(pthread_rwlock_t *restrict rwlock,
                        const pthread_rwlockattr_t
                        *restrict attr);
int pthread_rwlock_destroy(pthread_rwlock_t *rwlock);

int pthread_rwlock_rdlock(pthread_rwlock_t *rwlock);
int pthread_rwlock_tryrdlock(pthread_rwlock_t *rwlock);

int pthread_rwlock_trywrlock(pthread_rwlock_t *rwlock);
int pthread_rwlock_wrlock(pthread_rwlock_t *rwlock);

int pthread_rwlock_unlock(pthread_rwlock_t *rwlock);
```

The data race on the shared, global variable a in the a = a+27; example from above is correctly avoided by protecting this critical section by a mutex.

```
// A lock shared by all threads
pthread_mutex_t lock = PTHREAD_MUTEX_INITIALIZER;

pthread_mutex_lock(&lock);   // acquire lock
a = a+27;  // thread alone in critical section, no race
pthread_mutex_unlock(&lock); // release lock
```

Threads that try to execute the update concurrently will *serialize*: One thread after the other will be allowed to enter the critical section. If there is repeated competition for acquiring the lock, it may even happen (if the mutex is not starvation free) that some thread will never enter the critical section. If this happens, such a thread does not contribute to the parallel computation any more and the speed-up that might be possible is reduced accordingly. A lock for which many threads are competing is said to be *contended*. Contention is always a source of slowdown, since threads are waiting for their critical section instead of doing useful work.

To allow threads to do something useful in case of contention, many lock models offer a *try-lock* operation. Try-lock tries to acquire the lock, and if the lock is not already held by some other thread, it immediately acquires the lock. If the lock is held by another thread, try-lock returns with a condition code (**false**). It is, clearly, essential that try-lock acquires the lock immediately when the lock is free instead of returning with a condition code. This would be useless, since trying to acquire the lock after checking the condition code could well fail because of some other thread having taken the lock in-between. A great application of try-lock to the implementation of a concurrent priority queue with certain guarantees can be found in [122].

A means of alleviating lock serialization effects and slowdown takes advantage of the situation that accesses and updates to shared objects are often asymmetric. In some (many) critical sections, shared variables are only read, while in other (fewer) also actual updates (writes) have to be performed. All the threads that only need to read some shared object can do this concurrently, in parallel. For the writes, full mutual exclusion is needed, and all other reading as well as writing threads must be excluded from the critical section. *Readers-writer* locks that are found in many thread programming models, provide this functionality. Readers-writer locks have a lock acquire operation for threads that want to read (concurrently), and another lock acquire operation for threads that want to write under strict mutual exclusion. It is the programmer's responsibility to make sure that no updates (to shared variables) are performed in the critical sections when the lock is acquired for reading.

There are many ideas and algorithms for implementing locks (not treated in this lecture). An important pragmatic issue is how waiting for a lock is implemented and how waiting (blocking) interacts with the operating system (OS). In a *spin lock*, the processor-core executing the blocked thread actively keeps testing (spinning) for the lock to become free. That is, the processor-core is kept busy for as long as the thread is blocked on the lock acquire operation. Acquiring the lock is fast for spin locks, and this implementation is typically advantageous when the critical sections are short and there is no thread oversubscription. With a *blocking lock*, the thread that is waiting for the lock to become free is suspended by the OS, and the processor-core that was executing the thread is free to do something else. It could, for instance, wake up and run another thread. Blocking locks may be advantageous when the shared memory system is oversubscribed and the lock waiting times can be productively spent by the core doing something else. In **pthreads**, spinning behavior can be requested explicitly by using spin locks. This (strange) **pthreads** design decision means that code has to be rewritten, if spin locks are desired.

```
int pthread_spin_destroy(pthread_spinlock_t *lock);
int pthread_spin_init(pthread_spinlock_t *lock, int pshared);

int pthread_spin_lock(pthread_spinlock_t *lock);
int pthread_spin_trylock(pthread_spinlock_t *lock);
int pthread_spin_unlock(pthread_spinlock_t *lock);
```

2.2.7 Flexibility in Critical Sections with Condition Variables

Since **pthreads** programs must be data race free, locks (or other constructs, see the following) must be used when transferring information between threads. For instance, a value updated by a writing thread may be needed by several reading threads. The following first solution is obviously wrong since it easily leads to a deadlock: A reading thread entering its critical section before the writer thread will stay in the while-loop and prevent the writer thread from ever setting the **written** flag.

```
// reader threads
pthread_rwlock_rdlock(&lock);
while (!written);
a = b; // information transfer
pthread_rwlock_unlock(&lock);
```

2.2 Sixth Block (1–2 Lectures)

```
// writer thread
pthread_rwlock_wrlock(&lock);
b = ... ; // update
written = 1;
pthread_rwlock_unlock(&lock);
```

This situation is quite common. A thread having entered its critical section cannot proceed before some condition that involves other threads to enter their critical section is fulfilled. A solution that sometimes works is for the thread to leave the critical section and try again later, hoping for the condition to have been fulfilled. A more elegant solution involves so-called condition variables. A *condition variable* is an object associated with a mutex variable. A thread can perform a *wait* on a condition variable, meaning that the thread will be suspended and effectively out of the critical section (the lock is released) until some other thread performs a *signal* operation on the condition variable. When a waiting thread receives the signal and is woken up, the signalling thread will have left the critical section, such that mutual exclusion in the critical section is always maintained. More threads, for instance, readers as in the example above, can wait on the same condition variable. A single signal operation will wake up either of the threads: `pthreads` provides no fairness guarantee and no guarantee that a thread is not starved. To wake up all waiting threads, one after the other, a *broadcast* is also provided. Note that mutual exclusion is always maintained, one thread after the other will be in the critical section. A signal operation on a condition variable where no thread is suspended is lost. This is different from the *semaphore*, another well-known primitive synchronization mechanism, that is not natively supported with `pthreads`, though. The standard usage pattern for locks with condition variables is called a *monitor* [61]. Some thread models and interfaces support monitors directly, `pthreads` only indirectly via the condition variable mechanism.

```
int pthread_cond_destroy(pthread_cond_t *cond);
int pthread_cond_init(pthread_cond_t *restrict cond,
                      const pthread_condattr_t *restrict attr);

int pthread_cond_wait(pthread_cond_t *restrict cond,
                      pthread_mutex_t *restrict mutex);
int pthread_cond_signal(pthread_cond_t *cond);
int pthread_cond_broadcast(pthread_cond_t *cond);
```

The problem with the readers-writer lock transfer of information that we saw above can now be solved with condition variables.

```
pthread_cond_t data = PTHREAD_COND_INITIALIZER; // shared

// reader threads
pthread_lock(&lock);
while (!written) {
  pthread_cond_wait(&data,&lock);
}
a = b;
pthread_unlock(&lock);
```

```
// writer thread
pthread_lock(&lock);
b = ... ;
written = 1;
pthread_cond_broadcast(&data);
pthread_unlock(&lock);
```

Often the condition variable mechanism permits so-called *spurious signals* or *spurious wakeups*. These are false or outdated signals being sent to a waiting thread (this could be for implementation reasons). With `pthreads` this can indeed be the case. It is, therefore, good and common practice to always recheck the desired condition (in the example the `written` flag) when a sleeping thread is woken up. Condition variables can easily lead to deadlocks if used wrongly; there must always be a thread and a condition that signals and wakes up waiting threads.

2.2.8 Versatile Locks from Simpler Ones

The condition variable mechanism (monitor) is a powerful addition to the simple mutexes. For instance, the more versatile readers-writer locks can be constructed from simple locks using condition variables. Also, different priority schemes (writer or readers preferred) can be implemented. The following code example gives one such implementation.

```
typedef struct {
  int readers;   // count number of readers
  int waiting, writer;  // writers waiting
  pthread_cond_t read_ok, write_ok;
  pthread_mutex_t gatekeeper;
} rwlock_t;

void lock_read(rwlock_t *rwlock)
{
  pthread_mutex_lock(&rwlock->gatekeeper);
  while (rwlock->waiting>0||rwlock->writer) {
    pthread_cond_wait(&rwlock->read_ok,&rwlock->gatekeeper);
  }
  // acquired for read (possibly more than one)
  rwlock->readers++;
  pthread_mutex_unlock(&rwlock->gatekeeper);
  assert(rwlock->writer==0);  // at any time before unlock
}

void lock_write(rwlock_t *rwlock)
{
  pthread_mutex_lock(&rwlock->gatekeeper);
  rwlock->waiting++;
  while (rwlock->writer||rwlock->readers>0) {
    pthread_cond_wait(&rwlock->write_ok,&rwlock->gatekeeper);
  }
```

```
  // acquired for write (exactly one)
  rwlock->waiting--;
  rwlock->writer = 1;
  pthread_mutex_unlock(&rwlock->gatekeeper);
  assert(rwlock->readers==0); // at any time before unlock
}

void unlock_readwrite(rwlock_t *rwlock)
{
  pthread_mutex_lock(&rwlock->gatekeeper);
  if (rwlock->writer) rwlock->writer = 0; // done writing
  else rwlock->readers--; // one less reading
  pthread_mutex_unlock(&rwlock->gatekeeper);

  // resume possibly waiting threads
  if (rwlock->readers==0&&rwlock->waiting>0) {
    // wake up writer
    pthread_cond_signal(&rwlock->write_ok);
  } else {
    // wake up readers
    pthread_cond_broadcast(&rwlock->read_ok);
  }
}
```

The functions implement the functionality of the `pthreads` readers-writer lock. A simple `gatekeeper` lock is used to provide mutual exclusion when updating the variables that control the behavior of the readers-writer lock: A number of readers are allowed to acquire the lock for reading, but only one writing thread for writing. The number of threads waiting to acquire the lock for writing also needs to be kept track of. By unlock, if there are no readers and at least one thread waiting to acquire the lock for writing, one waiting thread is signalled. Otherwise, all possibly waiting reader threads are notified by a `pthread_cond_broadcast()`. When used in a multi-threaded application, our readers-writer lock is declared as a shared variable and the fields determining the waiting conditions initialized to zero. Also, the `gatekeeper` mutex must be initialized.

```
// declaration and initialization of readers-writer lock
rwlock_t lock;
lock.readers = 0;
lock.waiting = 0;
lock.writer  = 0;

pthread_mutex_init(&lock.gatekeeper,NULL);
pthread_cond_init(&lock.read_ok,NULL);
pthread_cond_init(&lock.write_ok,NULL);
```

A *thread barrier* is a construct which makes it possible for a thread to define a point in the execution beyond which it cannot progress before a certain number of other threads have reached the barrier synchronization point (see Sect. 1.3.14). `pthreads` defines function interfaces for such barriers; the count is the number of threads required to reach the barrier point. Each barrier (there can be several) is identified by an opaque `pthread_barrier_t` object, which needs to be shared among the threads.

```
int pthread_barrier_init(pthread_barrier_t *restrict barrier,
                         const pthread_barrierattr_t
                         *restrict attr,
                         unsigned count);
int pthread_barrier_destroy(pthread_barrier_t *barrier);

int pthread_barrier_wait(pthread_barrier_t *barrier);
```

Barriers can also trivially be constructed from mutexes with condition variables. Implementing efficient shared memory barriers is non-trivial, however, see for instance [78].

A final, common pattern is concurrent initialization, where one of the threads (the "first") should carry out some initialization code (function). This pattern can easily be implemented with mutexes, but `pthreads` provides a shorthand.

```
int pthread_once(pthread_once_t *once_control,
                 void (*init_func)(void));
```

2.2.9 Locks in Data Structures

Sequential data structures with their particular semantics and operations are often used in a parallel setting and this can make a lot of sense. Threads might want to share a linked list, for instance, used as the implementation of a set data structure with search, insert, and delete operations, or a stack, or a queue, or a hash map, or a priority queue, etc., and use the data structure operations as the means for communication and synchronization between threads. As long as the data structure does not become a sequential bottleneck (Amdahl's Law) by being too slow or by leading to thread serialization, shared data structures with sequential semantics can be helpful in the implementation of parallel algorithms.

The trivial way of making a(ny) sequential data structure useful in a parallel algorithm, is to use a global lock to protect all data structure operations. Mutual exclusion will ensure that the operations on the data structure are done one after the other. A collection of concurrent operations will thus execute in some order and behave according to the sequential semantics. The already available sequential implementation, perhaps complex and highly tuned, can be used right away, but the price is that all concurrent operations on the data structure will serialize. This can limit the possible speed-up of the algorithm. Thus, this solution is often not good enough. For data structures with read and write operations, like the set, which supports search (read) and insert/delete (write) operations, the more versatile readers-writer locks can alleviate some of the drawbacks. Concurrent, perhaps frequent read operations will have real parallelism, and only the write operations may be bottleneck operations.

When this is, for performance reasons or others, not acceptable, data structures and algorithms have to be rethought into more *concurrent data structures*.

Some data structures, like linked lists, easily allow for implementations with more "fine-grained", hands-over locking. The idea is to use a lock for each list element. As the list is being traversed, only the locks for the current element and its successor are acquired. Having locks on two successive elements makes it possible to link out an element or insert a new element between the two under mutual exclusion and thus without interference from other threads. For long lists, this makes it possible for many threads to perform operations on different parts of the lists. But since a thread having acquired the locks on elements at the front of the list will prevent other threads from traversing the list past this point, the improvement of this locking scheme is modest.

Developing data structures, even with the use of locks, that allow for a large amount of concurrent uses by many threads, is highly non-trivial and beyond the scope of these lectures. The point we make here is that locks can still be useful, but need to be used carefully (localized, short critical sections), and that in such cases a large number of locks will have to be used. Therefore, the (space) efficiency of the lock implementations provided by pthreads, OpenMP and other thread models is highly important.

2.2.10 Problems with Locks

Locks, semaphores, and similar constructs are Concurrent Computing constructs that were not designed for Parallel Computing with large numbers of active processing elements (threads, processes, processor-cores). The typically (inherently) limited scalability is a reason to use them sparingly. Locks have other problems:

- Deadlocks can easily be introduced by design (errors). For instance, in a program with two or more locks L_1 and L_2 (like the linked list with hands-over locking), one thread may try to acquire the locks in the order L_1, L_2 and some other thread in the order L_2, L_1. If the two threads execute roughly at the same time, they will both come to a point where they cannot proceed, because the lock that each is trying to acquire is already taken by the other thread. This sounds trivial to avoid, but it is not. The code for the two different threads may be in different parts of a large software package, may perhaps not written by the same people etc. Each of the code pieces can in itself be correct so that when tested in isolation, the deadlock situation does not occur. When the pieces of code are run together, the program deadlocks. In that sense, locks are not a mechanism that supports modular software development. A deadlock is always deadly, it proliferates and eventually the whole application cannot complete, because the deadlocked threads will not complete. To avoid deadlocks when using multiple locks, locks can be acquired in an agreed upon order (stratified locking) or release-temporary backoff-acquire techniques be used. With multiple locks, the try-lock operation can often be useful.
- A special case of deadlock can occur when a thread having acquired lock L tries to acquire L again. This may deadlock. So-called *recursive locks* (or

nested locks) explicitly allow a thread having a lock to acquire the lock again. The number of unlock calls may have to match the number of lock calls. `pthreads` makes it possible to initialize recursive locks by the use of a `PTHREAD_MUTEX_RECURSIVE` attribute.
- Locks that protect long critical sections lead to possibly harmful serialization which can severely degrade performance (Amdahl's Law).
- Infinitely long critical sections, for instance, a thread crashing in the critical section, lead to deadlocks. Locks are not fault-tolerant.
- Since locks are often not fair, threads can be starved and actually not be contributing to the progress of the parallel algorithm.
- When threads have priorities (possible with `pthreads`, but not covered in these lectures) locks can lead to the effect that a lower prioritized thread prevents a thread with high priority from running, even when this would have been possible. The phenomenon is called *priority inversion*.

2.2.11 Atomic Operations

The problem with the `a = a+27;` example was that the sequence of instructions in one thread's complex assignment operation (load, compute, store) could be interleaved with instructions executed by another thread as well as the data race on the final update to `a` (all threads writing). To prevent such interleavings, the assignment should be executed as an *atomic*, that is as an indivisible, unit. Mutual exclusion with locks is one way of guaranteeing atomic execution of the sequence of instructions. The drawback is that during the execution of the lock code, no other threads can do anything with the variables that are protected by the lock.

Another way of ensuring atomic execution of compound operations is offered by hardware implemented *atomic operations*. An atomic operation carries out a complex (but still relatively simple) compound instruction as a unit that cannot be interfered with by other threads or processor-cores. One kind of atomic operation is, for instance, the Fetch-And-Add (FAA) instruction which can implement the particular `a = a+27;` assignment as a single, indivisible instruction.

Special, *atomic instructions* for performing atomic operations are offered by all modern multi-core processors and systems. They operate on one or more memory words given by their memory addresses, sometimes with an additional value operand, and produce a result. Memory words that are operated on by atomic instructions are called and must be atomic. Typical atomic instructions are for instance:

1. *Test-And-Set* (TAS): On an atomic memory word, returns the contents of the address and updates the contents to 1 (**true**).
2. *Fetch-And-Add* (FAA), *Fetch-And-Increment* (FAI): On an atomic memory word, returns the contents of the address and updates the memory word by either adding a given value (FAA) or incrementing it by one (FAI). More

generally, *Fetch-And-Operate* (FAO) updates the memory word by a simple operation (logical, for instance) while returning the original contents of the address.
3. *Exchange* (EXCHG): On an atomic memory word, returns the contents of the address and replaces the contents with the given value.
4. *Swap* (SWAP): Swaps the contents of two atomic memory addresses.
5. *Compare-And-Swap* (CAS): On an atomic memory word, checks whether the contents equals a given expected value and if so replaces the contents with a new update value, and returns **true**. If the contents are not equal to the expected value, **false** is returned and the contents are not changed.
6. *Compare-Exchange* (CEX): On an atomic memory word, checks whether the contents equals a given expected value and if so replaces the contents with a new update value, and returns **true**. If the contents are not equal to the expected value, **false** is returned and the contents of the atomic memory word copied back to the expected value (given as a reference).

Beyond this lecture: These atomic operations form a hierarchy (hence the numbering, with CAS and CEX being what are called universal and the most powerful) characterized by the power of what they can do [59], more precisely for how many threads they can solve the so-called *consensus problem*. All these operations are quite natural and helpful in many contexts. For instance, the atomic Test-And-Set (TAS) instruction is exactly what is needed to implement a lock.

Atomic operations are indeed instructions like all other processor instructions, meaning that they complete in some finite, bounded number of clock cycles, regardless of what other processor-cores might be doing (even executing atomic operations). This essential property is called *wait-freeness*. This does not mean that atomic operations are always fast. Mostly, they are not when compared to other operations offered by the processor instruction set. On the contrary: Atomic operations are expensive, since they need to interact with the cache and memory system (locking and/or invalidating cache lines, flushing the write buffer). So, like locks, they should be used sparingly. But in contrast to locks, use of atomic operations cannot lead to deadlocks. A crashed (failed) thread will not affect the ability of the other threads to continue and make progress. Optimistically, we might assume that atomic operations are constant time $O(1)$ operations with relatively small constants, but bounded does not always mean constant.

In the `stdatomic.h` header for C, the following atomic operations are standardized for C; however, there is more to the C atomics than explained here (ordering guarantees and memory model, for instance). These operations work on atomic integer types. There is such an atomic integer type defined in the header for all C integer types, e.g., `atomic_bool`, `atomic_char`, `atomic_short`, `atomic_int`, `atomic_long`, etc. There is also a special, atomic flag type, `atomic_flag`. We list the operations as defined for atomic integers (they are also defined for other C word datatypes).

```
atomic_init(atomic_int *object, int value);
int atomic_load(atomic_int *object);
void atomic_store(atomic_int *object, int desired);

// TAS
_Bool atomic_flag_test_and_set(volatile atomic_flag* obj);
void atomic_flag_clear(volatile atomic_flag* obj);

// FAA, FAO
int atomic_fetch_add(atomic_int *object, int operand);
int atomic_fetch_and(atomic_int *object, int operand);
int atomic_fetch_or(atomic_int *object, int operand);
int atomic_fetch_sub(atomic_int *object, int operand);
int atomic_fetch_xor(atomic_int *object, int operand);
// EXCHG
int atomic_exchange(atomic_int *object, int desired);

// CEX
_Bool atomic_compare_exchange_strong(atomic_int *object,
                                     int *expected,
                                     int desired);
_Bool atomic_compare_exchange_weak(atomic_int *object,
                                   int *expected,
                                   int desired);

// Are operations lock-free?
_Bool atomic_is_lock_free(const volatile A* obj);

void atomic_thread_fence(memory_order order); // memory fence
```

Here is an interesting example: A number of threads update three counters stored in a global C structure. One counter is updated non-atomically, the two others with the `atomic_fetch_add` instruction. After execution, it will not necessarily hold that, for instance, `cnt0==cnt1` or `cnt0==cnt2`. And even if each of the two counters `cnt1` and `cnt2` are updated atomically, the compound update of both is not, therefore neither of the stated assertions will (always) hold.

```
typedef struct {
  int cnt0;
  atomic_int cnt1, cnt2;
} count3;

void *updates(void *arguments) {
  count3 *counters = (count3*)arguments;

  int i;
  int c1, c2;

  for (i=0; i<1000; i++) {
    counters->cnt0++;
    c1 = atomic_fetch_add(&(counters->cnt1), 1);
    c2 = atomic_fetch_add(&(counters->cnt2), 1);
    //assert(c1==c2); ?
```

```
        //assert(counters->cnt1==counters->cnt2); ?
    }
    pthread_exit(NULL);
}
```

It is a good exercise to try this example with varying numbers of threads.

Our final **pthreads** example is concerned with finding and listing all primes up to some upper `limit`. An obvious parallelization of this problem would be to use a loop of independent iterations and to check in each iteration whether the corresponding index is a prime by calling `isprime(i)`. This function more or less naively tries to find out whether `i` is divisible by some smaller number. Since it is called by all the threads, it must be thread safe. A little thought shows that this is not an efficient parallelization approach. First, the time for checking whether index `i` is prime depends strongly on `i` and is fast for numbers with small prime factors and slow for large primes. And second, primes are not uniformly distributed (Prime Number Theorem). For these two reasons, parallelization of the loop will have poor load balance. Some threads will finish their iterations fast and have to wait for other threads with many expensive checks. Load balancing by array compaction cannot help here: We neither know which indices are primes nor which indices are either fast or slow to check. Instead, we employ a shared counter which the threads can query and increment to get the next index to check for primality. A corresponding thread function is shown below.

```
typedef struct {
  int rank;
  int limit;
  int *next;  // shared counter
  int found;
} realargs;

void *primes_race(void *arguments)
{
  int i;
  realargs *next = (realargs*)arguments;

  do {
    i = (*(next->next))++;
    if (i<next->limit) {
      if (isprime(i)) { // prime found, take action
        next->found++;
      }
    } else break;
  } while (1);

  pthread_exit(NULL);
}
```

It is illustrative to try this code and check how many primes it reports per thread in field `found`. The problem is the non-atomic update of the shared counter `*next`, similarly to what we saw in the `a = a+27;` example. Since increments can easily be lost, the effect is that much primality checking is repeated by the threads of which there may be many. The solution is to use an atomic counter with a FAI instruction.

```
void *primes_atomic(void *arguments)
{
  int i;
  realargs *next = (realargs*)arguments;

  do {
    i = atomic_fetch_add(next->next,1);
    if (i<next->limit) {
      if (isprime(i)) { // prime found, take action
        next->found++;
      }
    } else break;
  } while (1);

  pthread_exit(NULL);
}
```

This is a simple example of a work pool with work-stealing (see Sect. 1.3.6). The work items are the indices to be checked and are maintained as a single, shared counter. Threads steal work atomically by reading the current value of the counter and incrementing it.

In general, an operation on a data structure is said to be *wait-free* if a thread executing the operation can always complete in a bounded amount of time, regardless of what the other threads are doing (including also performing the operation). An operation is said to be *lock-free*, if, when several threads are performing the operation, some thread will be able to complete in a bounded amount of time. Wait-freeness is the nonblocking analogy of starvation-freeness and *lock-freeness* the nonblocking analogy of deadlock-freeness. Like starvation freedom implies deadlock freedom, wait-freeness implies lock-freeness.

It can be shown that, with sufficiently strong atomic operations (CAS), it is possible to give a wait-free implementation of any sequential data structure [59]. This is a theoretically strong result, but does not mean that wait- and lock-free data structures also perform well in practical contexts. We have seen that a wait-free counter can be useful. Other lock- and wait-free algorithms and data structures are beyond these lectures.

2.3 Seventh Block (3 Lectures)

OpenMP ("Open Multi Processing"), a standard for C and Fortran dating back to around 1997, is our next example of a concrete programming interface that implements a shared memory programming model and is intended for running on parallel shared memory systems. Like pthreads, OpenMP is thread based, but offers much more and much stronger support for Parallel Computing. Historically, the main unit of parallelization in OpenMP was the loop of independent iterations, see Sect. 1.3.2. Around OpenMP 3.0, support for task parallelism was introduced, see Sect. 1.3.1. This lecture and the following ones give an introduction to parallel programming with OpenMP and cover the main features and constructs needed in Parallel Computing. There is more to OpenMP than we will cover here, though. In particular, thread teams and groups will be silently circumvented and also the recent support for accelerators like GPUs will not be treated. Some recommended and sometimes revealing books for OpenMP programmers are [27, 75, 87].

OpenMP is maintained and developed further by an *Architecture Review Board* (ARB) which includes academic institutions and industry in various roles. The OpenMP specification and additional information are freely available via www.openmp.org, including very helpful cheat-sheets, see for instance https://www.openmp.org/wp-content/uploads/OpenMPRefCard-5.1-web.pdf.

2.3.1 The OpenMP Programming Model

Like pthreads, OpenMP is a fork-join thread model but threads are less explicit than in pthreads. There is no specific object identifying a thread. A *master thread* can fork (activate) a consecutively numbered set of working threads that includes the master thread itself. The threads share in executing *work* as specified by a *work sharing construct*, e.g., a loop of independent iterations or a task graph. Upon completion, threads join, leaving the master thread to fork a next set of threads. An OpenMP program is a single program and all forked threads execute the same code (SPMD).

The main characteristics of the OpenMP programming model are:

1. Parallelism is (mostly) implicit through work sharing. All threads execute the same program (SPMD).
2. Fork-join parallelism: A master thread implicitly spawns threads through a parallel region construct. Threads join at the end of the parallel region.
3. Each thread in a parallel region has a unique integer thread identifier (id), and threads are consecutively numbered from 0 to the number of threads minus one in the region.
4. The number of threads can exceed the number of available processors/cores. Threads are intended to be executed in parallel by available processor-cores.
5. Constructs for sharing work across threads are provided. Work is expressed as loops of independent iterations and task graphs.
6. Threads can share variables; shared variables are shared among all threads. Threads can have private variables.
7. Unprotected, parallel updates of shared variables lead to data races and are illegal.
8. Synchronization constructs for preventing race conditions are provided.
9. Memory is in a consistent state after synchronization operations.

As for `pthreads`, OpenMP does not come with any performance model and gives neither guarantees nor prescriptions for the behavior and performance of compiler and runtime system. Different compilers and runtime systems for OpenMP sometimes deliver very different performance for the same code.

2.3.2 OpenMP in C

OpenMP requires compiler, library and runtime system support and must, therefore, be compiled with an OpenMP-capable compiler and linked against library and runtime system. Most C compilers are nowadays OpenMP-capable. For instance, OpenMP programs can be compiled with `gcc` by giving the `-fopenmp` option. C code using OpenMP must include the function prototype header file with the `#include <omp.h>` preprocessor directive. All OpenMP relevant functions and predefined objects are prefixed with `omp_`, which identifies the OpenMP "name space". Special OpenMP environment variables are prefixed with `OMP_`. OpenMP is not a language extension per se, but requires extensive compiler and runtime support for parsing and translating and efficiently executing the `#pragma omp`-directives. OpenMP programs are C programs, but constructs like `for`-loops and compound statements are given their OpenMP meaning by `#pragma omp` compiler directives. Pragmas are designations in the code that direct the compiler to handle the following (compound) statement in a certain way.

For the concrete explanations in the following sections, we use <...> as meta-language designation for statements and non-empty lists of names, [...] to denote zero or more (optionally comma-separated) repetitions of some pragma element (clause), and | for exclusive choice.

2.3.3 Fork-Join Parallelism with the Parallel Region

Threads are activated when the master thread reaches an OpenMP *parallel region* construct which is a structured C statement (simple statement or compound statement in curly brackets {...}) designated by the `omp parallel` pragma. Actual threads and their binding to processor-cores are managed by the OpenMP runtime system; threads may be created once and for all and put to sleep for later reactivation or may be created afresh at each parallel region. In the parallel region, a defined number of threads p will be active, all executing the structured statement (SPMD style). Variables declared in the parallel region exist per thread and are local to the threads while variables declared before and outside of the parallel region are per default shared by all the threads. Once started, the number of threads in the parallel region cannot be changed. The threads can, by suitable library function calls, look up their thread identifier (id) and the number of threads executing in the parallel region. The thread id is a C integer between 0 and $p-1$. That is, thread ids are consecutive $0, 1, \ldots, p-1$.

2.3 Seventh Block (3 Lectures)

Threads coming to the end of the parallel region *join* with the other threads by performing a barrier synchronization, leaving only the master thread active after the parallel region. The barrier synchronization operation is implicit with the end of a parallel region, and it is essential for the OpenMP fork-join model that this cannot be changed. The thread id of the master thread is always 0.

```
#pragma omp parallel [clauses]
<structured statement>
```

Activation and deactivation of threads takes place at entry and (syntactic) exit of the parallel region and entails a barrier synchronization where all threads of the region have to be involved. It is, therefore, not allowed to break into or out of a parallel region with `break;` or `goto` statements. Sometimes such requirements can be checked by the OpenMP compiler.

The number p of threads in a region can be controlled either by the runtime environment, by a library call, or by a `num_threads()` clause for the `omp parallel` pragma. The last takes priority over the library call, which takes priority over the environment setting. When controlled by the environment, either a default number of threads is used or p is determined by the environment variable `OMP_NUM_THREADS`. The `OMP_NUM_THREADS` variable can be set to a number of threads larger than the number of processor-cores; that is, it is possible to run OpenMP programs with *oversubscription*. This is often useful for debugging but rarely for performance. The default number of threads is typically the number of processor-cores of the system where the program is running but can also be the number of hardware supported threads: The CPU may support *hardware multithreading* where each core can execute a small number of unrelated instruction streams which makes it look as if a correspondingly larger number of cores are available.

Regardless of the number of threads that might be available for a parallel region, it can sometimes be useful and efficient not to employ any additional threads at all – the problem at hand could be so small that parallelization with any number of threads larger than one is actually detrimental. An `if (<expr>)` clause to a parallel region will only activate threads if the expression evaluates to **true**. Otherwise, the parallel region will be executed by the master thread alone.

An OpenMP program consists of a sequence of parallel regions and can be depicted as a fork-join task graph (see Sect. 1.3.1). The work of a parallel region executed with p threads is the sum of the sequential work done by the p threads, and the time of a parallel region is the time for the last thread to finish. Since threads are activated at the beginning of the region and the region is finished by a barrier synchronization, the time of a region can be defined as the time that can be measured by the master thread 0 from start to end. Note that this always entails at least $\Omega(\log p)$ time for the barrier synchronization, with hopefully small constants depending on the quality of the OpenMP implementation. The work of a complete OpenMP program is the total amount of work done by the threads over all parallel regions. Different parallel regions may use different numbers of threads. The parallel execution time of an OpenMP program is the time that can be measured by the master thread from start to the end of the computation, that is, the sum of the running times of the successive regions plus the time taken by the master thread when no parallel region is active. A good OpenMP program will have work proportional to the work of a best known sequential program for the given problem and has a small number of regions in each of which the work is well balanced over the threads executing in the regions. In particular, the number and total time of the regions will correspond to $T\infty(n)$, the longest path in the program and the dependent part of the work that has not been parallelized.

2.3.4 OpenMP Library Calls

By suitable OpenMP library calls, a thread can look up its non-negative integer thread number, determine the number of threads in a parallel region, get the maximum number of threads allowed by the environment, and set the number of threads for a parallel region.

```
int  omp_get_thread_num(void);
int  omp_get_num_threads(void);
int  omp_get_max_threads(void);
void omp_set_num_threads(int num_threads);
```

These OpenMP library calls are all *thread safe*, that is, they can be called concurrently, in parallel, without any risk of interference.

For measuring the time taken by the execution of a (sequence of) parallel region(s), OpenMP provides standardized access to a (stable, high precision) timer.

```
double omp_get_wtime(void);
double omp_get_wtick(void);
```

The library function `omp_get_wtime()` returns the *wall clock time* in seconds since some point in the past. To report the time in milliseconds or microseconds of a piece OpenMP code, read the time before and after the piece of code and multiply the difference by 1000.0 or 1.000.000, respectively. The `omp_get_wtick()` call returns the resolution (in seconds) of the timer.

2.3.5 Sharing Variables

Per default, all variables declared before a parallel region by the master thread are shared by the threads in the region. Variables declared in the structured statement (block) of the parallel region are *private* (local) to each thread which means that a private, local copy for each thread will be created by the OpenMP compiler.

2.3 Seventh Block (3 Lectures)

Sharing of variables can be controlled through sharing clauses to the `omp parallel` pragma directive.

```
private(<comma separated list of variables>)
firstprivate(<comma separated list of variables>)
shared(<comma separated list of variables>)
default(shared|none)
```

A list of variables declared by the master thread (before the parallel region), that will per default be shared in the parallel region, can be made private, which means that the compiler will generate a local copy for each thread. Variables declared private by the `private()` clause are *not* initialized. The `firstprivate()` clause additionally initializes each local copy to the value the variable had before the parallel region. Often, this is the desired and perhaps implicitly assumed behavior. Note, that this can be expensive if the variable denotes a large, statically (compiler) allocated array as in `int a[1000];`. In contrast, for pointers the value of the pointer is copied and not the object to which it points. There are many possibilities for making non-sharing mistakes with OpenMP.

It is good practice (many say) to explicitly not share any variables declared by the master thread before a parallel region by using the `default(none)` clause and to then explicitly list the variables to be shared in a `shared()` clause. Such discipline forces one to think about which variables need to be shared and which not.

Shared variables can be read concurrently by the threads in the parallel region, but an OpenMP program in which it can happen that a thread updates a shared variable concurrently with other threads reading (or writing) the shared variable is *incorrect*. This is a *data race* that may lead to a race condition and correct OpenMP programs must not exhibit data races. OpenMP provides different means for avoiding data races while still allowing to exchange information between threads via shared variables (see the following).

2.3.6 Work Sharing: Master and Single

The simplest work sharing OpenMP constructs designate work that is *not* to be shared among the threads but rather to be executed by only one thread.

```
#pragma omp master
<structured statement>
```

Here, the work of the structured statement is done by the master thread alone (the thread with `omp_get_thread_num()==0`). The other threads will skip the structured statement code and just continue execution. There is *no* barrier synchronization implied following the master thread code. Also, the code of the master thread is *not* executed under mutual exclusion. That is, the master thread must not update shared variables that can potentially be read or updated concurrently by the other threads that are not in the master statement.

```
#pragma omp single [clauses]
<structured statement>
```

With the `single` construct, the work of the structured statement is done by either one of the parallel, running threads, but it is not determined which of the threads; the OpenMP runtime system (or compiler) makes the decision. A parallel region can, of course, have several `single` statement blocks and each of the blocks may be executed by a different thread. The code executed by the chosen, single thread is, like for the `master` construct, not executed under mutual exclusion. So updates to shared variables possibly read or written by other threads are illegal. In contrast to the `master` construct, the `single` construct has an implied barrier at the end of the structured statement. A thread reaching this point, regardless of whether it was the thread executing the `single`-designated statement or one of the other threads, cannot proceed until all threads have reached this point. This implies that the number of encountered `single` statement blocks must be the same for all threads and so one must be careful with branches and loops in parallel regions.

The implied barrier at the end of the `single` block can be eliminated with the `nowait` clause. This can sometimes lead to better performance: A barrier can be expensive, and an OpenMP program should have no more barriers than absolutely necessary. On the other hand, a `nowait` clause can as easily make a correct program incorrect by introducing race conditions (data races). The `single` construct allows to make variables `private()` or `firstprivate()`; the `master` construct does not.

In the following example, the master thread reads input for a parallel computation to be done by private (non-parallelized) `for`-loops by the threads. Since there is no implied synchronization between the threads after the master has completed, an explicit OpenMP barrier (see next section) has been introduced, after which all threads can safely work on the input in the array `a`. The result in array `b` is written by some single thread. In order to ensure that all threads have completed their work before the array is written, again an explicit barrier is needed. The implicit barrier of this `single` construct is not needed here and is, therefore, eliminated by a `nowait` clause. Since there is always a barrier at the end of the parallel region, the extra barrier implied by the `single` work sharing construct would have been redundant here.

```
int n; // shared size
int *a, *b; // allocate somewhere
#pragma omp parallel if (n>10000) // only for large n
{
   int i; // private i for each thread
...
#pragma omp master
   readdata(a,n);
#pragma omp barrier
   // compute
   int i0, n0; // loop range for thread
   i0 = ...; n0 = ...n;
```

```
   for (i=i0; i<n0; i++) {
      b[i] = ...;  // per thread computation from a into b
   }
#pragma omp barrier
#pragma omp single nowait
   writedata(b,n);
}
```

If the explicit barriers were omitted, correctness could be guaranteed: There would be possible race conditions on both a and b arrays.

Code for single and master threads should be kept short, unless the other threads have sufficient other work to do. All threads in a parallel region should perform more or less the same amount of work.

2.3.7 The Explicit Barrier

An explicit barrier, a point in the code of a parallel region beyond which no thread shall continue before all other threads have reached this point, can be designated with the **barrier** construct as we saw in the previous section.

```
#pragma omp barrier
```

An explicit barrier is sometimes necessary, for instance, after a **master** construct or in situations where threads read values computed by other threads. Here, an explicit (or implicit) barrier can be necessary to ensure that the other threads have indeed completed the computation of the required values.

2.3.8 Work Sharing: Sections

The work to be done by some (part of an) algorithm can sometimes be statically expressed as some finite set of independent pieces that can potentially be executed in parallel by a set of available threads. In OpenMP, such work can be identified and the independent pieces can be designated as such. This work sharing construct is called **sections** with each independent piece forming a **section** of code.

```
#pragma omp sections [clauses]
<section block>
```

Each independent section of code in the section block (enclosed in {...}) is marked as such.

```
#pragma omp section
<structured statement>
```

A block of sections also ends with an implicit barrier synchronization point: No thread can continue beyond the sections code before all sections have been completed. This implicit barrier can be circumvented with the `nowait` clause. Before the block of sections, the sharing of variables can be restricted to either `private()` or `firstprivate()`.

In a parallel region with sections, the individual sections are assigned to the threads according to some schedule chosen by the OpenMP runtime system. Ideally, each thread will execute a section, and the threads will all run in parallel. If there are more sections than threads in the parallel region, some thread(s) will necessarily execute more than one section. Good OpenMP code will aim to make the amount of work in the sections balanced and, in particular, avoid having (too) few, very large sections that could lead to harmful load imbalance by many threads sitting idle at the barrier waiting for the other threads. OpenMP sections is a static division of work and normally the number of sections is small. We will see constructs for dynamic division of work in the following sections.

2.3.9 Work Sharing: Loops of Independent Iterations

Work is very often expressed as loops of independent iterations (see Sect. 1.3.2). This was and is still the basic, fundamental premise of OpenMP and the parallelized loop one of the basic, work sharing constructs. As we have seen, loops of independent iterations provide ample opportunity for keeping threads (processor-cores) busy by assigning (blocks of consecutive) loop iterations to threads. The assignment of particular iterations to threads is called *loop scheduling* in OpenMP and is expressed by a `for` work sharing pragma with clauses as part of a parallel region. Loop scheduling must at least fulfill that each iteration is executed exactly once by some thread as the sequential semantics of the loop require. By the independence condition, the iterations can be executed in any order and concurrently by the threads. Loops must take a specific, syntactic form called the *canonical form*.

```
#pragma omp for [clauses]
for (<canonical form loop range>)
<loop body>
```

In order that threads can independently of each other (perhaps supported by data structures in the OpenMP runtime system) schedule the iterations, the loop range must confirm to certain rules. The most important such rule is that all threads in the parallel region will be able to determine the *same* loop range. Thus, in a standard C for-loop

```
for (i=start; i<end; i+=inc)
<loop body>
```

all threads must compute the same values for the start and end iteration and must use the same increment (here, `i`, `start`, etc. are arbitrary variable names and expressions). These values must *not* change in any way during the execution of the loop. Also, loop ranges must be finite and determined; that is, the for loop must *not* be a camouflaged, open-ended while loop. Such a range can easily be split into blocks of iterations by the compiler.

Finally, OpenMP poses restrictions on the form of the loop bound condition, which must be of the form i<n, i<=n, i>n, i>=n, or i!=n only (i is an arbitrary variable and n an arbitrary expression). Also, increments must take either of the forms i++, i+=inc, or i=i+inc and similarly for decrements. Loops fulfilling such restrictions are said to be in *canonical form*. The loop variable, here i, is automatically made private for the loop body; otherwise, each iteration would be a race condition on i.

There is a composite, shorthand directive that combines the parallel region with one parallel loop. This is one of the most frequent directives in OpenMP programs.

```
#pragma omp parallel for [clauses]
for (<canonical form loop range>)
<loop body>
```

Inherited from the `omp parallel` construct, the composite loop directive does not allow the `nowait` clause since only the master thread 0 is to be active after the parallel for loop. For this reason, breaking into or out of a parallel loop is illegal. Such violations may sometimes be caught by the compiler. Try compiling the following loop.

```
#pragma omp parallel for
for (i=0; i<n; i++) {
  if (i==10) continue; // this is ok
  if (i==11) break;    // but not this!
}
```

The sharing clauses from the `omp parallel` region also apply to the `omp parallel for` loop. For parallel loop constructs there is a further sharing option which allows to transfer the value of a private variable to its shared counterpart, namely by capturing the value of the variable at the sequentially last iteration of the loop. Here is a handy use-case for capturing the value of the loop index variable after the last iteration as often used in sequential code.

```
int i; // shared i
#pragma omp parallel for lastprivate(i) // now private
for (i=0; i<n; i++) {
  a[i] = b[i];
}
assert(i==n);
```

The `lastprivate()` clause can also be used with parallel sections and will in that case capture the value of the variable in the syntactically last section.

In order for an OpenMP program to be correct, loop iterations, regardless of the order in which they are executed by the threads, must not cause data races by concurrent reads and writes to shared variables: The loop iterations must be independent and have neither forward, anti-, nor output dependencies. A simple, sufficient rule for independence of loops is the following. The loop does array updates only, each iteration updates at most one array element, and no iteration refers to an element updated by another iteration.

Some loop carried dependencies in simple, array only loops can be eliminated by transforming the loops. A loop like

```
for (i=k; i<n; i++) a[i] = a[i]+a[i+k];
```

where, sequentially, a[i] is updated in iteration i with the (sequentially) not yet updated, and, therefore, "old" value a[i + k], can equivalently be written as

```
for (i=k; i<n; i++) aa[i] = a[i]+a[i+k];
// swap
tmp = a; a = aa; aa = tmp;
```

by introducing a new array aa into which the updates are computed from the "old" values in array a and swapping the two arrays after the loop. A little care is required to allocate and free the extra arrays correctly.

The transformed loop is now a loop of independent iterations (also according to the simple rules for independent loops) and can, therefore, readily be parallelized with #pragma omp parallel for.

2.3.10 Loop Scheduling

Loop scheduling denotes the assignment of loop iterations to threads: How exactly is the work expressed by the loop of independent iterations shared across the threads of the parallel region?

For loop scheduling, the number of iterations in the loop range is divided into not necessarily same-sized *chunks* of consecutive iterations. Like the iterations, chunks are numbered consecutively such that they can be referred to by their number. The chunk numbering is for reference only and not something that has to be computed or maintained explicitly by the OpenMP runtime system.

OpenMP provides three basic types of loop schedules. In a *static* schedule, all chunks have (almost) the same size, and chunks are assigned in a round-robin fashion to the threads. For a loop range of n iterations, with chunksize c, and p threads, there are $k = \lceil n/c \rceil$ chunks, and the iterations of chunk $i, 0 \leq i < \lceil n/c \rceil$ are executed (one after another) by thread i mod p. That is, thread 0 executes the iterations of chunk 0, thread 1 the iterations of chunk 1, thread 2 the iterations of chunk 2, and so on. If there are more than p chunks, again thread 0 executes the iterations of chunk p, thread 1 the iterations of chunk $p + 1$, and so on, until all iterations of all chunks have been executed. If the loop range has been divided into at least p chunks, all threads can be kept busy, but not necessarily all of the time: That depends on the exact number of chunks and on the time that each iteration takes, which may be different for different iteration indices. For instance, if the work per iteration in chunks $0, p, 2p$ etc. is very small, which could be the case if a condition on the loop iteration index fails for these chunks, there is nothing to do for these chunks except for going through the iterations and checking the condition. So, thread 0 might be able to finish much faster than the other threads.

In a *dynamic* schedule all chunks also have the same size c, but the chunks are not assigned to the threads in any predetermined, static fashion. Chunks are executed by the threads in increasing order. Instead of a fixed assignment, each thread *dynamically* grabs the next not yet assigned chunk as soon as it has finished its previous chunk. With a dynamic schedule, the situation sketched above will not happen. As soon as thread 0 finishes (fast) with chunk 0 it will grab the next unassigned chunk and, thus, help with finishing the loop iterations faster than the static schedule could.

Like in a dynamic schedule, a *guided* schedule assigns chunks to threads dynamically as the threads become available. Unlike both static and dynamic schedules, the chunk size is no longer fixed. Instead, when a thread has finished executing an earlier, smaller numbered chunk, it grabs a chunk for the next iteration that has not yet been executed. Instead, the size (number of iterations) of the chunk is computed dynamically as the number of remaining, not yet executed or assigned iterations divided by the number of threads p.

The advantage of the static schedule is that computation of chunk numbers and assignment to threads can be done very fast and efficiently. Essentially, each thread decides for itself which chunks it will have to execute, which is possible due to the restrictions on parallelizable loops by the canonical form. Thus, static schedules have low scheduling overhead. A static schedule can be expected to give good performance when the work per iteration or per chunk is more or less the same for all iterations or chunks. Many, but certainly not all loops have this property. The time per iteration can be influenced heavily by the memory and cache access patterns even for code where the iterations incur the same number of instructions to be executed. Dynamic and guided schedules can be preferable for loops with conditions depending on the iteration index or otherwise varying amounts of work per iteration. The guided schedule is motivated by the assumption that, when a thread becomes ready to execute the next chunk, the work in the remaining iterations will be more or less the same per iteration, in which case it makes sense to divide these remaining iterations evenly into p chunks. This assumption may or may not hold, and a guided schedule may or may not perform better than a static or dynamic schedule. Dynamic and guided schedules both have a higher scheduling overhead than static schedules. For instance, dynamic scheduling could be implemented by the OpenMP runtime system with a simple *work pool* (see Sect. 1.3.6) that maintains the next, not yet executed loop iteration index. Implementing such a work pool would require just one *atomic counter*:

```
do {
  start = atomic_fetch_and_add(&i, chunksize);
  if (start>=n) break;
  end = min(start+chunksize,n);
```

```
      for (j = start; j<end; j++) {
          // execute iterations in chunk
      }
} while (1);
```

Here, `chunksize` is the fixed chunksize c. It was tacitly assumed that the loop increment was 1.

In OpenMP, the particular schedule type for a parallel for loop can be determined by an explicit *schedule clause* that takes an optional, explicit chunk size parameter. For static and dynamic schedules this optional chunk size is the exact size in number of iterations of the chunk (except possibly for the last chunk), whereas for guided schedules, the chunk size parameter is a lower limit on the gradually decreasing number of iterations for the chunks.

```
schedule(static[,chunksize])
schedule(dynamic[,chunksize])
schedule(guided[,chunksize])
```

If a schedule clause is not given, a static schedule with default chunk size is used. If no chunk size is given, a default chunk size is used. For a static schedule for a loop range with n iterations, this is approximately n/p, such that there are exactly p chunks, one for each thread. If p does not divide n, the smallest chunk size is $\lfloor n/p \rfloor$ and one or more chunks will have one or more extra iterations. The OpenMP specification deliberately does not specify which chunks will get the extra iterations. For dynamic and guided schedules, the default chunk size is 1.

For simple loops over arrays, it can sometimes make sense to let the chunksize c be some multiple of the *cache line* (block) size in order to avoid *false sharing*.

There are two additional schedule types that can be given with the schedule clause.

```
schedule(runtime)
schedule(auto)
```

With the `runtime` type schedule, the schedule can be set externally through the `OMP_SCHEDULE` environment variable, which can be very useful for tuning and experimenting with different schedules. Here are some examples of `OMP_SCHEDULE` settings.

```
"static,1"
"static,8"
"dynamic"
"guided,100"
```

With the `auto` type schedule, the choice of "best" schedule is left to the OpenMP compiler and runtime system.

To understand and memorize the OpenMP loop schedules, it is a good exercise to implement code that for a (large) iteration range n in a parallel for loop records for each iteration which thread was responsible for that iteration and as well makes sure that all iterations were indeed executed (and no more).

```
int t = omp_get_max_threads();
int iter[t]; // iterations per thread
int loop[n]; // who did iteration i; careful for large n

for (i=0; i<t; i++) iter[i] = 0;
#pragma omp parallel for schedule(runtime)
for (i=0; i<n; i++) {
  loop[i] = omp_get_thread_num();
  iter[omp_get_thread_num()]++; // problematic for large n
}
int nn = 0;
for (i=0; i<t; i++) nn += iter[i];
assert(nn==n); // all iterations done
```

Running the code with $n = 20$ (small) and printing out the values `loop[i]` and `iter[i]` with `OMP_SCHEDULE` set to "static,2" and seven threads (`OMP_NUM_THREADS` set to 7) gives the following output.

Thread for iteration $i, 0 \leq i < n$.

0	1	2	3	4	5	6	7	8	9	10	11	12	13	14	15	16	17	18	19
0	0	1	1	2	2	3	3	4	4	5	5	6	6	0	0	1	1	2	2

Iterations per thread.

0	1	2	3	4	5	6
4	4	4	2	2	2	2

2.3.11 Collapsing Nested Loops

Many computations, for instance, computations involving matrices, are often expressed with (doubly, triply) nested, parallelizable loops. If all the loops in the loop nests are loops of independent iterations, either of them can be parallelized with the parallel for directive. Deciding which one to parallelize may not be obvious and depends (among other things) on the amount of work per iteration and the number of iterations per loop. Often it makes sense to parallelize the loop with the largest number of iterations, but since that could be either of the loop nests, attempting this could lead to code blow up by having to maintain parallelizations for all the possibilities. A sometimes good solution is to treat the nested loops as one single larger loop; that is, to transform code of the form

```
for (i=0; i<m; i++) {     // parallelize this loop?
  for (j=0; j<n; j++) {   // or this loop?
    x[i][j] = f(i,j);
  }
}
```

into

```
for (ij=0; ij<m*n; ij++) {
  i = ij/n; j = ij%n;
  x[i][j] = f(i,j);
}
```

This transformation is valid in the sense that each iteration of the nested loop is performed exactly once by the transformed loop. This is possible under the condition that all loop bounds can be computed before the two loops and do not change during the iterations.

By adding the `collapse(<nesting depth>)` clause to the `parallel for` directive, the outlined transformation can be performed automatically (to any nesting depth) by the OpenMP compiler. The loops must be perfectly nested. This means that the body of an outer loop must consist of only the next inner loop (no extra statements). As for all OpenMP parallelizable loops, the iteration ranges must be in canonical form prescribed by OpenMP. The two nested loops can then be parallelized as follows:

```
#pragma omp parallel for collapse(2)
for (i=0; i<m; i++) {    // OpenMP makes one loop out of two
  for (j=0; j<n; j++) {
    x[i][j] = f(i,j);
  }
}
```

If the loop has this form instead,

```
#pragma omp parallel for collapse(2)
for (i=0; i<m; i++) {
  for (j=i; j<n; j++) {
    x[i][j] = f(i,j); // upper matrix triangle
  }
}
```

where the start index of the inner loop depends on the outer loop, the loops cannot be collapsed automatically and the OpenMP compiler will (most likely) complain.

The `schedule()` clause and all other clauses allowed for parallel for loops can be used and will be interpreted as if the loop had been transformed (collapsed, flattened) as outlined. According to the OpenMP specification, the sequential execution order of the iterations in uncollapsed loops determines the order of the iterations of the collapsed iteration range.

2.3.12 Reductions

Two frequently occurring loop patterns are the following:

- Prefix sums

2.3 Seventh Block (3 Lectures)

```
for (i=1; i<n; i++) {
  a[i] = a[i-1]+a[i]; // the classic flow dependency
}
```

- Reduction

```
sum = a[0];
for (i=1; i<n; i++) {
  sum += a[i]; // data race on sum
}
```

Both of these loop patterns are loops of dependent iterations. Therefore, they cannot be correctly parallelized with the OpenMP constructs for loop parallelization seen so far. The computations expressed by the two loops (parallel prefix sums and simple reductions) require different, parallel algorithms in order to be performed with any speed-up by a set of threads working together (see Sect. 1.4.7 and Sect. 1.4.9). Thus, either non-trivial compiler transformations of the loop patterns into better, parallel algorithms or the execution of preimplemented algorithms at runtime is required to handle such loop patterns well. Good parallel algorithms require the binary operator used in the patterns (here: +) to be at least associative.

The reduction pattern loop can be handled, i.e., parallelized efficiently, with OpenMP by using the `reduction()` clause with the parallel `for` directive. How well the parallelization works will depend on the OpenMP compiler and runtime system among other things.

The `reduction()` clause is quite flexible. It takes an associative, binary reduction operator and a list of reduction variables on which a reduction with this operator is to be performed in the loop. The order of the reductions follow the loop iteration order, but it is not defined where brackets are put: The associativity of the reduction operator is exploited to allow efficient reduction in parallel. Different reduction operators can be used in the same loop by giving a reduction clause for each.

```
reduction(<reduction operator>:<reduction variables>)
```

The allowed operators in C are +, -, *, &, |, ^, &&, ||, as well as special `min` and `max` operators. Minimum and maximum operations are expressed either with special operators or with code patterns like

```
mi = (x<mi) ? x : mi;
if (x>ma) ma = x;
```

that will be recognized by the compiler as minimum or maximum computations, respectively. Here `mi` and `ma` are global variables declared by the programmer.

The reduction clause can also be used with parallel regions and the sections work sharing construct. In such cases, the reduction will be performed in thread or section order.

Since OpenMP 5.0 the scan/prefix sum pattern can be also handled. This is expressed by modifying a reduction in a parallelizable loop to "capture" the reduced result for the current iteration, i.e., the prefix sum for that iteration. Prefix sums reductions are called **inscan** reductions and are expressed as follows:

```
reduction(inscan,<reduction operator>:<reduction variables>)
```

A reduction is performed with the reduction operator on the reduction variables in the sequential loop order. The corresponding prefix sum is captured with either a

```
#pragma omp scan exclusive(<reduction variables>)
<structured statement>
```

directive for a structured statement (for the exclusive prefix sums), or a

```
#pragma omp scan inclusive(<reduction variables>)
<structured statement>
```

directive for a structured statement (for the inclusive prefix sums). For the inclusive prefix sums computation, the reduction variables can be used in the block of the **scan** directive and will contain the result of applying the reduction operator up to and including the current iteration of the parallel loop; conversely, the result of the reduction for the current iteration used before the **scan** directive will be the exclusive prefix sum up to but not including the current iteration. There can be only one **scan** directive in a parallel loop, and in such a loop scheduling clauses cannot be used. The following example shows how to compute inclusive and exclusive prefix sums for an input array **a** with the result stored in **b**.

```
x = 0;
#pragma omp parallel for reduction(inscan,+:x)
for (i=0; i<n; i++) {
  x += a[i]; // reduce
#pragma omp scan inclusive(x)
  b[i] = x;  // and save the prefix (current value)
}

x = 10;
#pragma omp parallel for reduction(inscan,+:x)
for (i=0; i<n; i++) {
  b[i] = x;  // save the prefix
#pragma omp scan exclusive(x)
  x += a[i]; // and reduce for next iteration
}
```

A natural application of reduction with the scan directive is array compaction as discussed in Sect. 1.3.11 and Sect. 1.4.6. The marked elements of an input array **a** have to be compacted into a shorter output array **b**. To do this, a running index for each marked element is needed. Exactly this can be captured by the running, exclusive prefix sum.

```
int mark[n];
// mark[i] == 0/1 for input elements a[i]

int j = 0;
```

```
#pragma omp parallel for reduction(inscan,+:j)
for (i=0; i<n; i++) {
  if (mark[i]) b[j] = a[i];
#pragma omp scan exclusive(j)
  j += mark[i];
}
```

2.3.13 Work Sharing: Tasks and Task Graphs

Dynamically evolving work can often be expressed as a Directed Acyclic task Graph (DAG) as discussed in Sect. 1.3.1. The OpenMP *task* work sharing constructs make it possible to express such computations.

Consider the recursive Quicksort algorithm as discussed in Sect. 1.3.1. In each Quicksort invocation, the input array is partitioned into two roughly equally large parts, each of which can be Quicksorted independently of the other. With several threads activated like in an OpenMP parallel region, each Quicksort call can be wrapped as a *task* to be executed by a thread that may happen to be available and has no other work to do. In an OpenMP parallel region, any piece of code, like a procedure call (Quicksort), a function call, or even a structured block, can be marked as a *task* with the following work sharing construct.

```
#pragma omp task [clauses]
<structured statement>
```

During execution, the code designated as a task will be prepared and wrapped by the thread executing the omp task pragma (with compile time help from the OpenMP compiler). A created task may be executed by any (other) thread in the parallel region, possibly at a later time. All created tasks will be executed and completed at the latest at a point in the code where completion is requested. One such point of completion is the implicit barrier at the end of the parallel region. All generated tasks can also be completed by an explicit #pragma omp barrier construct. In the terminology of Sect. 1.3.1, the tasks being wrapped by a thread are *ready*, but they can have dependencies on (private and shared) variables of the thread that generated the task. Thus, in a correct OpenMP task program, the generating task shall not update any variable that can be referred to by the generated tasks. If it does, data races, which are illegal in OpenMP, may arise.

A thread that generates one or more tasks may depend on these tasks to be executed and completed before it can continue its computation. A thread can have many omp task directives, for instance, through recursive calls, and it may need results from these tasks in order to continue. Waiting for completion of the tasks directly generated by the thread can be enforced by the taskwait construct.

```
#pragma omp taskwait [clauses]
```

The only allowed clauses are `depend()` clauses. They express input and output dependencies on other tasks in the same lexical scope. Dependencies are not treated in this lecture. It is also possible to generate tasks in such a way that waiting is done not only for the directly generated (children) tasks, but also for all tasks descending from these tasks (for instance, from recursive calls). This is done by generating the tasks in a `taskgroup` region which we shall not describe further in these lectures.

Here is a standard example of an algorithm that can be parallelized with tasks. The problem is to count the number of occurrences of some value x in an unordered array a of n elements. The algorithm is recursive. If $n = 1$ the problem is trivial: There is an occurrence if $a[0] = x$, otherwise not. If $n > 0$ the array is split into two halves, the number of occurrences in both halves counted and added together. This idea can obviously be formulated as a computation on a task graph and be implemented in OpenMP as shown below.

```
int occurs(int x, int a[], int n)
{
   if (n==1) {
     return (a[0]==x) ? 1 : 0;
   } else {
     int s0, s1; // private variables for executing thread
#pragma omp task shared(s0,a)
     s0 = occurs(x,a,n/2);
#pragma omp task shared(s1,a)
     s1 = occurs(x,a+n/2,n-n/2);
#pragma omp taskwait

     return s0+s1;
   }
}

int main(...)
{
   int a[n];
   int x;
   int s;

#pragma omp parallel shared(x) shared(s) shared(a)
   {
#pragma omp single
     s = occurs(x,a,n);
   }
}
```

Each recursive call is marked as an `omp task`. In order to sum the number of occurrences for each half of the array, an explicit `omp taskwait` is necessary. The computed results (and the array pointer) must be classified as `shared`. This is crucial, since each task can be executed by any thread of the parallel region, in particular by a thread that is different from the one that allocated the variable. The thread that executes a task must be able to update the variable that was

2.3 Seventh Block (3 Lectures)

possibly allocated by another thread. This is possible only if the variable is shared among the two different threads.

In the main program, the threads are activated by the **parallel** region. One of the threads, arbitrarily chosen by the **single** work sharing construct, shall initiate the search. If **single** (or **master**) is forgotten, all threads will start performing the search operation, which leads to superfluous work (by a factor of the number of threads p) and (in this example) definitely to data races.

In the example, the recursion is done all the way down to the bottom $n = 1$ condition. This is rarely a good choice, neither sequentially, nor in parallel. Generally, finding a good cut-off for recursive algorithms is a difficult problem, which we will not solve here. In order to prevent too many, too small tasks, a task can be designated as **final**, meaning that the task will generate no additional tasks. **untied** tasks are tasks that may be suspended and are allowed to be resumed by any other thread. Together with a conditional **if**-clause this can possibly be used as a substitute for an explicit cut-off programmed into the recursive task.

The **omp task** work sharing construct offers further possibilities for controlling when a task will be ready for execution. Input-output dependencies can be expressed with **depend()** clauses. With the **priority()** clause, tasks can be prioritized as hints to the OpenMP runtime system in which order the tasks should preferably be executed.

Here is a(n almost) complete parallelization of the sequential Quicksort algorithm with OpenMP tasks.

```
void Quicksort(int a[], int n)
{
  int i, j;
  int aa;

  if (n<2) return; // recursion all the way down

  // partition
  int pivot = a[0]; // choose an element (non-randomly...)
  i = 0; j = n;
  while (1) {
    while (++i<j&&a[i]<pivot);
    while (a[--j]>pivot);
    if (i>=j) break;
    aa = a[i]; a[i] = a[j]; a[j] = aa;
  }
  // swap pivot
  aa = a[0]; a[0] = a[j]; a[j] = aa;

#pragma omp task shared(a)
  Quicksort(a,j);
#pragma omp task shared(a)
  Quicksort(a+j+1,n-j-1);
  // #pragma omp taskwait - not needed
}
```

```
int main(int argc, char *argv[])
{
   ...

   start = omp_get_wtime();
#pragma omp parallel
   {
#pragma omp single nowait
      Quicksort(a,n);
      //#pragma omp taskwait
   }
   stop = omp_get_wtime();
}
```

Assuming that a good pivot can be selected that divides the input array into two approximately equally large halves, the recurrence relations for work and parallel time complexity are as follows.

$$W(n) = 2W(n/2) + O(n)$$
$$W(1) = O(1)$$
$$T(n) = T(n/2) + O(n)$$
$$T(1) = O(1)$$

Using the Master Theorem 9, the solutions are $W(n) = O(n \log n)$ and $T(n) = O(n)$, respectively (Case 2 and Case 1). This tells us that the parallelism of this implementation is low, in $(n \log n/n) = O(\log n)$ and that only up to that number of processors can be employed with linear speed-up. The culprit is obviously the linear-time partition step which we have to pay for in full at the first recursive call. In order to improve, a parallel algorithm, for instance by using prefix sums and compaction as discussed, would have to be employed. Unfortunately, in this lecture, we will have no means for using task and loop parallelism (with scan-reduction) together. We could use our own parallel prefix sums algorithms parallelized with omp taskloop (see Sect. 2.3.16).

2.3.14 Mutual Exclusion Constructs and Atomic Operations

In order to prevent data races in parallel regions, OpenMP provides direct support for *mutual exclusion* through named critical sections.

```
#pragma omp critical [(name)]
<structured statement>
```

Threads that encounter a (named) *critical section* will all execute the code in the critical section, but under mutual exclusion; that is, at most one thread at a time can execute the code for its critical section. In a critical section, one or more shared variables can be updated, shared variables can be read, and the thread can make decisions based on the read values. Since no other threads will be executing code for the named critical section at the same time, such updates are technically not data races, and it is possible to ensure a definite outcome of the parallel execution of the threads. The order in which the threads enter the critical section is nondeterministic. It depends on the relative speeds of the threads, on when they encounter the critical section, on how many threads arrive "at the same time" and compete for the critical section, and on how the mutual exclusion (locking) algorithms of the runtime system resolve the conflicts. Thus, relying on some specific behavior of the critical section construct will lead to incorrect programs. A concrete case is the implementation of reduction like operations: Implementations with critical sections will be correct only when the reduction operators being used are commutative.

Critical sections are always (relatively) expensive constructs and will, therefore, have an impact on the overall performance of a parallel program. In particular, they lead to serialization between the threads and should, therefore, be used sparingly and with care.

Here is a technically incorrect, but defensible use of mutual exclusion to find the maximum of values computed by threads in a parallel region which can, dependent on the input, alleviate some serialization penalties.

```
int max = 0;
#pragma omp parallel shared(max)
{
   int x;
   x = ...;

   if (x>max) {
#pragma omp critical
      if (x>max) max = x;
   }
}
```

With this kind of speculative, *optimistic locking* the critical section is entered only by threads whose x value could potentially be the maximum. Under mutual exclusion in the critical section where the thread is alone, it must then be rechecked that the x value is still a candidate in which case the shared max variable is updated. The recheck is necessary since another thread could have updated max before the thread enters the critical section such that x is after all not the largest value seen "so far" (try this). The pattern is also known as test-and-test-and-set (TATAS). The advantage is that values that definitely cannot be the maximum do not cause serialization. Technically, such code is not data race free; but as argued it leads to correct results: not all race conditions are harmful.

In case the update and work to be done in a critical section has a particularly simple form, it may be possible to use a hardware assisted atomic operation instead. OpenMP provides access to certain types of *atomic operations* by the following construct.

```
#pragma omp atomic [read|write|update|capture|compare]
<atomic statement>
```

The atomic operations ensure that operations by different threads are performed in order without interruption (atomically). Atomic operations also lead to serialization, but since they are performed directly by the hardware they can be considerably less costly than explicit critical sections. Atomic **update** and **capture** operations allow the use of Fetch-And-Add (FAA) type atomic operations. For the **atomic update** construct, the atomic statement is restricted to be of the form `x++;`, `++x;`, `x--;`, `--x;`, and `x = x binop expr;` and similar syntactically equivalent forms that will result in a simple arithmetic-logical update to x. For the **atomic capture** construct, the atomic statement must have the form `y = x++;` or a similar syntactically equivalent form that will read the previous content of x and cause an update. Here x and y are variables where the C operators apply and **binop** one of the word wise C operations +, *, -, /, &, ^, |, « or ».

A frequent, convenient use case for captured atomic operations is the computation of unique indices for threads that execute a parallel loop.

```
int ixc = 0; // shared index counter
#pragma omp parallel for shared(ixc)
for (i=0; i<n; i++) {
  int ix; // next index

  if (need) { // index needed
#pragma omp atomic capture
    ix = ixc++;
  }
}
```

Each time a thread performing some iteration of the loop needs an index, it performs the atomic Fetch-And-Increment (FAI) to retrieve the current value of the shared counter and atomically increments this to the next index.

After OpenMP 5.0, also the (hardware) Compare-And-Swap (CAS) operation is supported as an OpenMP atomic operation. For this, **compare** or **capture compare** clauses have to be given. A conditional statement like `if (x==e) x = d;` is executed atomically, preferably by a corresponding hardware atomic operation. This makes it possible to implement certain lock- and wait-free algorithms directly in OpenMP, but this is beyond this lecture, see for instance [59]. The atomic operations defined for C in the `stdatomic.h` library can also be used with OpenMP.

2.3.15 Locks

Sometimes named critical sections are insufficient, for instance, for implementing list-based algorithms with hands-over locking, where a lock (critical section) is needed for each element of the list. For that reason, OpenMP provides locks that can be allocated dynamically similarly to the `pthreads` locks. Locking and unlocking a lock is called set and unset in OpenMP.

```
void omp_init_lock(omp_lock_t *lock);
void omp_init_nest_lock(omp_nest_lock_t *lock);
void omp_destroy_lock(omp_lock_t *lock);
void omp_destroy_nest_lock(omp_nest_lock_t *lock);
void omp_set_lock(omp_lock_t *lock);
void omp_set_nest_lock(omp_nest_lock_t *lock);
void omp_unset_lock(omp_lock_t *lock);
void omp_unset_nest_lock(omp_nest_lock_t *lock);
int  omp_test_lock(omp_lock_t * lock);
int  omp_test_nest_lock(omp_nest_lock_t * lock);
```

Locks in OpenMP do not have condition variables. OpenMP provides nested (recursive) locks. Nested (recursive) locks in OpenMP must be unlocked as many times as locked by the threads having successfully acquired the lock. Locks (now) also have a try-lock operation called `omp_test_lock`. OpenMP does not provide readers-and-writer locks. Thus, OpenMP is not intended for involved programming with locks the same way that `pthreads` and other thread interfaces are.

2.3.16 Special Loops

Loops of independent iterations where the operation(s) per iteration have a particularly simple form, for instance, expressing a simple n-element vector addition like:

```
for (i=0; i<n; i++)
   c[i] = a[i]+b[i];
```

can benefit from hardware capabilities for operating on small vectors with a single instruction. Modern processors typically have such capabilities in the form of extended vector instructions for operating on $2, 4, 8, 16$ float or double elements with one single instruction (SSE, AVX instructions). Such instructions are called SIMD instructions. With OpenMP, the compiler can be instructed to try to use SIMD instructions with the following three loop parallelization constructs. A sequential loop to be executed by one thread with SIMD instructions is designated with the `simd` pragma.

```
#pragma omp simd [clauses]
for (<canonical form loop range>)
<loop body>
```

A loop within a parallel region to be shared among the threads of the region with each chunk executed with SIMD instructions is designated with the `for simd` pragma.

```
#pragma omp for simd [clauses]
for (<canonical form loop range>)
<loop body>
```

A parallel region with SIMD loop sharing can be written with a shorthand, composite construct.

```
#pragma omp parallel for simd [clauses]
for (<canonical form loop range>)
<loop body>
```

For the compiler to be able to exploit SIMD instructions, often certain preconditions must be fulfilled, for instance, on alignment and size of loop ranges. Hints and assertions that this is the case can be expressed by additional clauses. Such hints and conditions are beyond these OpenMP lectures.

A different way of parallelizing a loop of independent iterations is to (recursively) break the iteration range into smaller ranges that are executed as tasks. While such a loop parallelization can easily be done by hand (give a recipe for this), OpenMP provides a construct for automatically performing the transformation into and execution of a loop as tasks.

```
#pragma omp taskloop [clauses]
for (<canonical form loop range>)
<loop body>
```

A taskloop is initiated by a single thread in a parallel region. A taskloop does not take a `schedule()` clause (scheduling is done by the task scheduling algorithm); instead, the size of the parts of the iteration range can be controlled by a `grainsize()` clause. Alternatively, the number of tasks across which the loop is split can be set with the `num_tasks()` clause. Nested loops can be collapsed with the `collapse()` clause. Also, reductions can be performed over task loops by adding a `reduction()` clause. Prefix sums `inscan`-reductions are not allowed, though (think about reasons why).

2.3.17 Parallelizing Loops with Hopeless Dependencies

Instead of completely giving up on (not parallelizing) loops with dependency patterns that cannot be handled by reductions, scans or any other of the means that we have seen, OpenMP makes it possible as a last resort to mark a part of the loop code as having to be executed in the sequential iteration order. This is done by the `ordered` parallel for loop clause and by marking the section of code that has to be done in the sequential iteration order with the corresponding OpenMP construct.

```
#pragma omp ordered
<structured statement>
```

2.3.18 Example: Parallelizing a Sequential Algorithm with Dependencies

The prime sieve of Erathostenes is an amazing recipe for listing all prime numbers in increasing order starting at 2 (the first prime) up to some given n. The idea is this. Write all (optimization: odd) numbers in a list. Start going through the list. The first number (2) is a prime, write this down, and cross out all multiples of this prime on the list. Go to the next number still on the list (3). This must be a prime, otherwise it would have been crossed out by being a multiple of some earlier found prime. Write it down, and cross out all multiples of this prime. Continue like this $(5, 7, 11, \ldots)$ until all numbers have been considered.

The function `primesieve()` below implements Erathostenes prime sieve with a few clever, well-known optimizations worth pondering.

```
int primesieve (int n, int primes [])
{
  int i, j, k;
  unsigned char *mark;
  mark = (unsigned char*) malloc (n*sizeof (unsigned char));

  for (i=2; i<n; i++) mark[i] = 1; // possibly prime
  k = 0;
  for (i=2; i*i<n; i++) {
    if (mark[i]) {
      primes[k++] = i; // list prime
      for (j=i*i; j<n; j+=i) mark[j] = 0; // j not prime
    }
  }
  for (; i<n; i++) {
    if (mark[i]) primes[k++] = i; // list remaining primes
  }
  free (mark);

  return k;
}
```

First, if some i is composite, $i = pq$, then either $p \leq \sqrt{i}$ or $q \leq \sqrt{i}$. Therefore, it suffices to examine the list only up to $\sqrt{n}$. At iteration i, multiples of all $j < i$ have been crossed out, therefore if i is found to be prime and is still on the list (`mark[i]` is **true**), then $2i, 3i, \ldots (i-1)i$ have already been crossed out. It is, therefore, correct to start the crossing out from index i^2. The algorithm performs $O((n-i)/i)$ operations for the crossing out for each found prime $i, 2 \leq i < n$. The complexity is thus bounded by $O(\sum_{i=2,\text{isprime}(i)}^{n} n/i)$ which is in $O(n \log \log n)$ by a result from number theory, see for instance [56, Theorem 427].

Proposition 1. *The prime sieve algorithm lists all primes in the range from 2 to n in $O(n \log \log n)$ operations.*

The idea and the program above have obvious room for parallelization and some obstacles in the form of dependencies. The inner loop for crossing out multiples is a standard loop of independent iterations and straightforwardly parallelizable with OpenMP. It is performed $\sqrt{n}$ times. However, the final loop for compacting the marked primes has our now well-known dependency on k. Still, we might just give up and mark these iterations as to be performed in the sequential order.

```
#pragma omp parallel for private(i) schedule(static,1024)
for (i=2; i<n; i++) mark[i] = 1; // possibly prime

k = 0;
for (i=2; i*i<n; i++) {
  if (mark[i]) {
    primes[k++] = i;
#pragma omp parallel for private(j) schedule(static)
    for (j=i*i; j<n; j+=i) mark[j] = 0; // j not prime
  }
}
j = i;
#pragma omp parallel for ordered
for (i=j; i<n; i++) {
  if (mark[i])
#pragma omp ordered
    primes[k++] = i;
}
```

Performance will likely not be any better than just performing the loop sequentially: there is no other work in the loop that could benefit from being executed in parallel. It is instructive to try it out. However, by now, the array compaction pattern is familiar (see Sect. 1.3.11). We can parallelize either manually by using any one of our algorithms for computing the prefix sums of the mark array or by using the OpenMP construct for capturing the prefix sums in a loop reduction.

```
#pragma omp parallel for private(i) schedule(static,1024)
for (i=2; i<n; i++) mark[i] = 1; // possibly prime

k = 0;
for (i=2; i*i<n; i++) {
  if (mark[i]) {
    primes[k++] = i;
#pragma omp parallel for private(j) schedule(static)
    for (j=i*i; j<n; j+=i) mark[j] = 0; // j not prime
  }
}
```

```
j = i;
#pragma omp parallel for reduction(inscan,+:k)
for (i=j; i<n; i++) {
  if (mark[i]) primes[k] = i;
#pragma omp scan exclusive(k)
  if (mark[i]) k = k+1;
}
```

Again, it is instructive to compare this parallelization against the version with the `omp ordered` clause and, of course, against the sequential implementation to estimate the achievable (relative) speed-up. The parallel time complexity of the parallel solution with prefix sums compaction is clearly $O(\sqrt{n})$ (why?).

2.3.19 Cilk: A Task Parallel C Extension

Cilk (alluding to "silk", C and "ilk") is (was) a C language extension for task parallel programming originally developed at MIT in the mid-1990ties. It focuses on provably efficient execution by the runtime system of dynamically generated acyclic task graphs [8,20,21,73]. Cilk was supported by `gcc` and other compilers for a number of years, but is unfortunately being deprecated since 2018. However, there is recently an open version called OpenCilk, see www.opencilk.org [73,95]. The OpenMP task model has surely been inspired by Cilk, and Cilk programs can easily be reimplemented with OpenMP. Cilk provides three new keywords to C.

```
cilk_spawn <function call>
cilk_sync
cilk_for (<canonical form iteration space>) <loop body>
```

Generation of tasks is called *spawning* in Cilk, and the `cilk_spawn` keyword indicates to compiler and run-time system that a function or procedure call can be executed as a task, thus, concurrently with other tasks on processor-cores that may be available. This corresponds to the `omp task` construct, which is, however, more general: With OpenMP an arbitrary structured statement can be wrapped as a task. Directly spawned children tasks will be waited for at the end of the statement block doing the task spawns. If waiting for the directly spawned tasks to complete is required (as in the search program discussed in Sect. 2.3.13) the keyword `cilk_sync` can be used, much in the same way as `omp taskwait`. Finally, the `cilk_for` keyword is used as a shorthand for parallelizing loops as collections of tasks, much in the same way as `omp taskloop`.

Cilk has no explicit concept of threads. The `cilk_spawn` construct indicates that a function or procedure call *may* be executed concurrently with the code following the spawn (called the *continuation*); but not *how* or by which processor-core or thread this is done. The `cilk_sync` construct introduces a dependency point where the execution must wait for the spawned calls to complete. By removing the `cilk_spawn` and `cilk_sync` keywords and replacing `cilk_for` with a C for a Cilk program should be a correct, sequential C program. This is sometimes helpful for understanding the Cilk semantics (a similar observation, btw., should hold for OpenMP programs). The Cilk runtime system executes spawned threads with a clever *work-stealing* algorithm. In the multi-threaded runtime system, threads execute spawned tasks from a local task-queue and, when running out of local tasks, *steal* tasks from other runtime threads. They continue this until there are no more tasks to be executed. The Cilk constructs give rise to highly structured, acyclic tasks graphs, so-called *(fully) strict computations*. For (fully) strict computations with $T_1(n)$ total work and $T\infty(n)$ work on the longest path, it can be shown that the computation can be completed in $O(T_1(n)/p + T\infty(n))$ expected time steps by the work-stealing runtime system on a dedicated parallel shared memory computing system with p processors running the p worker threads [8]. This is within a constant factor of optimal. In this sense, Cilk comes with a provably efficient runtime system. The Cilk runtime work-stealing algorithm implements a randomized, greedy scheduling strategy. A similar work-stealing algorithm is most likely also the basis for the OpenMP runtime system for executing OpenMP tasks. This, however, is deliberately not specified by the OpenMP standard.

As seen with the OpenMP examples, task parallel programs often follow from recursive, divide-and-conquer algorithms where the recursive calls are independent of each other. This was the case with the search algorithm and the Quicksort example. Also, sorting by merging can be expressed in this way. Runtime bounds for recursive algorithms, both with regard to the total number of work, and the work of a single path of recursive calls down to the base case, can often be expressed as recurrence relations. In many cases, the solutions follow directly from the Master Theorem 9; if not, the recurrence must be solved (by induction) by hand.

For standard implementations of Quicksort and mergesort such analyses reveal that $T_1(n) = O(n \log n)$ and $T\infty(n) = O(n)$. The parallelism is modest in $O(\log n)$, meaning that linear speed-up can be achieved only for a modest range of processor-cores and threads. In the two cases, the bottlenecks were the sequential partitioning step and the sequential merge operation. To achieve more parallelism, parallel algorithms for the bottleneck operations must be found.

In Sect. 1.4.1, several parallel approaches were given for merging in parallel in $O(n/p + \log n)$ time steps. A drawback of these algorithms for implementation as task parallel algorithms (with no explicit notion of threads) is that the number of processors p is used and must be known. The final algorithm in this part of the lecture notes is a different, recursive divide-and-conquer merging algorithm that addresses these issues and can readily be implemented with Cilk and OpenMP tasks.

```
void par_merge(int A[], int n, int B[], int m, int C[])
{
  if (n<m) { // for the bounds, it must hold that n>=m
    int k;
    int *X;
    k = n; n = m; m = k;
```

2.3 Seventh Block (3 Lectures)

```
    X = A; A = B; B = X;
  }
  if (m==0) {
    par_copy(C,A,n); // copy in parallel
    return;
  }
  int r = n/2; // it holds that n>=m
  int s = rank(A[r],B,m); // determine rank of A[m] in B
  C[r+s] = A[r];
  cilk_spawn par_merge(A,r,B,s,C);
  cilk_spawn par_merge(A+r+1,n-r-1,B+s,m-s,C+r+s+1);
  cilk_sync; // not necessary, implicit in Cilk
}
```

The algorithm ranks the middle element of the larger of the arrays in the other array. It computes rank(A[$\lfloor n/2 \rfloor$], B) by binary search (see Sect. 1.4.2), which gives two pairs of sufficiently smaller subarrays that can be (recursively) merged together. In case a pair has an array without any elements, a (task) parallel copy operation is used to copy the other array to the output array. For the parallel recursion to terminate (and to avoid redundant computation on an element whose position in the output is now known), the element A[$\lfloor n/2 \rfloor$], which is larger than or equal to all previous elements in A and larger than or equal to B[s] and all previous elements in B, is written immediately to its correct position in the output array.

The recurrences, assuming for the two arrays that $n = m$ with total input size $2n$, are as follows.

- Work $T_1(n)$:

$$T_1(2n) = T_1(n/2 + \alpha n) + T_1(n/2 + (1-\alpha)n) + O(\log n)$$

for some $\alpha, 0 \le \alpha \le 1$ that can vary throughout the evaluation of the recurrence and corresponds to where the rank in the smaller array is found. The $n/2$ term is the split index of the larger array.

- Time $T\infty(n)$:

$$T\infty(2n) \le T\infty(3/2n) + O(\log n)$$

since the larger of the input arrays is always halved and in the worst case merged (recursively) with the other array of (at most) n elements.

The second recurrence can be solved by the Master Theorem (Case 2 with $a = 1, b = \frac{2n}{3/2n} = 4/3, d = 0, e = 1$) to give $T\infty(n) = O(\log^2 n)$, whereas the first requires a direct induction proof to give $T_1(n) = O(n)$. To see this, conjecture the solution to be

$$T_1(n) \le Cn - c\log_2 n$$

for constants C and c where the time to rank an element in a sequence of length n is at most $c \log n$. Using this as induction hypothesis, the recurrence relation now gives

$$\begin{aligned} T_1(2n) &\leq T_1(n/2 + \alpha n) + T_1(n/2 + (1-\alpha)n) + c \log_2 n \\ &= C(n/2 + \alpha n) - c \log_2(n/2 + \alpha n) + \\ &\quad C(n/2 + (1-\alpha)n) - c \log_2(n/2 + (1-\alpha)n) + c \log_2 n \\ &= C2n + c \log_2(n/2 + \alpha n) + c \log_2(n/2 + (1-\alpha)n) + c \log_2 n \ . \end{aligned}$$

Assuming the worst case in both logarithmic terms; that is, $\alpha = 1$ and $\alpha = 0$, respectively, gives

$$\begin{aligned} &C2n + c \log_2(n/2 + \alpha n) + c \log_2(n/2 + (1-\alpha)n) + c \log_2 n \\ &= C2n - 2c \log_2(3/2n) + c \log_2 n \\ &= C2n - 2c \log_2(3/2) - 2c \log_2 n + c \log_2 n \\ &= C2n - 2c \log_2 3 + 2c - c \log_2 n \\ &= C2n - 2c \log_2 3 + 2c - c(\log_2 2n - 1) \\ &= C2n - 2c \log_2 3 + c - c \log_2 2n \\ &= C2n - c(2 \log_2 3 - 1) - c \log_2 2n \\ &\leq C2n - c \log_2 2n \end{aligned}$$

using $\log_2 2n = \log_2 2 + \log_2 n = 1 + \log_2 n$ and $2 \log_2 3 - 1 > 0$ which then establishes the induction hypothesis.

We summarize in the following proposition.

Proposition 2. *The merging problem can be solved work-optimally with* $T_1(n) = O(n)$ *and* $T\infty(n) = O(\log^2 n)$.

By computing instead the co-ranks in A and B for index for $\lfloor (n+m)/2 \rfloor$ as described in Sect. 1.4.3 we could have found the exact indices of the parts of A and B to merge to get exactly the two halves of the resulting C array. This would have saved us from the (nevertheless illustrative) induction proof and would give a possibly faster algorithm. It is a good exercise to implement and compare the two possibilities.

The Cilk merge algorithm can now be plugged into another recursive, task parallel algorithm for sorting by merging. The algorithm works top-down by first mergesorting the two halves of the input array, and then merging together the two sorted halves.

```
void par_mergesort(int A[], int n)
{
  if (n==1) return;

  cilk_spawn par_mergesort(A,n/2);
  cilk_spawn par_mergesort(A+n/2,n-n/2);
```

```
  cilk_sync; // necessary
  // allocate temporary array B
  par_merge(A,n/2,A+n/2,n-n/2,B);
  par_copy(A,B,n);
  // free B again (possibly inefficient)
}
```

The algorithm is not an *in-place* algorithm since the additional n-element work array B is needed for the merging. which also necessitates an explicit copy back operation. For the complexity we have

$$T_1(n) = 2T_1(n/2) + O(n)$$

for the work and

$$T\infty(n) = T\infty(n/2) + O(\log^2 n)$$

for the parallel time complexity. Both recurrences can be solved by Case 2 of the Master Theorem 9 to give

$$T_1(n) = O(n \log n)$$
$$T\infty(n) = O(\log^3 n) \quad .$$

We summarize in the following proposition.

Proposition 3. *An n-element array can be sorted work-optimally by mergesort in $O(\frac{n \log n}{p} + \log^3 n)$ parallel time.*

2.4 Exercises

1. Does the PRAM model presuppose a cache-memory?
2. Implement the six sequential, loop based matrix–matrix multiplication algorithms corresponding to the six possible permutations of the order of the loops. Write each variant as a function with function prototype

    ```
    void mmm(const base_t* A, const base_t* B, base_t* C,
             const int m, const int l, const int n)
    ```

    for $m \times l$ matrix A, for $l \times n$ matrix B and for $m \times n$ matrix C represented as one-dimensional arrays with a user defined base type `typedef int base_t;` (where `int` can be replaced by any other arithmetic C datatype). Time the six variants for matrices of total size around $m \approx n \approx l \approx 1200$ elements with and without compiler optimization (-O3 and other). Investigate the effects of different base datatypes, e.g., `char`, `int`, `float` and `double`. Allocate the matrices in linear storage with `malloc(m*n*sizeof(base_t))` and typecast into a pointer to rows of the given number of elements.
3. Give and analyze a recursive algorithm for adding two $m \times n$ matrices.

4. Implement the recursive, divide-and conquer matrix–matrix multiplication algorithm for multiplying two $m \times l$ and $l \times n$ matrices into an $m \times n$-matrix. Represent submatrices as blocks inside larger $m \times n$-matrices defined by row start index and number of rows $i0, m0$ and column start index and number of columns $j0, n0$. Use, for instance, a C structure as shown below together with an iterative submatrix–submatrix multiplication procedure to handle the base cases.

```
typedef struct {
  int m, n;      // rows and columns of matrix
  int i0, m0;    // block start and size
  int j0, n0;
  base_t *M;     // the matrix elements, row after row
} matblk;

void mmmite(matblk A, matblk B, matblk C)
{
  // the matrices
  base_t (*a)[A.n] = (base_t(*)[A.n])A.M;  // the matrix
  base_t (*b)[B.n] = (base_t(*)[B.n])B.M;
  base_t (*c)[C.n] = (base_t(*)[C.n])C.M;
  int i, j, k;

  for (i=0; i<A.m0; i++) {
    for (j=0; j<B.n0; j++) {
      c[C.i0+i][C.j0+j] = 0;
    }
  }
  for (i=0; i<A.m0; i++) {
    for (k=0; k<A.n0; k++) {
      for (j=0; j<B.n0; j++) {
        c[C.i0+i][C.j0+j] +=
          a[A.i0+i][A.j0+k]*b[B.i0+k][B.j0+j];
      }
    }
  }
}
```

The recursive algorithm can then be written as outlined below (with only two of the 8 submatrix multiplications shown) where the matrices are cut roughly in half along the two dimensions. Complete the code and make sure it works (correctly) for any non-negative m, l, n.

```
void mmmrec(matblk A, matblk B, matblk C)
{
  if (A.m0<=CUT || A.n0<=CUT || B.n0<=CUT) {
    mmmite(A,B,C);
  } else {
    matblk A00;
    matblk A01;
    matblk A10;
    matblk A11;
```

2.4 Exercises

```
    matblk B00;
    matblk B01;
    matblk B10;
    matblk B11;

    // 4 intermediate results in two matrices
    matblk C0;
    matblk C1;

    C0.m = A.m0;
    C0.n = B.n0;
    C0.M = (base_t*)malloc(C0.m*C0.n*sizeof(base_t));
    base_t (*c0)[C0.n] = (base_t(*)[C0.n])C0.M;

    C1.m = C0.m;
    C1.n = C0.n;
    C1.M = (base_t*)malloc(C1.m*C1.n*sizeof(base_t));
    base_t (*c1)[C1.n] = (base_t(*)[C1.n])C1.M;

    A00.m = A.m;
    A00.n = A.n;
    A00.M = A.M;
    A00.i0 = A.i0;
    A00.m0 = A.m0/2;
    A00.j0 = A.j0;
    A00.n0 = A.n0/2;

    A01.m = A.m;
    A01.n = A.n;
    A01.M = A.M;
    A01.i0 = A.i0;
    A01.m0 = A.m0/2;
    A01.j0 = A.j0+A.n0/2;
    A01.n0 = A.n0-A.n0/2;

    B00.m = B.m;
    B00.n = B.n;
    B00.M = B.M;
    B00.i0 = B.i0;
    B00.m0 = B.m0/2;
    B00.j0 = B.j0;
    B00.n0 = B.n0/2;

    B10.m = B.m;
    B10.n = B.n;
    B10.M = B.M;
    B10.i0 = B.i0+B.m0/2;
    B10.m0 = B.m0-B.m0/2;
    B10.j0 = B.j0;
    B10.n0 = B.n0/2;

    C0.i0 = 0;
    C0.m0 = A00.m0;
    C0.j0 = 0;
```

```
        C0.n0 = B00.n0;

        C1.i0 = C0.i0;
        C1.m0 = C0.m0;
        C1.j0 = C0.j0;
        C1.n0 = C0.n0;

        mmmrec(A00,B00,C0);
        mmmrec(A01,B10,C1);

        // Analogous cases for the remaining C submatrices
        ...

        int i, j;
        base_t (*c)[C.n] = (base_t(*)[C.n])C.M;
        for (i=0; i<C.m0; i++) {
          for (j=0; j<C.n0; j++) {
            c[C.i0+i][C.j0+j] = c0[i][j]+c1[i][j];
          }
        }

        free(C0.M);
        free(C1.M);
    }
}
```

Experiment with good cut-off criteria for the base case; the one shown is just one, simple possibility. The explicit submatrix representations come with a certain redundancy: Improve the argument structure to get a leaner, more economic implementation; for instance, the matrix orders m, l, n can be factored out and kept constant throughout the recursions. Do your code improvements make any difference in running time?

5. Implement the blocked, 6-fold nested loop matrix–matrix multiplication algorithm. Compare to the best and worst of the 3-fold nested loop implementations. Estimate a good block (tile) size based on the size of the last-level cache in your processor. Conduct experiments using different base datatypes.
6. Implement in the style of the previous exercises a *fused-multiply-add* matrix–matrix operation working on $m \times l$ and $l \times n$ input matrices A and B and an input-output $m \times n$ result matrix C. The function shall compute $C \leftarrow C+AB$.

```
void fmma(const base_t* C,
          const base_t* A, const base_t* B,
          const int m, const int l, const int n) {
  base_t (*a)[l] = (base_t(*)[l])A;  // the matrices
  base_t (*b)[n] = (base_t(*)[n])B;
  base_t (*c)[n] = (base_t(*)[n])C;

  int i, j, k;
  // ... best three loops for multiply-add
}
```

2.4 Exercises

The code will be useful as a building block in both shared- and distributed matrix–multiplication implementations.

7. Complete and implement the false sharing example (update to elements in a C structure or a C array) from Sect. 2.1.6 with either OpenMP or pthreads. Compile with different optimization options (also entirely without optimization) and time the duration for a medium large number of updates, say r=10000. What is the observed difference in running time between the three cases (false sharing with struct, false sharing with array, no false sharing via local variables)? Why is the difference between the optimized and non-optimized versions so large? Explain by studying the generated assembler code.

8. Consider the false sharing example from Sect. 2.1.6 with an array of 16 int elements. Eliminate false sharing by padding each element of the array to occupy a full cache line (on your processor), either by accessing the array in strides of 16 (your cache line size) or by declaring the array to be of structures occupying a full cache line. Benchmark against the original version and comment on the results (performance differences).

9. The following two code snippets compute the in- and out-degree of a graph with n vertices represented by its 0/1 adjacency matrix A[n][n].

```
int A[n][n]; // sample declaration; beware for large n
int indeg[n];

for (i=0; i<n; i++) indeg[i] = 0;
for (j=0; j<n; j++) {
  for (i=0; i<n; i++) indeg[j] += A[i][j];
}
```

```
int A[n][n]; // sample declaration; beware for large n
int outdeg[n];

for (i=0; i<n; i++) outdeg[i] = 0;
for (i=0; i<n; i++) {
  for (j=0; j<n; j++) outdeg[i] += A[i][j];
}
```

The code snippets are executed on a single processor with a directly mapped cache and the adjacency matrix and all n element arrays are larger than what can fit in the cache, i.e., the cache capacity is less than n integers. The block size (cache line size) of the cache is 64 bytes and a cache block can therefore hold 16 consecutive integers starting from a block-aligned address in memory. For simplicity, we assume that there is no hardware prefetching mechanism. What is the read cache miss rate of the in-degree computation? What is the read cache miss rate of the out-degree computation? Can the read cache miss rate of either of the codes be improved? Write the corresponding program(s). Run the code snippets on your own computer and check for differences in running time. Consider how to ensure that the matrix and the arrays are not in cache prior to each time measurement.

10. Consider the following standard, sequential merging algorithm to be run on a single core with a certain cache system.

```
i = 0; j = 0; k = 0;
while (i<n&&j<m) {
    C[k++] = (A[i]<=B[j]) ? A[i++] : B[j++];
}
while (i<n) C[k++] = A[i++];
while (j<m) C[k++] = B[j++];
```

The inputs are two large arrays of integers, A and B, with n elements each (both n and m set to n). The output array C contains $2n$ elements. All arrays are much larger than can fit in the small cache. The block size of the cache is 16 integers à 4 bytes each, thus, 64 bytes in total. The cache is directly mapped. The cache is write non-allocate, so that writes to C will not cause cache misses (this may not correspond to any existing cache system). It also means you can ignore array C when investigating the cache behavior. The input arrays have been allocated in such a way that each element $A[i]$ goes to the same cache line as element $B[i]$ (same index i).
In order to analyze the cache behavior of the merging algorithm under these conditions, consider different possible inputs. Assume that once an array element (e.g., A[i]) has been accessed (loaded from cache, with or without a cache miss), it will be in a register. Thus, subsequent accesses to a specific element (e.g., A[i]) will not be counted as a cache access.

a. Construct a *best case* input for A and B leading to the smallest number of cache misses (hint: how should the values inside A and B look like to avoid cache misses?). Give the cache miss rate for the $2n$ iterations of the merging algorithm.

b. Construct a *worst case* input for A and B with the largest number of cache misses. Give the cache miss rate for the $2n$ iterations of the merging algorithm.

11. In Sect. 2.2.3, creating and starting p **pthreads** threads was done in a sequential loop and, thus, in $\Omega(p)$ time steps; this may be too slow for a large numbers of threads on a multi-core processor with a larger number of processor-cores, especially if the p threads are started and stopped many times. Show how to create the p threads in $\Omega(\log p)$ operations by using a recursive, tree-like algorithm where each started thread creates two (or some other, constant number of) new threads. Do this in such a way that the threads are also correctly terminated (with **pthread_exit()** and **pthread_join()**.

12. Remove the race condition from the **pthreads hello()** program by protecting the **printf()** call with a **pthreads** mutex that can be accessed through the arguments to the **hello()** function. Enhance the program with a shared counter to let the threads also print the order in which they are activated (the order in which they acquire the mutex).

13. Implement a linear pipeline with **pthreads**. Each stage in the pipeline is represented by a thread that gets a work item from the predecessor thread in

the pipeline, processes the item and signals the successor thread that a work item is ready. The first and last threads are special. The first thread generates the work items or reads them from the input. The last thread consumes the final output for each work item and stores it as output. Termination can be detected by sending a null-work item through the pipeline. For inspiration, see, for instance, the book by Rauber and Rünger [90].

14. Complete and implement the `primes_race()` and `primes_atomic()` examples with `pthreads` and a simple, pseudo-polynomial `isprime(i)` predicate-function (check for divisibility from $2, 3, \ldots, \lfloor\sqrt{i}\rfloor$). Run on your system with different (larger) numbers of threads and compare the outcome in number of primes found. You may verify against the OpenMP prime-sieve implementation.

15. Implement a simple mutex (lock) with lock and unlock operations using the C TAS instruction `atomic_flag_test_and_set()`. Define a worst-case throughput benchmark in which threads for some allocated time slot aggressively lock and unlock but actually do nothing useful in their critical section. Compare the implemented lock to the `pthreads` mutex with this benchmark for varying numbers of threads.

16. A potential improvement of the TAS lock from the previous exercise is to use the possibly expensive TAS operation speculatively by first checking with a non-atomic read operation whether the lock may be free:

```
volatile int locked;
while (!locked) {
   while (locked); // spin on lock non-atomically
   locked = atomic_flag_test_and_set(*lock);
}
```

Compare the performance of this lock against the TAS implementation with increasing numbers of threads. Are there notable performance differences? This technique is known as Test-And-Test-And-Set (TATAS). Why is the `volatile` qualifier needed? What is it supposed to ensure or prevent?

17. Compare the predefined `pthreads` readers-writer mutex against the one written using condition variables by creating a suitable benchmark: What is the mutex throughput in number of operations over a certain time slice? Do the implementations behave similarly (number and distribution of threads being granted mutual exclusion)? Make it possible to configure the number of readers and writers and the distribution of the operations to be performed.

18. In an OpenMP program, some unit of `work()` per thread is to be repeated a number of times until some `done` condition is fulfilled. The `work()` function is assumed to be thread-safe. There are (at least) two natural ways of expressing this.
The first alternative is:

```
while (!done) {
#pragma omp parallel
  {
    int t = omp_get_thread_num();
```

```
      work(t); // do some work
   }
   done = ...;
}
```

The second alternative is:

```
#pragma omp parallel
  {
    int t = omp_get_thread_num();
    while (!done) {
      work(t); // do some work

      done = ...;
    }
  }
```

Explain the differences and discuss advantages and pitfalls of one alternative over the other (performance, thread activation, coordination between threads, barrier synchronization, race conditions, etc.). What are requirements for the work() function for either of the two styles?

19. Consider the following two OpenMP program snippets:

```
int i, n;
#pragma omp parallel private(i)
{
   for (i=0; i<n; i++) {
     // loop work O(1) per iteration
   }
}
```

and

```
int i, n;
#pragma omp parallel for
for (i=0; i<n; i++) {
  // loop work O(1) per iteration
}
```

Explain the differences between the two cases, in particular which work is performed and how it is shared. Why is the `private(i)` declaration in the first case needed? Assume that the work per iteration is constant $O(1)$. When executed with p threads, what is the total work as a function of n and p?

20. Consider the following OpenMP program fragment.

```
int b[2000];
int i;
for (i=0; i<2000; i++) b[i] = -i;
#pragma omp parallel private(b)
{
   int i;
   for (i=0; i<2000; i++)
     assert(b[i]==-i); // first assertion
   for (i=0; i<2000; i++) b[i] = omp_get_thread_num();
```

2.4 Exercises

```
        for (i=0; i<2000; i++)
            assert(b[i]==omp_get_thread_num()); // second
    }
```

Assume the program is executed with $p, p > 1$ threads. Explain which threads execute which iterations of the inner loops in the `parallel` region. Explain why there are no race conditions on the updates to `b[i]`. Explain why the first assertion is violated. Propose, using only OpenMP clauses, a way to make the assertion hold, regardless of p. Explain why the second assertion holds. Assume now, after your repair, that the initialization in the master thread is changed to

```
    int c[2000];
    int *b = c;
```

Will the repair still work? Why is the second assertion violated?

21. Let two ordered arrays of n and m elements, respectively, that have to be merged stably into a single, ordered array of $n+m$ elements be given. Use the co-ranking preprocessing idea of Sect. 1.4.3 to let each of p threads compute the co-rank indices in the two input arrays needed to make it possible for each thread to sequentially merge segments of the input arrays into a unique segment of the output array of roughly $\frac{n+m}{p}$ elements. Implement this algorithm with OpenMP using a single, parallel region. How does the parallel implementation compare against your best known, sequential merge implementation (which should be used as a subroutine in the parallel implementation)? How many co-rank computations are needed per thread? How many explicit (and implicit) barrier synchronization operations do your implementation need? With how few can you do?

22. In order to understand the OpenMP schedules and the way loop iterations are assigned to threads, it is instructive to run the example code from the end of Sect. 2.3.10 that records the threads assigned to iterations and counts the number of iterations performed per thread. Run the example with $1, 2, 4, 5, 9, 30$ threads by setting the `OMP_NUM_THREADS` environment variable. Run with schedules like "static", "static,1", "static,4", "static,7", "static,9", "dynamic", "dynamic,2", "dynamic,3", "dynamic,10", "guided" and "guided,10" by setting the `OMP_SCHEDULE` environment variable, and check the outcome to verify your understanding. Try with a larger loop range; experiment with different loop starts and loop increments.

23. In order to understand the scheduling of sections, complete and run the following code with varying numbers of threads.

```
    int s, a, b, c, d; // shared counters
    int i, r = 10; // some number of repetitions
    int t = omp_get_max_threads();
    int thrd[t];
    for (i=0; i<t; i++) thrd[i] = 0;
    s = 0; a = 0; b = 0; c = 0; d = 0;
    #pragma omp parallel private(i)
    {
```

```
         for (i=0; i<r; i++) {
#pragma omp sections reduction(+:s) nowait
         {
#pragma omp section
           {
             a++; s++;
             thrd[omp_get_thread_num()]++;
           }
#pragma omp section
           {
             b++; s++;
             thrd[omp_get_thread_num()]++;
           }
#pragma omp section
           {
             c++; s++;
             thrd[omp_get_thread_num()]++;
           }
#pragma omp section
           {
             d++; s++;
             thrd[omp_get_thread_num()]++;
           }
         }
      }
   }
   for (i=0; i<t; i++) {
      printf("Thread %d: %d\n",i,thrd[i]);
   }
   assert(a+b+c+d==s);
   assert(s==4*r);
```

24. Implement a parallel copy function `par_copy(int a[], int b[], int n)` as a simple parallel loop.

```
#pragma omp parallel for schedule(runtime)
for (i=0; i<n; i++) {
   a[i] = b[i];
}
```

Benchmark the performance with medium large, dynamically allocated arrays of, say, $n = 1\,000\,000$, $n = 10\,000\,000$ and $n = 100\,000\,000$ elements for different number of threads. Experiment with "static" and "static,1" schedules (and other schedules) by setting the OMP_SCHEDULE environment variable accordingly. Compare outcomes and explain the differences. Run with $1, 48, 10, 100$ threads by setting the OMP_NUM_THREADS environment variable accordingly. Try replacing the parallel for directive with taskloop. What is a suitable grainsize? Does the performance depend on how the arrays are initialized (first touched)? More concretely, initialize both arrays before the copy loop each with a simple, sequential loop setting each entry to, say, -1. Experiment with sequential initialization and with parallel initialization with

2.4 Exercises

both the same as well as with a different schedule from the one used in the copy loop. Explain possible performance differences.

25. Repeat the previous exercise with gather and scatter loops (see Sect. 1.3.11) instead of the simple copy loop. Use different index arrays, first permutations (identity permutation, pairwise swaps, reverse order, random permutation, etc.), second possibly surjective index arrays (all indices to 0, etc.).

```
#pragma omp parallel for schedule(runtime)
for (i=0; i<n; i++) {
  a[i] = b[ix[i]];
}
#pragma omp parallel for schedule(runtime)
for (i=0; i<n; i++) {
  a[ix[i]] = b[i];
}
```

26. Consider the following C function for doing matrix–vector multiplication.

```
int seq_mv(int *A, int m, int n, int *b, int *c)
{
  int (*a)[n] = (int(*)[n])A; // the matrix
  int i, j;

  for (i=0; i<m; i++) c[i] = 0;
  for (i=0; i<m; i++) {
    for (j=0; j<n; j++) {
      c[i] += a[i][j]*b[j];
    }
  }
}
```

Parallelize this function with straightforward omp parallel for loop parallelization. Which loops can be parallelized and which cannot? Can both loops be parallelized and collapsed? Run the code for a number of repetitions, say 30, determine the time per repetition (use omp_get_wtime() to time all repetitions, compute average), for inputs with $m = 600$ and $n = 1003$ for different numbers of threads. Try different loop schedules. Compute the speed-ups relative to the sequential (best known?) code.

27. In oder to understand the transformation and scheduling of collapsed OpenMP loops, write a simple, three-loop nest

```
#pragma omp parallel for collapse(3) schedule(runtime)
for (i=0; i<m; i++) {
  for (j=0; j<n; j++) {
    for (k=0; k<l; k++) {
      // keep book: which thread did what?
    }
  }
}
```

Insert code to record which thread did which iteration (i, j, k) and how many iterations each thread performed. Experiment with different schedules similarly to Exercise 22.

28. Use OpenMP to parallelize the six different, three loop algorithms for matrix–matrix multiplication from Exercise 2 with m, n and l as the input size parameters. Which loops can be immediately parallelized? Which loops can be collapsed? Benchmark the six implementations for input sizes around 1000 and 2000 elements per matrix dimension for varying numbers of threads. Perform a number of repetitions and compute either average or median values for the running times. Compute the speed-up relative to the best performing sequential variant (loop order). Is the cache-behavior different in the sequential and parallel cases?

29. Consider again the straightforward, two loop implementation of matrix–vector multiplication. In contrast to the parallelization in Exercise 26, now parallelize the innermost loop doing the actual product summations by using a parallel reduction() clause. Verify correctness against the sequential code. Benchmark and compare the running time for different numbers of threads and total matrix sizes to both the sequential implementation and the simple loops parallelizations from Exercise 26.

30. Run and benchmark the matrix–vector multiplication code of Exercise 26 again with small m (on the order of the number of threads) and (very) large n and different numbers of threads. Try with schedules "static,1", "static,16" and the default "static". Explain the observed differences in performance.

31. Implement a function par_max(a,n) for finding the maximum of n numbers (of some C base type, int, double, ...) in an array a. Use a simple loop and parallelize it with a reduction(max:...) clause. Is this better than the simple, sequential loop? For specific input sizes, what is the number of threads leading to the smallest running time?

32. Implement the task parallel function occurs(x,a,n) of Sect. 2.3.13 for counting the number of occurrences of x in array a as a parallel loop. Which OpenMP constructs can be used to do this correctly? Which of the two implementations can give speed-up with more than one thread, which is faster? What can you do to improve the performance of the task parallel implementation?

33. Show how a simple for-loop for (i=0; i<n; i++) ... in canonical form can be parallelized with OpenMP tasks by giving a recursive algorithm for splitting the iteration range into single iterations each to be performed as a task. The algorithm is analogous to the task parallel occurs(x,a,n) function of Sect. 2.3.13. As in Exercise 22, instrument the loop body with performance counters that record for each iteration executed as a task which thread executed that iteration/task and for each thread how many tasks (iterations) that thread executed. What is the overhead in terms of number of tasks that split loop ranges and do not directly perform iterations? Run with very large iteration ranges (be careful with output) and with different numbers of threads. Is the work well divided between the threads?

34. Consider the following OpenMP program fragment.

```
int cc;
#pragma omp parallel default(none) shared(cc)
{
   int r;
#pragma omp single
   cc = 0;
   for (r=0; r<omp_get_num_threads(); r++) {
#pragma omp single
      cc++;
      assert(cc-1==r);
#pragma omp barrier
   }
}
```

Explain why the assertion will hold. Explain why the explicit omp barrier is necessary. What happens if the omp single inside the loop is given a nowait clause? What about the omp single before the loop? What happens if omp single is replaced with omp master? What happens if omp single is replaced with omp critical? Formulate an assertion that will always hold for this last case.

35. The following program fragment uses single work sharing, critical sections, atomic updates and reductions to update global variables. Apparently, the intention is that a is incremented once per iteration up to iter, and that the other variables are incremented by each thread in each iteration.

```
int a, b, c, d;
a = 0; b = 0; c = 0; d = 0;
#pragma omp parallel private(a) reduction(+:d)
{
   int i;
   for (i=0; i<iter; i++) {
#pragma omp single
      a++;
#pragma omp critical
      b++;
#pragma omp atomic update
      c++;
      d++;
   }
}
```

What is are the values of a, b, c and d after iter=20 iterations and $p = 8$ threads? Does this correspond to the intended outcome? Is the program correct (no race conditions)? What must possibly be done to make the program correct with the intended outcome?

36. Define for a not necessarily ordered n-element array A rank(i, A, n) to be the number of elements $A[j]$ with either $A[j] < A[i]$ or $A[j] = A[i] \vee j < i$. Suggest an OpenMP loop to parallelize the computation of rank(i, A, n). The function (not necessarily the parallelized version) is used in the following piece of code that moves the elements of A to a new array B.

```
for (i=0; i<n; i++) {
  int j = rank(i,A,n);
  B[j] = A[i];
}
```

What does this code accomplish? What is the sequential complexity as a function of n? Find another parallelization instead of relying on the parallelized rank(i, A, n) function. Compare the resulting two parallel implementations, also in terms of performance. Can loop collapsing be used?

37. Implement the shared-counter solution for prime finding discussed in Sect. 2.2.11 in OpenMP in two versions. Let `int next;` be a shared counter initialized to 2, the first prime. Activate the threads in a parallel region where they each iterate until the counter has reached the upper limit for the range to be checked. In the first solution, use `#pragma omp atomic capture` to atomically read and update `next`. In the second solution, read and update under mutual exclusion using `#pragma omp critical` to declare the critical section. Time the prime checking for up to 1 000 000, 10 000 000 and, if possible 100 000 000 integers (repeat a number of times) for different numbers of threads, and compare the two solutions.

38. Implement the recursive inclusive prefix sums algorithm described in Sect. 1.4.7 as a C program with OpenMP. Benchmark against a best known sequential implementation with arrays of $n = 100\,000$, $n = 1\,000\,000$, and $n = 10\,000\,000$ elements (of C `int` and/or `double` type), respectively. What speed-up can be achieved? In case speed-up is below expectation, explain possible reasons.

39. Implement the iterative inclusive prefix sums algorithm described in Sect. 1.4.9 as a C program with OpenMP. Benchmark against a best known sequential implementation with arrays of $n = 100\,000$, $n = 1\,000\,000$, and $n = 10\,000\,000$ elements (of C `int` and/or `double` type), respectively. What speed-up can be achieved? In case speed-up is below expectation, explain possible reasons.

40. Implement the Hillis–Steele inclusive prefix sums described in Sect. 1.4.10 as a C program with OpenMP. Benchmark against a best known sequential implementation with arrays of $n = 100\,000$, $n = 1\,000\,000$, and $n = 10\,000\,000$ elements (of C `int` and/or `double` type), respectively. What speed-up can be achieved? In case speed-up is below expectation, explain possible reasons.

41. Improve either of the prefix sums algorithms implemented in the previous exercises by using blocking (see Sect. 1.4.11). This means that the p threads first sequentially, but in parallel, preprocess disjoint parts of the input array to arrive at a prefix sums problem of size $O(p)$ which is then solved by a parallel prefix sums function. Post-processing will in addition be necessary to arrive at the solution to the original prefix sums problem. Benchmark against a best known sequential implementation with arrays of $n = 100\,000$, $n = 1\,000\,000$, and $n = 10\,000\,000$ elements (of C `int` and/or `double` type), respectively. Compared to your direct implementation from either of the previous exercises, how much improvement could be achieved?

42. Discuss algorithmic and architectural obstacles to obtaining full linear (perfect) speed-up for the OpenMP prefix sums implementations.

2.4 Exercises

43. Implement recursive matrix–matrix multiplication based on your sequential implementation from Exercise 4 using OpenMP tasks to handle the recursive calls. Compare the results to the best parallelization of standard, three-loop matrix–matrix multiplication of Exercise 28. Experiment with different matrix sizes (say, order $n = 850, n = 1000, n = 1200$). Experiment with different recursion cut-off strategies and values. How good performance can you achieve in comparison? You can try to implement the cut-off with OpenMP clauses.
44. The following part of a larger program shows two ways of computing the sum of n numbers (here the numbers are all just ones).

```
int jr = 0; // reduction-sum
int ja = 0; // atomic-sum
#pragma omp parallel for reduction(+:jr)
for (i=0; i<n; i++) {
  jr += 1;
#pragma omp atomic update
  ja += 1;
}
assert(jr==ja);
```

Explain the differences between the two possibilities. Run the code and look for differences in performance.

45. The following part of a larger program shows two ways of computing indices and storing them in an array.

```
int us[n];
int ua[n];
int jr = 0;
int ja = 0;
#pragma omp parallel for reduction(inscan,+:jr)
for (i=0; i<n; i++) {
  us[i] = jr;
#pragma omp scan exclusive(jr)
  jr += 1;
}
#pragma omp parallel for
for (i=0; i<n; i++) {
#pragma omp atomic capture
  ua[i] = ja++;
}
assert(jr==ja);
```

Explain the differences between the two possibilities. Run the code and look for differences in performance and outcome. You will be surprised.

46. Consider the following loop which is parallelized as a collection of tasks. The loop computes for each thread how many of the iterations were done by that thread (as done in Sect. 2.3.10).

```
int t = omp_get_max_threads();
int iter[t];
for (i=0; i<t; i++) iter[i] = 0;
// with task loop
```

```
#pragma omp parallel
{
#pragma omp taskloop
  for (i=0; i<n; i++) {
    iter[omp_get_thread_num()]++; // careful with large n
  }
}
nn = 0;
for (i=0; i<t; i++) nn += iter[i];
assert(nn==n); // all iterations done
```

Run with a large n. Why does the assertion not hold? Repair the program by inserting an additional, single work sharing construct. Compare the running time, now with large n, of the loop against the running time of the same loop parallelized with a standard, scheduled omp parallel for pragma. Try computing the total number of operations performed into a new, shared variable int nnn = 0; by using a reduction() clause. Does this have an effect on the time taken to execute the loop?

47. Use the Master Theorem 9 as a blueprint for a program to explore the way OpenMP may schedule tasks to threads. The following function follows the pattern of the recurrences covered by the Master Theorem for integer coefficients a, b, d, e and generates tasks for each evaluation of the recurrence.

```
void masterrecursion(const int a, const int b,
                     const int d, const int e,
                     int n, int ops[], int tsk[])
{
  tsk[omp_get_thread_num()]++;
  if (n<=1) {
    ops[omp_get_thread_num()]++;
  } else {
    int i, k;
    int nn = 1;
    for (i=0; i<d; i++)   nn *= n;
    k = 0;
    for (i=1; i<=n; i*=2) k++;
    int nk = 1;
    for (i=0; i<e; i++)   nk *= k;
    ops[omp_get_thread_num()] += nn*nk;

    for (i=0; i<a; i++) {
#pragma omp task
      masterrecursion(a,b,d,e,n/b,ops,tsk);
    }
  }
}
```

Execute the function in a parallel region with values for a, b, d, e as found in the different recurrences discussed in the text, and different values for n.

```
#pragma omp parallel
  {
#pragma omp single
```

```
        masterrecursion(a,b,d,e,n,ops,tsk);
    }
```

Extend the `masterrecursion` function to count atomically the total number of tasks and the total number of operations, and verify against the per thread computed values.

48. Is the usage of the `omp barrier` construct below legal? State and add an assertion on j that will hold after the loop.

```
int i, j;
j = 0;
#pragma omp parallel private(i) firstprivate(j)
{
    for (i=0; i<500; i++) {
        if (i%2==0) {
#pragma omp barrier
            j++;
        } else {
            j--;
#pragma omp barrier
        }
    }
}
```

49. Complete the implementation of the task parallel Quicksort using OpenMP tasks from Sect. 2.3.13. Benchmark your implementation with inputs of $n = 1\,000\,000$, $n = 10\,000\,000$ and $n = 100\,000\,000$ elements of basetype `int` and `double` and different input permutations. What is the largest speed-up achievable compared to a sequential Quicksort implementation on your system and with how many threads? Experiment with different cut-off value to end the recursion earlier. Time permitting, experiment with different pivot-selection strategies.

50. Analyze the following variant of Quicksort where the partitioning step is done by a work-optimal parallel algorithm and the recursive calls are done sequentially. What cut-off is required in order to make the algorithm work- and cost-optimal?

```
void Quicksort(int a[], int n)
{
    int i, j;
    int aa;

    if (n<CUTOFF) {
        // sort sequentially
        return;
    }

    // pivot selection
    int pivot = a[0]; // choose an element (non-randomly...)

    // partition
    // mark elements <pivot
```

```
// array compaction, last index j
// mark elements >=pivot
// array compaction with pivot in a[j]

Quicksort(a,j);
Quicksort(a+j+1,n-j-1);
}
```

Assume that an exclusive prefix sums computation can be done in parallel by a function `exclusive_prefix_sums(a,n)` (see previous exercises) and use this to write out the Quicksort code. Implement and benchmark this algorithm against the task parallel formulation.

51. Implement a parallel partition function which partitions the elements into three blocks: those smaller than the given pivot, those equal to the pivot, and those larger than the pivot.
52. Show how to implement Quicksort partitioning in parallel using OpenMP scan-reduction. Benchmark your implementation against a best known sequential implementation and a parallel implementation using a hand-written, adapted prefix sums computation.
53. Show how to implement Quicksort partitioning in parallel using OpenMP atomics to maintain indices in arrays for the classes of elements (smaller than, larger than the pivot). Compare the performance of this implementation against an implementation with OpenMP scan-reductions (previous exercise).
54. Use the parallel merge function implemented in Exercise 21 to devise a parallel mergesort algorithm based on co-ranking. What is the asymptotic running time of your implementation? How does the cut-off at which we change to sequential merging have to be chosen in order to arrive at a cost-optimal algorithm? What is the parallel running time of your algorithm including (or excluding) the time for explicit and implicit barrier synchronizations? Implement your algorithm in OpenMP and benchmark against your best, sequential mergesort implementation.
55. Implement the prime sieve algorithm in the three versions discussed in Sect. 2.3.18 in OpenMP: sequential, parallelization with ordered and parallelization with inscan-reduction. You may consider also using an own, hand-written prefix sums algorithm for the array compaction. Benchmark and compute the (relative) speed-up for finding primes up to $n = 1\,000\,000$, $n = 10\,000\,000$ and $n = 100\,000\,000$ with different number of threads. Verify correctness by comparing to the work pool shared counter implementation.
56. Implement the task parallel merge from Sect. 2.3.19 with OpenMP (or with Cilk). Benchmark against a sequential merge implementation (see Sect. 1.4.1). How large speed-up can be achieved with how many threads? What is a good cut-off for the recursion/task spawning?
57. Improve the task parallel merge implementation from the previous exercise by using the co-ranks of $n/2$ to split exactly the input arrays A and B. Benchmark against your previous implementation. Are there any improvements? How can an input arrays be constructed that lead to a worst possible behavior for the simple, rank-based algorithm?

2.4 Exercises

58. Implement a mergesort algorithm as in Sect. 2.3.19 using the task parallel merge algorithm in either OpenMP or Cilk. Benchmark against a standard, sequential, recursive, top-down mergesort. How large speed-up can be achieved with how many threads? What is a good cut-off for the recursion/task spawning? Can you eliminate or minimize the use of the additional array B with the extra, parallel copy operations? Hint: It is possible to do with at most one copy operation.

59. Here is an implementation sketch of a sequential Breadth-First Search (BFS) algorithm that assigns distance labels to all vertices of a graph starting from some given start vertex. The graph is represented by a collection of adjacency lists written in matrix notation. We assume that a FIFO `queue` data structure with `init()`, `enq()` and `deq()` operations and a `nonempty()` predicate has been implemented.

```
int n, m;   // size of graph
int i;

int s;   // start vertex
int u, v;

int deg[n];      // vertex degrees
// space inefficient adjacency list representation
int adj[n][n];

int dist[n];   // labels to be assigned

queue Q, N;   // queues to be implemented, simple array FIFO
init(Q);
init(N);

int l;   // level

l = 0; dist[s] = l;
enq(s,Q);   // put start vertex in Q

do {
  // process level l;
  do {
    deq(u,Q);   // Q is not empty
    for (i=0; i<deg[u]; i++) {
      v = adj[u][i];
      if (dist[v]==-1) {
        dist[v] = l+1;
        enq(v,N);
      }
    }
  } while (nonempty(Q));
  Q = N; l++;   // next level
} while (nonempty(Q));
```

Complete the sequential implementation. Write a simple graph generator in order to test and benchmark your implementation. Now, parallelize the sequential code with OpenMP. Use critical sections or atomic operations in order to dequeue and/or enqueue elements from your FIFO queue. Consider allowing optimistic/speculative updates of the distance labels. Where is the parallelism in your parallel algorithm/implementation? What is the time complexity of your algorithm? Which properties of the input graph do you use in your complexity statement? Is your algorithm work-optimal? Benchmark with simple graphs using your input generator and compute the achieved speed-ups compared to your sequential (best possible) implementation.

60. The Floyd–Warshall algorithm [39] for the *all-pairs shortest path problem* (APSP) is based on the following observation. Let a weighted, directed graph $G = (V, E)$ be represented by an $n \times n$ distance matrix $W[n, n]$ where $W[i, j]$ is the weight of the edge between vertex i and vertex j, $i, j \in V$. Let $W^k[i, j]$ be the distance (weight of a shortest path) between vertices i and j using only paths with vertices $0, 1, \ldots, k-1$. It then holds that $W^0[i, j] = W[i, j]$, $W^k[i, j] = \min(W^{k-1}[i, j], W^{k-1}[i, k] + W^{k-1}[k, j])$ for $0 < k < n$ and that $W^{n-1}[i, j]$ is the length of a shortest path in G between i and j. The following function computes $W^{n-1}[i, j]$.

```
int *fw_apsp2(int *w, int *wnext, int n) {
  int (*W)[n]     = (int(*)[n])w;
  int (*Wnext)[n] = (int(*)[n])(wnext);
  int (*WW)[n];

  int i, j, k;
  for (k=0; k<n; k++) {
    for (i=0; i<n; i++) {
      for (j=0; j<n; j++) {
        Wnext[i][j] = (W[i][j]>W[i][k]+W[k][j]) ?
                      W[i][k]+W[k][j] : W[i][j];
      }
    }
    WW = W; W = Wnext; Wnext = WW;
  }

  return (int*)W;
}
```

What is the complexity of the Floyd–Warshall algorithm on graphs G with n vertices, $n = |V|$? Which of the loops can possibly be parallelized? Give a parallelization with OpenMP.

An alternative, less obvious implementation using only one matrix is given below.

```
void fw_apsp(int *w, int n) {
  int (*W)[n] = (int(*)[n])w;

  int i, j, k;
  for (k=0; k<n; k++) {
    for (i=0; i<n; i++) {
      for (j=0; j<n; j++) {
        if (W[i][j]>W[i][k]+W[k][j]) {
```

```
                W[i][j] = W[i][k]+W[k][j];
            }
        }
      }
    }
}
```

Argue that the two implementations compute the same correct distance matrix. The second implementation has subtle dependencies and is less obviously parallelized. What can be done? Which of the implementations perform best? What is the speed-up that can be achieved on matrices of order $n = 1000$ with different numbers of threads on your system?

Chapter 3
Distributed Memory Parallel Systems and MPI

Abstract. The last third of the lectures on Parallel Computing deals with characteristics of distributed memory systems where processors or shared memory processor-nodes are interconnected via a communication network which has to be used to exchange information between processes or threads in any non-trivial parallel algorithm. The concept of message-passing as a means to structure and implement algorithms is introduced, concretely via MPI which is dealt with extensively, regarding both fundamental message-passing ideas but also the way these have been realized in a concrete specification.

3.1 Eighth Block (1 Lecture)

This lecture block is an introduction to performance relevant aspects of real, parallel, distributed memory systems.

A naïve, parallel distributed memory system model consists of a set of p processors each with local memory for program and data (MIMD architecture). Processors execute independently and asynchronously and exchange information through explicit communication through an *interconnection network*. Communication is (significantly) more expensive than accessing data in local memory and may be subject to additional constraints. The network may provide means for synchronizing the processors.

In a corresponding distributed memory programming model, processes (or threads) communicate explicitly by executing communication operations, either pairwise, or in more complex collective patterns. Distributed memory programming models also offer means for synchronizing processes.

The concrete, distributed memory programming interface will be MPI, the *Message-Passing Interface*, which is treated in depth in the following parts of the lecture notes.

3.1.1 Network Properties: Structure and Topology

The distinguishing, new feature of distributed memory systems is the interconnection network (sometimes just called the *interconnect*) needed for communication between processors, which can be individual cores, multi-core CPUs, or larger entities consisting of many multi-core CPUs, nowadays often enhanced with GPUs and other accelerators, see Sect. 3.1.5. These entities are physically connected (electric or optical cables, or other, often just called *links*), and not all of these entities may be immediately, directly connected with each other; typically, they are *not*! Also, some elements in the network may not be processors used for computation, but simply *network switches* serving communication with other network elements. It is clear that both the physical and the topological properties of the network (speed of the connections, processing capabilities, the composition and structure of the network) play a decisive role for the performance of algorithms and programs running on distributed memory systems. It is also clear that without a powerful interconnect, there can be no Parallel Computing: We are interested in non-trivial problems requiring non-trivial communication and interaction between processors.

An interconnect where the processors are also the communication elements and in which there are no switches is called a *direct network*. An interconnect, in which there are also special switch elements (special communication processors with connections to other elements) is, on the other hand, called an *indirect network*.

First, we are interested in investigating how structural properties of the network influence the communication performance and the capability to solve problems that we are interested in.

The structure or *topology* of a communication network, both direct and indirect, can be modeled as a(n un)directed, (un)weighted graph $G = (V, E)$, where the vertices (nodes) V denote processors or network communication elements, and the edges E model the immediate connections or links between communication elements. Two elements (processors or switches) $u, v \in V$ are immediately connected (adjacent neighbors) if there is a (directed) edge (arc) $(u, v) \in E$. For most communication networks, if network element u can send data directly to network element v via a link (u, v), then also v can send data directly to u; that is, communication networks are most often undirected (or bidirected) graphs, and the link (u, v) can be used in both directions. It can nevertheless sometimes be relevant to consider directed graphs; and indeed there have been (few) examples of real, parallel distributed memory systems built on directed interconnection networks.

When two processors u and v are not adjacent in the network, a path between u and v must be found along which u and v can then communicate. Let a path between nodes u and v have length l. Communicating some data from u to v along this path will take at least l successive communication operations.

Recall that the *diameter* of a graph $G = (V, E)$ is the maximum over all shortest paths between pairs of nodes $u, v \in V$.

3.1 Eighth Block (1 Lecture)

$$\mathrm{diam}(G) = \max\{\mathrm{dist}(u,v) | u, v \in V\}$$

Here, $\mathrm{dist}(u,v)$ denotes the distance in number of links that have to be traversed to get from u to v in G. It is defined as the length of a shortest path from u to v measured in number of edges to traverse. The diameter is a lower bound on the number of communication steps for communication operations and algorithms that involve message transmission between nodes u and v which have the longest distance in the communication network. Note that we always take the diameter to be finite: Disconnected networks cannot be used for Parallel Computing.

The out-degree(G) of a graph $G = (V, E)$ is the largest number of outgoing edges from a node in G; that is, the largest node degree of a node in G.

$$\mathrm{degree}(G) = \max\{\mathrm{degree}(u) | u \in V\}$$

where the node degree of $u \in V$ is given by $\mathrm{degree}(u) = |\{v \in V | (u,v) \in E\}|$. The in-degree is defined analogously.

The *bisection width* of a graph $G = (V, E)$ is the smallest number of edges that must be removed in order for the graph to fall apart into two roughly equal-sized parts (in number of vertices); that is, to partition the vertices of G into two disjoint subsets with no edges between pairs of vertices in the two subsets. If two vertex subsets V', V'' are roughly equal when $||V'| - |V''|| \leq 1$, the formal definition is as follows.

$$\mathrm{bisec}(G) = \min_{\substack{V', V'' \subset V \\ V' \cup V'' = V \\ V' \cap V'' = \emptyset \\ ||V'| - |V''|| \leq 1}} |\{(u,v) \in E, u \in V', v \in V''\}|$$

While both $\mathrm{diam}(G)$ and $\mathrm{degree}(G)$ can be easily computed in polynomial time for any given network topology graph G, $\mathrm{bisec}(G)$ can (most likely) not. The problem of finding $\mathrm{bisec}(G)$ is essentially the *Graph Partitioning* problem, one of the classical, standard NP-complete problems [44, ND14].

The best possible communication network in terms of diameter and bisection width is the *fully connected network* $G = (V, E)$, where $(u, v) \in E$ for all $u, v \in V$ (assume either $(u, u) \in E$ or $(u, u) \notin E$ as convenient). For a fully connected network G, $\mathrm{diam}(G) = 1$ and $\mathrm{bisec}(G) = |V|^2/4$ (for $|V|$ even). The significant drawbacks of the fully connected network are the large number of links, namely $|V|(|V|-1)$ and the high degree (number of links per node), namely $\mathrm{degree}(G) = |V| - 1$.

The worst possible communication networks that can support Parallel Computing are the *linear processor array* and the *processor ring*, which are graphs

$A, R = (V, E)$ consisting of either a single path from vertex $u \in V$ to vertex $v \in V$ both having degree 1 with all other vertices in-between having degree 2 (processor array $A = (G, V)$), or a single cycle spanning all vertices $v \in V$ each of which have degree 2 (processor ring R). For the linear array, $|E| = |V| - 1$, $\text{diam}(A) = |V| - 1$ and $\text{bisec}(A) = 1$, and for the ring $|E| = |V|$ (for $|V| > 2$), $\text{diam}(R) = \lfloor |V|/2 \rfloor$ and $\text{bisec}(R) = 2$. A significant advantage of linear arrays and rings is the small(est possible) number of links (to keep the graph connected) and the low degree. A *tree network* $T = (V, E)$ likewise has $|E| = |V| - 1$, $\text{bisec}(T) = 1$, but typically $\text{diam}(T) = O(\log |V|)$.

Number of communication edges (links) and node degrees entail concrete, physical costs (money and space for cables and network connections) when building Parallel Computing systems with given network properties, as do other factors like, for instance, the necessary physical lengths of cables. It is, therefore, interesting, relevant, and highly challenging to find good compromises between costs and structural network properties desirable for supporting non-trivial Parallel Computing. Many different (with and with no commercial potential) solutions have been given, see, for instance, the aforementioned http://www.top500.org.

Numerous networks between the two extremes have been proposed and studied, see, for instance [72], and are not the topic of these lectures. Only three classes of such communication networks shall be mentioned, namely *trees*, *d-dimensional tori/meshes*, and *hypercubes*.

In a *tree network*, the topology graph $T = (V, E)$ is a tree (minimally connected graph over the nodes in V), most often with logarithmic diameter as in balanced binary or k-ary trees, binomial trees, etc. (the linear array is a special case). Being minimally connected, tree networks have $\text{bisec}(T) = 1$, since removing any one link will make the network fall apart.

In a d-dimensional *mesh network* with dimension sizes (or orders) $r_0, \ldots, r_{d-1}$, the processors are identified with the set of d-element integer vectors $V = \{(x_0, \ldots, x_{d-1}) | x_i \in \{0, 1, \ldots, r_i - 1\}\}$. The number of processors in such a d-dimensional mesh is therefore $|V| = \prod_{i=0}^{d-1} r_i$. There is a bidirected link (u, v) between two processors $u = (x_0, \ldots, x_{d-1})$ and $v = (y_0, \ldots, y_{d-1})$ if $|x_i - y_i| = 1$ for some coordinate $i, 0 \leq i < d$ and $x_j = y_j$ for all other coordinates. A *torus network* or *torus* is a mesh network with additional "wrap-around" edges between processors at the "borders" of the mesh. That is, between two processors $u = (x_0, \ldots, x_{d-1})$ and $v = (y_0, \ldots, y_{d-1})$ if $x_i = 0$ and $y_i = d_i - 1$ for some ith coordinate and $x_j = y_j$ for all other coordinates $j \neq i$. The diameter of a mesh $M = (V, E)$ is $\text{diam}(M) = \sum_{i=0}^{d-1} (r_i - 1)$, and the degree is $\text{degree}(M) = 2d$. The diameter of a torus $S = (V, E)$ is $\text{diam}(S) = \sum_{i=0}^{d-1} \lfloor r_i/2 \rfloor$, and the degree is likewise $\text{degree}(S) = 2d$.

A uniform (symmetric, homogeneous) mesh or torus network has the same order for all dimensions, $r = \sqrt[d]{p}$. The bisection width of a symmetric mesh is $\text{bisec}(M) = p^{\frac{d-1}{d}} = p/\sqrt[d]{p} = p/r$ and of a symmetric torus $\text{bisec}(S) = 2p^{\frac{d-1}{d}} = 2p/r$ (for r even).

A *hypercube network* $H = (V, E)$ is a special case of a uniform torus (or mesh) network in which all coordinates are either $x_i = 0$ or $x_i = 1$; note that in this case, mesh and torus coincide, the hypercube torus has no more edges than the hypercube mesh. Thus, the number of processors is $p = 2^d$ for some d, that is, a power of 2, or the other way around, the dimension of a p-processor hypercube is $d = \log_2 p$. Each processor has d neighboring processors which for processor $u = (x_0, \ldots, x_{d-1})$ are found by changing one of the i coordinates from x_i to $1 - x_i$. This is the same as flipping the ith bit in u when viewed as a binary number. Both the degree and the diameter of a hypercube are degree$(H) = $ diam$(H) = d = \log_2 p$. The bisection width is bisec$(H) = p/2$.

Modern high-performance systems are often built as torus networks of $d = 3, 5, 6$ dimensions or as indirect networks with multiple switches of small, fully connected networks, often called *multi-stage networks* of which there are many examples (e.g., InfiniBand). Hypercube networks were once popular, but are currently not built (what could some reasons be?).

3.1.2 Communication Algorithms in Networks

Communication from a processor u to another processor v in a given network $G = (V, E)$ requires at least dist(u, v) communication steps, in which processor u sends data to a neighboring processor that is closer to v (along an edge in E). This neighbor, in turn, sends data to a neighboring processor that is closer to v (along an edge in E), etc., until the data reaches v. This is independent of the amount of data to be transferred and the concrete costs incurred by sending and receiving some amount of data (see later). It is relevant to study the number of such communication steps that may be required for other, more complex communication operations, apart from just the transmission of information from one processor to another. We, therefore, first assume that data to be communicated are all of some small unit and that each communication step takes the same unit of time.

In a communication step, a processor $u \in V$ can communicate with a neighbor in the communication network $G = (V, E)$. What exactly a processor can do in a communication step depends on the *capabilities* of the communication system. We say that a communication system is *one-ported* (or *single-ported*) if a processor can engage in at most one communication operation in a step. A communication system where a processor can be involved in up to k communication operations in the same step (that is, concurrently) is called *k-ported* or just *multi-ported*.

If communication in a step between neighboring processors $u \in V$ and $v \in V$ with $(u, v) \in E$ is only in one direction from u to v or from v to u, it is *unidirectional*. The communication system is said to be unidirectional if it can support only unidirectional communication in a step. Communication in both directions, from u to v and from v to u, is bidirectional, telephone-like (in an old sense of "telephone" where only two parties could speak at the same time). A communication system that can support such communication is said to be *bidirectional*. Communication where a processor u receives from a processor w and sends to a processor v is said to be general, send-receive, bidirectional communication. A system that can support such communication in a step is said to be *bidirectional* in the general, *send-receive* sense.

Most modern communication systems and networks can, approximately, support general, send-receive, bidirectional communication. It can be measured to what extent this is the case by system benchmarking, in itself an interesting activity. Systems with indirect, multi-stage communication networks are often one-ported, whereas torus-based systems are most often $2d$-ported and can approximately support communication with all torus neighbors in a step.

Processors in a communication network can work independently and concurrently. For the analysis of communication algorithms, we count the total number of steps in which processors are communicating that are required for solving the given problem, i.e., for the last processor to finish. In each step, some or all of the processors in the network may be involved. Sometimes, actually often, steps where many pairs of processors are communicating are called *rounds*.

Interesting communication problems often correspond to parallelization patterns that are useful in complex algorithms and applications, for instance, broadcasting data from one processor to other processors, exchanging information between all processors, etc. (see Sect. 1.3.12). In any such communication pattern that involves transmission of data from a processor u to a processor v where $\text{dist}(u,v) = \text{diam}(G)$, an obvious lower bound on the number of steps (rounds) required to complete the pattern operation is $\text{diam}(G)$. One such pattern is the broadcast operation which we formalize as the following communication problem.

Definition 12 (Broadcast Problem). Let $G = (V, E)$ be a communication network, and $r \in V$ a given *root processor* which has some indivisible unit of data that needs to be transmitted to all other processors $u \in V$. The *broadcast problem* is to devise for a given network $G = (V, E)$ and any root $r \in V$ an algorithm with the smallest possible number of communication rounds that transmits the data unit from r to the other processors of G.

Both the (structure and capabilities of the) network G and the chosen root processor r are known to all processors and can be used in the algorithm. A solution to the broadcast problem for a given class of communication networks is an algorithm that solves the problem for the given root r and describes the communication steps for each processor, together with a proof that the algorithm completes in the claimed number of communication rounds. In particular, G is not part of the input but fixed (given) and can be used in the algorithm design, whereas the root processor r is usually taken to be an input parameter which is, however, known to all processors.

In tree, torus, and hypercube networks, the diameter lower bound argument gives a non-constant bound on the number of rounds needed to solve the broadcast problem. It depends on the number of processors in the network. But even

3.1 Eighth Block (1 Lecture)

in a fully connected network with constant diameter one, the number of communication rounds is non-constant, as captured by the following, important statement.

Theorem 14. *In a fully connected, p-processor network $G = (V, E), p = |V|$ with k-ported, unidirectional communication capabilities for $k \geq 1$, the number of communication rounds necessary and sufficient for solving the broadcast problem is $\lceil \log_{k+1} p \rceil$.*

The proof for the lower bound part of the claim is the following information-theoretic argument. The best that an algorithm that solves the broadcast problem can do is the following. In the first communication step, only the root processor has the data which it can disseminate to at most k new processors that so far did not have the data. In the next round, the best that each of the $k+1$ processors that now have the data can do is to disseminate the data to k new processors that so far did not have the data. Therefore, after the second round, the number of processors that have the data are at most $(k+1)+(k+1)k = (k+1)^2$. In summary, from one communication round to the next, the best that an algorithm can achieve is that a factor of $k+1$ more processors will have the data. The smallest number of communication rounds i that are required for all processors to eventually receive the data is found by solving $(k+1)^i \geq p$ which by taking the logarithm on both sides gives $i \log(k+1) \geq \log p$. Thus, $i \geq \lceil \log_{k+1} p \rceil$ since the solution (number of rounds) must be integral.

The argument almost immediately leads to an algorithm that matches this lower bound. Partition the communication network into $k+1$ pieces of roughly the same number of processors. The root processor r belongs to one of these pieces; for the other pieces, a virtual root processor is chosen. The processors must be able to do this with no communication, based on the information they have on the identity of r and the fact that G is fully connected. The root r sends the data it has to the k virtual roots. The broadcast problem has now been reduced to $k+1$ proportionally smaller broadcast problems, still on fully connected networks, which can be solved recursively, in parallel. The number of recursive steps needed for all pieces to have been reduced to a single processor is $\lceil \log_{k+1} p \rceil$ after which the data have been broadcast.

Good and even optimal solutions for the broadcast problem, in the sense of matching a known lower bound, are known for many types and classes of networks, like trees, tori, and hypercubes (and many, many others), but not always trivial. Efficient algorithms for broadcast and other collective communication problems are not the subject of these lectures.

The broadcast problem for an arbitrary graph G, now as part of the input, is NP-complete [44, ND49].

The bisection width of a communication network gives a lower bound on the number of communication rounds required for another important communication problem.

Definition 13 (All-to-All Problem). Let $G = (V, E)$ be a communication network and assume that each processor $i \in V$ has, for each other processor in

G, specific data that have to be sent to that processor. The *all-to-all problem* is to devise for a given network $G = (V, E)$ an algorithm with the smallest possible number of communication rounds that transmits all data from all processors to all other processors.

The *all-to-all problem*, also called personalized or individual exchange, is a highly communication intensive problem. All processors have distinct data for each of the other processors, so for each processor $|V| - 1$ data have to be sent and received. The total communication volume is thus $|V|(|V| - 1)$ data. What is the smallest number of communication rounds required to handle this volume? Partition the set of processors into two roughly equal sized sets of $|V|/2$ processors (for simplicity, we assume that $|V|$ is even). The volume of data to be exchanged between the two sets is $|V|^2/4$, independent of how the processors were partitioned, since the all-to-all problem is symmetric. Now let the partition of the processors be the partition that corresponds to the bisection width $\text{bisec}(G)$ of the network G. Since there are only $\text{bisec}(G)$ links connecting the two parts, each of which can carry data in a communicating step, the required number of steps for any algorithm solving the all-to-all problem is $\frac{|V|^2}{4\text{bisec}(G)}$.

Theorem 15. *Let G be a direct communication network with bisection width $\text{bisec}(G)$. The number of communication rounds necessary to solve the all-to-all problem is at least $\frac{|V|^2}{4\text{bisec}(G)}$.*

For the fully connected network with the highest possible bisection width, the all-to-all problem could possibly be solved in a single communication round. This would, on the other hand, require that each processor can communicate with all other processors in a single step, which is not realistic. The bisection width lower bound alone is most often too optimistic and not the sole limiting factor on achievable all-to-all communication performance. On the other hand, a poor network with constant bisection width (independent of the number of processors) like a ring or a tree would need a quadratic number of communication steps (in the number of processors) for all-to-all communication, and there is nothing that can be done about that.

3.1.3 Concrete Communication Costs

Communication mostly involves not only small, indivisible units of information, but (complex) data of some size m (bytes, integers, other relevant, but explicitly stated unit). What is the concrete, real cost (in time) of transmitting such data between processors in the network?

As a first shot, often a simple linear-affine time cost model is adopted. The *linear-affine transmission cost model* states that transmitting m units, $m \geq 0$, from u to v in $G = (V, E)$ along a communication edge $(u, v) \in E$ takes

$$\alpha + \beta m$$

time units, where α is a fixed, *start-up latency* independent of m (for the given network, so perhaps dependent on p) and β a *time per unit* of data transmitted.

The linear-affine time cost model is a crude, first, and perhaps even misleading approximation of the cost of communication between processors in a network or distributed memory Parallel Computing system. Nevertheless, for lack of better, such a model is (tacitly) assumed in the analysis of the distributed memory algorithms in these lecture notes. The model correctly emphasizes that communication takes time, both in terms of cost per transmitted unit and latency and reminds us that both of these terms can be considerable. In particular, since

$$\alpha + \beta m \leq \sum_{i=0}^{k-1}(\alpha + \beta m_i) = k\alpha + \beta m$$

the model stresses that splitting a message of m units into k smaller messages of sizes $m_i, 0 \leq i < k$ with $m = \sum_{i=0}^{k-1} m_i$ can be detrimental to communication performance. Conversely, combining smaller messages into larger ones can, whenever this is possible, be of advantage by saving latencies (see next section). The linear-affine transmission cost model is a *homogeneous* model and treats all pairs of processors the same by ignoring their placement in the network (distance in network, placement in shared memory compute node). It also abstracts away routing and overall traffic (contention, congestion) in the network, which will be treated next.

3.1.4 Routing and Switching

In a not fully connected network $G = (V, E)$ where not every processor can communicate directly with any other processor, a general purpose *routing system* (*routing algorithm*, *routing protocol*) makes it possible for any processor $u \in V$ to send data to any other processor $v \in V$ via some path of intermediate processors in V. In a sense, the routing system turns a not fully connected network of processors into a virtually fully connected network, where any processor can communicate directly with any other other processor, however, not necessarily at the same cost of communication (see Sect. 3.1.3). A routing algorithm could be *centralized*, but, typically, routing is thought of as a Distributed Computing problem. A routing protocol or scheme consists of a set of local, per processor/switch algorithms, each making decisions on what to do with a received message based on its own state and possibly the state of some of the immediately adjacent processors or switches (local information). Sometimes, parallel algorithms are designed without a routing system by explicitly describing how processors communicate with each other and along which paths. Such an approach can make it possible to give more precise, better bounds on the expected running time but is not general purpose and comes with a high design cost (specialized algorithm). A routing system may be realized in hardware, in software, or in a combination of hard- and software. This is why the term routing *system* is used. Designing and analyzing routing algorithms for different types of graphs is a typical Distributed Computing topic (recall Definition 2), but routing systems and algorithms are not a topic of these lectures. A few terms are useful, though.

The most important requirement to a routing system (algorithm, protocol), is *deadlock freedom*: A message sent from a processor u to a processor v must eventually arrive correctly (and uncorrupted) at processor v, *regardless* of any other traffic in the communication network. A deadlock could arise when two processors or network elements at the same time require a certain resource and mutually block each other. For instance, they could want to send data to the same processor or switch element, possibly over the same network edge resulting in a conflict that cannot be resolved. It may also be seen as the task of the routing system to ensure *reliable communication* meaning that no data or parts of messages are lost, no data are corrupted, and perhaps even that data are delivered in some specific order (as must, for instance, be guaranteed by MPI, see Sect. 3.2.11). This is important, since network hardware does not always guarantee such properties (think of reasons why this may be the case).

A routing system should be (as) fast (as possible). In the linear-affine time cost model, routing data of m units from processor u to processor v along a path of length l would take $l(\alpha+\beta m)$ time units. For sufficiently large numbers of data units, this can be improved by *pipelining* as follows: The m units are separated into smaller *packets* of some maximum size of b units (assuming $m > b$) that are sent one after the other. The time for the last packet to arrive at the destination processor v would then be

$$l(\alpha + \beta b) + (\lceil m/b \rceil - 1)(\alpha + \beta b) = (l + \lceil m/b \rceil - 1)\alpha + \beta(l-1)b + \beta m$$
$$= (l-1)\alpha + \lceil m/b \rceil \alpha + \beta(l-1)b + \beta m$$

The first $l(\alpha + \beta b)$ term on the left-hand side is the time for the first packet to arrive at v. The $(\alpha + \beta b)$ factor in the second term is the time for each following packet, of which there are $\lceil m/b \rceil - 1$ remaining in total. Simplifying the expression, we see that sending all $\lceil m/b \rceil$ packets has a cost of βm since the last packet may be smaller than b units. If the packet size b can be chosen freely, a best possible packet size that minimizes the total transmission time can be found by (calculus or) balancing the terms $\lceil m/b \rceil \alpha$ and $\beta(l-1)b$, which both depend on b, against each other. That is, we solve $\lceil m/b \rceil \alpha = \beta(l-1)b$ for b which yields a best packet size b of

$$b = \sqrt{\frac{m}{l-1}}\sqrt{\frac{\alpha}{\beta}}$$

and a shortest transmission time of

$$(l-1)\alpha + 2\sqrt{(l-1)m}\sqrt{\alpha\beta} + \beta m$$

provided that $l > 1$. The important result is that with pipelining the βm term is not linearly dependent on the path length l. Furthermore, the constant factor in the βm term is one, meaning that the network is utilized to the fullest extent: We pay only the cost per unit β per data element of m once. Routing with pipelining is sometimes called *packet switching*, whereas routing without pipelining is called *store-and-forward*. Both store-and-forward and packet switching routing require some intermediate buffer space in the routing system, either for all m data units or for a block of up to b units. These and other terms are used somewhat differently in different fields, depending on the level at which the network is examined, the use (internet computing is different from Parallel Computing!), tradition, and many other factors [110].

In a communication network there may be several, partially different paths from a processor u to a processor v. When data are sent from processor u to processor v, the routing system chooses an appropriate path. This choice, of course, depends on u and v and the network topology $G = (V, E)$, but may also depend on the current traffic in the system; that is, concurrent communication between other processors.

With *deterministic (oblivious) routing*, the route (path to be taken) is determined solely by the endpoints u and v and the structure of the network G, whereas network traffic plays no role. With *adaptive routing*, the routing system takes other, concurrent communication into account. Thus, the route from u to v can be different from time to time. A routing algorithm is said to use *minimal routing* when routing from u to v is always along a shortest path (of length $\text{dist}(u, v)$). When several paths are possible and pipelining (packet-switching) is employed, it may be that different blocks (packets) are taking different routes. In such cases, packets could potentially arrive at the destination processor v in a different order than the order in which they were sent from the source processor u. It is then the task of the routing system to assemble the packets in the right order at the destination.

In the presence of traffic in the communication network due to many pairs of processors communicating at the same time, the optimistic, model based estimate of the transmission time from u to v will most likely not hold, and data communication times will (for some pairs of processors) be much higher. This can be due to network *contention*, for instance, on edges (u, v) that occur in many paths and are needed by several pairs of processors or to resource *congestion* because of a too high load on intermediate buffers or processors. The best that can be hoped for in such cases is a serialization slowdown proportional to the contention or congestion. The routing system can apply different strategies to alleviate and control contention and congestion, typically some form of *flow control*.

3.1.5 Hierarchical, Distributed Memory Systems

In modern Parallel Computing systems, the communication system has a more complex, hybrid structure, consisting of communication networks at different levels. Thus, a single, unweighted graph that alone describes the topology of the whole system may not be adequate or helpful.

A two-level hierarchical system, for instance, could consist of a number of shared memory compute nodes, interconnected by a, typically, indirect network. Thus, processor-cores within the same shared memory compute node may communicate with different communication characteristics than processor-cores residing on different compute nodes. In particular, if several processor-cores on the same compute node at the same time need to communicate with processor-cores on other compute nodes, they will have to share the network that connects the compute nodes. There will be congestion on the node (connection to the network) and possibly contention in the network as well. The simple, latency α and cost per unit β, linear-affine transmission cost model breaks down.

3.1.6 Programming Models for Distributed Memory Systems

Programming models for distributed memory systems usually abstract away from concrete network properties as discussed in the previous sections. They assume that the active entities of the model (processes, threads, ...) can freely communicate as in a fully connected network. Often, they come without a concrete cost model that says at what cost (in time) active entities can communication with each other. Processes are usually not synchronized, operate on local data that are invisible to other processes (shared nothing) and follow instructions of local programs (SPMD or MIMD). They cooperate (exchange data, synchronize) with the other processes by explicit or implicit communication. Distributed memory programming models usually assume that message transmission between processes and threads is deadlock free, reliable and correct and typically ordered according to certain constraints and rules. For the implementation of such programming models, the runtime system and routing algorithms need to ensure reliable message delivery between any processes in the model. Distributed memory programming models sometimes provide means for reflecting and exploiting properties of the underlying, hierarchical communication system. The programming model underlying MPI is a good example.

A concrete implementation of a distributed memory programming model like MPI, has concrete costs in time (and in other factors) for the different types of communication offered by the model. If one benchmarks communication performance (time) under different loads and between processes residing in different parts of the system, network and system properties will become manifest. Concrete communications do usually not obey a simple, homogeneous cost model. Such system and implementation dependent differences may or, rather, may not be reflected in the cost models for the programming model. MPI does not come with a cost model at all. Therefore, strictly speaking, all cost analysis of MPI programs will be based on external assumptions, benchmarking results, and known system properties. Simplified assumptions are often made (homogeneous communication with linear transmission costs), though, and can be indicative of actual performance, but can also be off target or even grossly misleading.

Distributed system programming models can be classified as either *data distribution centric* or *communication centric*. In a data distribution centric model, the data structures defined by the model (arrays, multi-dimensional arrays, vectors, matrices, tensors, complex objects, ...) are distributed according to given rules across the processes (see Sect. 1.3.10). When a process accesses or updates a part of a distributed data structure that resides with another process, communication and possibly "remote" computation is implied. When a process on the other hand accesses local data "owned" by itself, it can itself perform the specified computation. This is often called the *owner computes* rule. A communication centric model on the other hand usually does not define distributed data structures. Such models instead focus on properties of explicit communication and synchronization operations.

Examples of data distribution centric models are so-called *Partitioned Global Address Space* models (*PGAS*). In such models, data structures, typically simple $1, 2, 3, \ldots$-dimensional arrays, can be distributed across threads (processes), and access to non thread-local parts of arrays implicitly leads to communication. Otherwise, computations are done following the owner computes rule. An example implementation of a PGAS model is *Unified Parallel C* (*UPC*) [37]. PGAS models and languages will not be treated further in these lectures. MPI is, on the other hand, a communication centric model.

3.2 Ninth Block (3–4 Lectures)

Our concrete example of a distributed memory programming interface implementing a communication centric, distributed memory programming model is *MPI*, the *Message-Passing Interface* [52, 82, 83, 105, 106]. MPI is an older interface dating back to around 1992. It is (nevertheless, still) widely used, especially in HPC, and relevant to study and learn because of the concepts it introduces. MPI is an interface for C and Fortran (still an important programming language in HPC). MPI is maintained and developed further by the so-called MPI Forum, an open forum of academic institutions, laboratories, compute centers and industry. Incidentally, many of the MPI Forum members are or were also part of the OpenMP ARB. The standard is freely available and can be found via www.mpi-forum.org. These pages also give information on the standardization process (currently towards MPI 4.2).

The reference for programming (and learning) MPI is the latest version of the standard [83]. Some helpful reading are the series of books on "Using MPI" [51,

53, 54]. Many elementary textbooks on parallel programming, e.g., the books by Rauber and Rünger and Schmidt et al. [90, 96] deal extensively with aspects of MPI.

This block of lectures gives an introduction to MPI for Parallel Computing, covering all its fundamental concepts and features. Some aspects of MPI will not be dealt with, most notably support for I/O, dynamic management of processes (spawning and joining communication domains, see later), and tool building (highly important, though).

3.2.1 The Message-Passing Programming Model

The message-passing programming model goes way back, at least to papers by Dijkstra and Hoare in the 1960s and 1970s. The idea is that parallel computations can be structured as sequential processes with no shared information that communicate explicitly by sending and receiving *messages* between each other [62, 63]. Restricting interaction between the sequential processes to explicitly specified (synchronous) communication operations was seen as a means to develop provably correct, parallel and concurrent programs. This message-passing model was called *Communicating Sequential Processes (CSP)*. CSP programs, in particular, cannot have data races. The programming model that is implicitly behind MPI is much wider in scope than CSP. It incorporates both synchronous and asynchronous point-to-point communication (CSP focussed on synchronous, handshaking communication), one-sided communication and collective communication. It also provides features for data layout description and interaction with the communication system and the external environment (file system I/O).

Some main characteristics of the MPI message-passing programming model are:

1. Finite sets of processes in immutable *communication domains*. Processes in the same domain can communicate with each other. The same process can belong to several communication domains.
2. In communication domains, processes are identified by their rank. Ranks are consecutive $0, \ldots p-1$, with p being the number of processes in the communication domain (size). A process can have different ranks in different communication domains.
3. New communication domain are created from existing communication domains and a default domain consisting of all externally started processes.
4. Processes operate exclusively on local data. All communication between processes is explicit.
5. Communication is reliable and ordered.
6. Communication is network oblivious and possible between all processes.
7. Three basic communication models:

a. Point-to-point communication between pairs of processes with different modes, non-local and local completion semantics.
　　b. One-sided communication between one process and another with different synchronization mechanisms, local and non-local completion mechanisms.
　　c. Collective communication, many different operations, non-local (and local) completion semantics.
8. Structure of communicated data is orthogonal to communication model and mode.
9. Communication domains may reflect the physical network topology and communication system.

MPI has no performance model, and there are no prescriptions in the MPI standard on how the many, many different MPI constructs are to be implemented nor on which algorithms are to be used, in particular, not for the collective communication operations. Thus, detailed (asymptotic) performance analysis of MPI programs must make external assumptions and informed guesses on how specific features are implemented and how they perform.

However, MPI is designed with the intention of making high-performance implementations possible on wide ranges of Parallel Computing systems, meaning that the functionality and semantics are close to what an underlying communication system will offer, that preprocessing and communication of metainformation is not necessary (or strictly confined), and that memory required by library internals is bounded and/or can be controlled. These design objectives explain the concrete "look-and-feel" of the many MPI functions.

3.2.2 The MPI Standard

The MPI standard is largely a well-reasoned, semantic specification of the large set of MPI operations. The MPI standard is an open standard maintained by the so-called MPI Forum, which in principle anybody can join; see `mpi-forum.org` for the rules and current discussions on the standard. The current version of the standard is MPI-4.1 [83]. The standardization efforts over the past 20 years have mostly resulted in extensions, additions, and clarifications that maintain backward compatibility to the original standard published in 1993. This may change.

3.2.3 MPI in C

MPI is defined and implemented as a library, and MPI functionality can be used by linking code against a concrete MPI library. There are several such libraries available, notably the open source libraries `mpich`, `mvapich`, and OpenMPI as well as vendor libraries, often for specific High-Performance Computing systems. C code using MPI must include the MPI function prototype header with the `#include <mpi.h>` preprocessor directive. All MPI relevant functions and predefined objects are prefixed with `MPI_`, which identifies the MPI "name space". It is considered illegal and is in any case very bad practice to use the `MPI_` prefix for own functions or objects in the code. MPI programs are usually compiled with a special compiler (wrapper, mostly) that takes care of proper linking against the MPI library. The typical example is `mpicc`, which will also accept standard optimization options and arguments.

We explain the MPI functions by listing the C prototypes with the C types for all arguments and explain the outcome for given inputs, loosely a before-after explanation.

MPI functions return an error code, and it is good practice to check the error code (which is very often not done). The error code `MPI_SUCCESS` indicates success of the call.

3.2.4 Compiling and Running MPI Programs

An MPI program is, unlike an OpenMP program, simply a C (or Fortran) program with library calls to MPI functions. MPI programs can therefore be compiled with a standard C compiler. Usually, *an MPI program* means a single program that will be run by all started processes: Mostly, MPI code follows the SPMD paradigm. It is possible, though, to let different MPI processes run different programs. To facilitate linking against the MPI library, normally an `mpicc` compiler command that is just a wrapper around the C compiler command is provided. It takes the standard C compiler flags and options.

Running an MPI program with a desired number of processes is somewhat complex. Resources, cores and compute nodes, for the processes must be allocated and the processes started at the allocated compute resources. For small, stand-alone systems (say, laptop or workstation, small server) this is often done from the command line with a command like `mpirun`. More commonly, and on larger systems, a batch scheduling system like `slurm` is used.

When processes have been started, they become MPI processes after having initialized the MPI library. In the MPI context, processes are most often bound ("pinned") to specific processor-cores or at least to compute nodes. This binding is outside the control of MPI.

It is usually possible to start more MPI processes than there are physical processor-cores in the system (which can be useful when developing programs on a small system). But as with OpenMP and `pthreads`, such *oversubscription* must be used with care and is often detrimental to performance.

3.2.5 Initializing the MPI Library

After the processes are started on the system, the internal data structures of the MPI library must be initialized. This done by the `MPI_Init` call which takes the standard C argument count and argument array as arguments. The C argument count and full argument array are normally copied to all started MPI processes, so there is no need to transfer this information explicitly with MPI communication operations. After use, all activity of the MPI library is completed and resources freed with an `MPI_Finalize` call, which should not be forgotten: The program may otherwise terminate improperly. Prior to `MPI_Init` and after `MPI_Finalize`, no MPI calls can be performed, except for the two check calls `MPI_Initialized` and `MPI_Finalized` that tell the caller (perhaps an application specific library written with MPI with its own initialization function) whether MPI has been initialized or completed. When the MPI library has been finalized, it cannot be initialized again within the same program.

```
int MPI_Init(int *argc, char ***argv);
int MPI_Finalize(void);
int MPI_Finalized(int *flag);
int MPI_Initialized(int *flag);

int MPI_Abort(MPI_Comm comm, int errorcode);
```

The `MPI_Abort` call can be used to force termination of the running MPI program in an emergency situation.

An MPI library can provide limited information about itself and its environment by the following operations.

```
int MPI_Get_version(int *version, int *subversion);
int MPI_Get_library_version(char *version, int *resultlen);

int MPI_Get_processor_name(char *name, int *resultlen);
```

These calls illustrate the tediousness of MPI being a library (and the shortcomings of C for manipulating strings): For the strings `version` and `name`, the user must reserve space of at least `MPI_MAX_LIBRARY_VERSION_STRING` and `MPI_MAX_PROCESSOR_NAME` characters, respectively. The strings are copied into these arrays, in C properly terminated by a null character. The number of actual characters, excluding the trailing null character, will be stored in `resultlen`. Thus, in C, output arguments (result values) are always of pointer type.

A process can read the wall-clock time from some time point in the past (in seconds). The timers are local and (usually) not synchronized across processes and processor-cores. The call can be used to time process-local operations and is heavily used for this.

```
double MPI_Wtime(void);
double MPI_Wtick(void);
```

Whether the timers are synchronized (global) can be queried by reading an attribute. The attribute mechanism of MPI is not covered in these lectures, although it is important for library building with MPI [117]. The existence of the attribute mechanism in the MPI standard illustrates how MPI is intended to and can support portable, application specific library building; but also the tediousness of MPI being a library and not an integrated part of a programming language: Information must flow in and out of the MPI library to and from the application (specific library) explicitly through the MPI functions. Information that the compiler possesses is not known to the MPI library, but must be explicitly transferred.

3.2.6 Failures and Error Checking in MPI

The MPI functions are, at first sight (do not be scared!), often quite involved and sometimes take confusingly long lists of arguments that all must be used correctly. If an argument is not as specified according to the precondition of the operation, there is no guarantee that the function will have the specified effect and produce the desired outcome or any useful outcome at all! MPI libraries perform only rudimentary argument checks (whether preconditions are fulfilled), but the extent of this is not specified in the standard and MPI libraries differ in the amount and kinds of such checks done. Sometimes, tools or options can be used to perform more extensive checking, which can, of course, be helpful in the development phase of an application. But the programmer can most surely not rely on the MPI library to catch mistakes and errors. The MPI standard specifically states [82, page 340] that "An MPI implementation cannot or may choose not to handle some errors that occur during MPI calls. [...] The set of errors that are handled by MPI is implementation dependent. [...] Specifically, text [in the standard] that states that errors *will* be handled should be read as *may* be handled" (emphasis original). The most recent MPI 4.1 standard takes the same stance [83, Page 449].

As mentioned, almost all MPI functions return error codes. It can make sense to check those and try to take action on certain error return codes. But there is no guarantee that this will be possible: The application may have crashed before returning from the operation and no error code will ever be seen by the user. This is still the most typical MPI behavior on (programming) errors and (hardware) faults. Therefore, MPI programs typically do only a limited amount of error code checking and only of certain functions. In particular, communication failures due to processor/node crashes or failures in the communication system are typically not handled and will in most cases cause the whole application to abort. The own, self-inflicted, and most common reason for an application to crash is memory corruption through wrong use of MPI functions leading to memory being overwritten and/or wrongly addressed. Here, memory diagnostic tools that check bounds and accesses can be most helpful.

Part of the reason for MPI not doing extensive error checking and handling is that MPI is designed to allow for high-performance implementations. It, therefore, does not impose (expensive, extensive) checks for errors and wrong usage of the MPI functions.

MPI aims to make it possible to control the response of the library in case of failures. This is accomplished through *error handlers* which are special functions that can be attached to communicator objects (see next section) and are invoked by the MPI library when an error condition occurs in an MPI call on that communicator object. Error handlers are beyond the scope of these lectures. The quotes from the MPI standard cited above still apply.

```
int MPI_Errhandler_create(MPI_Handler_function *function,
                          MPI_Errhandler *errhandler);
int MPI_Errhandler_set(MPI_Comm comm,
                       MPI_Errhandler errhandler);
int MPI_Errhandler_get(MPI_Comm comm,
                       MPI_Errhandler *errhandler);
int MPI_Errhandler_free(MPI_Errhandler *errhandler);
```

3.2.7 MPI Concept: Communicators

After processes have been started and the MPI library initialized with `MPI_Init`, the started processes are put into a *communication domain* called `MPI_COMM_WORLD`. In addition, each process is also put into a domain by itself called `MPI_COMM_SELF`. A communication domain represents an ordered set of processes that can communicate with each other, each process with any other process in the domain, and only in that domain. A domain has a *size*, which we often denote by p, which is the number of processes it contains. Each process has a unique, non-negative *rank* r in the domain, $0 \leq r < p$. In MPI, communication domains are called *communicators*. A *communicator* is a distributed object that can be operated on by all processes belonging to the communicator. A communicator is referenced by a *handle* of type `MPI_Comm`. In particular, processes can look up the *size*, i.e., the number of processes, in any communicator `comm` to which they belong, and determine their own *rank* in the communicator by the following functions.

```
int MPI_Comm_rank(MPI_Comm comm, int *rank);
int MPI_Comm_size(MPI_Comm comm, int *size);
```

Thus, the code snippet

```
int rank, size;

MPI_Comm_rank(MPI_COMM_WORLD ,&rank);
MPI_Comm_size(MPI_COMM_WORLD ,&size);
assert(0<=rank&&rank<size); // the condition on ranks
```

will, when executed by any of the started MPI processes, identify the process relative to all other started processes by its serial number (rank) in the MPI_-COMM_WORLD communicator. As good as any MPI program will have such a code sequence somewhere after the MPI_Init call, and the processes decide what to do based on rank and size. Here, as almost always, the error return codes of the two function calls are ignored.

This trivial piece of code illustrates a number of important MPI concepts and principles.

- Processes belong to communication domains, which are called communicators in MPI. In particular, they belong to the MPI_COMM_WORLD communicator consisting of all externally started processes created by the MPI_Init call.
- Processes have a *rank* (serial number) in a communicator. Ranks are consecutive from 0 to $p-1$ where p is the size of the communicator ($0 \leq$ rank $<$ size).
- The rank of a given process in MPI_COMM_WORLD is determined by external factors like how the processes were started and where in the system the process is placed (processor-core, shared memory compute node, particular processor in the network) relative to the other processes. However, the rank of a process in a communicator will never change.
- Communicators are identified by handles of type MPI_Comm, which are *opaque* objects on which certain operations are defined (that is, their internal composition is not specified and cannot be know to the application programmer).
- There can be several communicators in an application, and the same process can belong to many communicators, possibly with a different rank in each.
- The communicator is the most fundamental object/concept in MPI: All communication is relative to a communicator, all collective operations (see later) are relative to a communicator. In particular, processes in different communicators cannot communicate, and simultaneous communication on different communicators can never interfere.
- MPI objects are static objects. They cannot be changed (only freed when no longer of use). New objects, for instance communicators, can be created from already existing ones by appropriate functionality.

For any communicator there is a special process rank MPI_PROC_NULL outside the range from 0 to $p-1$ that can actually be referenced and used for non-communication: Communication with MPI_PROC_NULL has no effect (see later).

The principle that communication is always with respect to a communicator and that communication between processes in one communicator can never interfere with communication between processes in another is fundamental. It is what allows construction of *safe, parallel libraries*. If each library used in an application uses its own communicator(s), communication going on in different libraries can never interfere.

For library construction, the essential operation on communicators is the creation of a duplicate communicator. The duplicate represents a communication domain with the same set of processes in the same order, but is nevertheless a different domain. Thus, communication on a communicator and its duplicate can never interfere. The MPI_Comm_dup operation is shown below. It is the first example of a so-called *collective operation*, meaning that it has to be called by all processes in the communicator comm.

```
int MPI_Comm_dup(MPI_Comm comm, MPI_Comm *newcomm);
int MPI_Comm_dup_with_info(MPI_Comm comm, MPI_Info info,
                           MPI_Comm *newcomm);

int MPI_Comm_split(MPI_Comm comm, int color, int key,
                   MPI_Comm *newcomm);
int MPI_Comm_create(MPI_Comm comm, MPI_Group group,
                    MPI_Comm *newcomm);
int MPI_Comm_create_group(MPI_Comm comm, MPI_Group group,
                          int tag, MPI_Comm *newcomm);
int MPI_Comm_split_type(MPI_Comm comm,
                        int split_type, int key,
                        MPI_Info info, MPI_Comm *newcomm);
```

The MPI_Comm_split and MPI_Comm_create functions allow to create new communicators from existing ones, possibly with fewer processes and possibly with a different order. Both calls are collective: All processes belonging to the communicator in the comm argument have to make the call. When the calls return, each calling process will be part of the new communicator newcomm and still also of the old comm communicator which was used to coordinate all the processes making the call. The MPI_Comm_split operation takes an integer color argument and all processes with the same color (argument) will end up in the same newcomm communicator. The key argument can be used to control the numbering of the processes in the new communicator in the following way. Processes with the same color are sorted non-increasingly by the key argument and this determines their ranks in newcomm. The process with the smallest key will become rank 0, the process with the next smallest key rank 1, and so on. Processes with equal key arguments are kept in their rank order in the comm communicator. The special MPI_UNDEFINED argument as color, indicates that a process calling with this color is not going to belong to any communicator. Thus, MPI_Comm_split is a very flexible operation for partitioning an existing communication domain into new communication domains. The discussion again illustrates some fundamental principles.

- MPI functions have input and output arguments. Output arguments in C have pointer type (we already saw this with MPI_Comm_rank and MPI_Comm_size).
- There are functions in MPI that are *collective*, meaning that they have to be called eventually by all the processes belonging to the input communicator. In MPI, collective functions are *always* called symmetrically: All processes (in the communicator) make the same call with possibly different argument values. The input arguments given by a process determine the role of that process in the call.
- On return from an MPI_Comm_split call, each calling process will have, in addition to the still existing, unchanged input communicator comm, a new communicator newcomm to which it belongs together with all the other processes that called with the same color argument. The rank is given by the

position in the list of processes with the same color sorted by the key argument.
- After completion of a communicator creating operation, each calling process will (in case of MPI_Comm_split) belong to two communicators, comm and newcomm, possibly of different sizes and possibly with a different rank in each.
- New processes are neither created nor started by these functionalities. The communicator creating functions operate on a given set of processes represented by their ranks in an input communicator. Only ranks and sizes may be different in the created communicators. The calling processes remain alive as they are, and retain their rank in the input communicator used in the call.

The MPI_Comm_create call likewise allows to create arbitrary new communicators from existing ones. This is based on process groups, a new MPI concept that will be explained briefly later. As with MPI_Comm_split, the newcomm returned to some processes can be an invalid MPI_COMM_NULL communicator, a communicator on which no (communication) operations can be performed, and that can mostly not be used as input argument to MPI functions. The last two operations, MPI_Comm_create_group and MPI_Comm_split_type, albeit both useful and semantically interesting, are not treated in these lectures.

In the following, we will see concrete examples of the use of MPI_Comm_split and MPI_Comm_create, for instance, in the implementations of Quicksort-like algorithms, stencil computations and matrix–matrix multiplication.

After use, a communicator is freed by the MPI_Comm_free call. Since communicators are distributed objects, all processes in the communicator have to eventually call MPI_Comm_free on the communicator. That is, this is also a collective operation. A program where some processes are correctly calling MPI_Comm_free on a created communicator and others not may not be able to complete properly.

```
int MPI_Comm_free(MPI_Comm *comm);
```

A communicator typically is a "costly object" in terms of required memory space (depending on the quality of the MPI library implementation). Also for that reason it is *always* good practice to free MPI objects that are no longer going to be used.

There is sometimes helpful functionality in MPI for comparing two communicators.

```
int MPI_Comm_compare(MPI_Comm comm1, MPI_Comm comm2,
                     int *result);
```

The possible outcomes are MPI_IDENT, meaning that the two input communicators are indeed referring to the same object, MPI_CONGRUENT, meaning that the two input communicators represent the same processes in the same rank order, MPI_SIMILAR, meaning that the two input communicators represent the same processes but not necessarily in the same order, and MPI_UNEQUAL for anything else. A communicator and its duplicate would, thus, be MPI_CONGRUENT, but not MPI_IDENT. This functionality is typically for use in application specific libraries and more seldomly directly used in applications.

3.2 Ninth Block (3–4 Lectures)

To illustrate the concepts and functionality introduced so far, a first (and our only) full-fledged MPI program follows below; in the following examples we will skip header-files, `main()`-function definitions, mostly also `rank`- and `size`-lookup, etc. The program creates a duplicate of the `MPI_COMM_WORLD` communicator from which it splits off a communicator with processes ranked in reverse order. It next partitions the `comm` communicator into communicators containing the processes with even rank (in `comm`) and the processes with odd rank. All processes at this point belong to three new communicators (plus `MPI_COMM_WORLD` and `MPI_COMM_SELF`), partly with different ranks. Finally, the program creates a subcommunicator in which the process with the highest rank in the calling communicator has been excluded by giving this process (which has rank equal to `size-1`) the special color `MPI_UNDEFINED`. This type of subcommunicator can be useful for master-worker applications (see Sect. 1.3.7), in which the worker processes need to communicate between themselves, for instance, by collective operations (see Sect. 3.2.28), without involving the excluded "master" process. Note that the program, including the assertions, is constructed in such a way that it can run for any number of started MPI processes.

```
#include <stdio.h>
#include <stdlib.h>
#include <string.h>

#include <assert.h>

#include <mpi.h>

int main(int argc, char *argv[])
{
  int rank, size;
  MPI_Comm comm, mmoc, evodcomm, workcomm;
  int result;

  MPI_Init(&argc,&argv);

  comm = MPI_COMM_WORLD;
  MPI_Comm_compare(comm,MPI_COMM_WORLD,&result);
  assert(result==MPI_IDENT);

  MPI_Comm_dup(MPI_COMM_WORLD,&comm);
  MPI_Comm_compare(MPI_COMM_WORLD,comm,&result);
  assert(result==MPI_CONGRUENT);

  MPI_Comm_rank(comm,&rank);
  MPI_Comm_size(comm,&size);

  MPI_Comm_split(comm,0,size-rank,&mmoc);
  MPI_Comm_compare(comm,mmoc,&result);
  assert(size==1||result==MPI_SIMILAR);

  MPI_Comm_split(comm,rank%2,0,&evodcomm);
  MPI_Comm_compare(comm,evodcomm,&result);
  assert(size==1||result==MPI_UNEQUAL);
```

```
  MPI_Comm_free(&mmoc);
  MPI_Comm_free(&evodcomm);

  MPI_Comm_split(comm,(rank==size-1 ? MPI_UNDEFINED : 1),0,
                 &workcomm);
  if (workcomm!=MPI_COMM_NULL) {
    MPI_Comm_compare(comm,workcomm,&result);
    assert(result==MPI_UNEQUAL);

    MPI_Comm_free(&workcomm);
  }

  MPI_Comm_free(&comm);

  MPI_Finalize();

  return 0;
}
```

3.2.8 Organizing Processes*

We touch briefly on convenient functionality to give more structure to the organization of MPI processes than just the rank in a communicator.

A running example through this part of the lectures is the stencil computation (see Sect. 1.3.5). In a large d-dimensional matrix, all entries have to be updated according to the same stencil rule for each entry, for instance, an average over neighboring elements "up, down, left, right, front, rear" (3-dimensional example) [36]. This update is iterated a large number of times, until some convergence criterion is met. In a distributed memory, message-passing setting, the matrix is conveniently cut into rectangular block-submatrices, one submatrix for each process, with all submatrices being of roughly the same size. We will return to this example shortly (in Sect. 3.2.14 and 3.2.24).

For the communication that is needed for a parallel implementation of the stencil update, it can be convenient to be able to think of the processes as points in a d-dimensional grid with integer coordinates. Dedicated MPI communicator creation functionality makes it possible to organize the processes into such a d-dimensional grid with sizes d_i such that $d_0 \times d_1 \times \cdots d_{d-1} = p$ by giving each process a d-dimensional coordinate vector describing its position in the grid. A communicator with an imposed grid structure is called a *Cartesian communicator*. Cartesian communicators are created and used with the functionality listed below.

```
int MPI_Cart_create(MPI_Comm comm, int ndims,
                    const int dims[], const int periods[],
                    int reorder, MPI_Comm *cartcomm);
```

3.2 Ninth Block (3–4 Lectures)

```
int MPI_Cartdim_get(MPI_Comm cartcomm, int *ndims);
int MPI_Cart_get(MPI_Comm cartcomm, int maxdims,
                 int dims[], int periods[], int coords[]);
int MPI_Cart_sub(MPI_Comm cartcomm, const int remain_dims[],
                 MPI_Comm *subcomm);

int MPI_Topo_test(MPI_Comm comm, int *status);

int MPI_Dims_create(int nnodes, int ndims, int dims[]);
```

A Cartesian communicator, the cartcomm returned by the MPI_Cart_create call, is like any other communicator and can be used wherever a "normal" communicator could, but carries additional information about the size of the grid, namely the number of dimensions d and the size along each dimension. The number of dimensions d is given as input ndims. The sizes of the dimensions are stored in the input array dims[] with d entries and dims[i] = d_i. It must hold that $\prod_{i=0}^{d-1}$ dims[i] $\leq p$ where p is the size of the old communicator comm. If $\prod_{i=0}^{d-1}$ dims[i] $< p$ some processes in comm will not be part of the new cartcomm communicator. These processes will be returned the value MPI_COMM_NULL. Thus, in the newcomm communicator, the product of the dimension sizes equals the size of the communicator. The Cartesian grid is the set of integer vector coordinates

$$\{(c_0, c_1, \ldots, c_{d-1}) | 0 \leq c_i < \text{dims}[i], 0 \leq i < d\}$$

and each process in cartcomm is uniquely associated with one such vector. The association of processes with coordinate vectors is by *row-major* assignment ("last coordinate changes the fastest"). More precisely, a process with coordinates $(c_0, c_1, \ldots, c_{d-1})$ has rank r with

$$r = \sum_{i=0}^{d-1} c_i \prod_{j=i+1}^{d-1} d_j$$

where d_j = dims[j] and the empty product $\prod_{j=i+1}^{d-1} d_j$ for $i = d-1$ being 1. The rank r can, of course, be computed in $O(d)$ steps (better than the $O(d^2)$ steps implied by the formula). When stored in a C array coords[], the coordinates are stored as coords[i] = c_i for $0 \leq i < d$, and coords[$d-1$] is the fastest changing coordinate for $r = 0, 1, \ldots, p-1$.

The periods array is a Boolean (0/1) array indicating whether the grid is periodic in the ith dimension, $0 \leq i < d$. Periodic in the ith dimension means that a coordinate vector $(c_0, \ldots, d_i, \ldots, c_{d-1})$ is treated as $(c_0, \ldots, 0, \ldots, c_{d-1})$, and $(c_0, \ldots, -1, \ldots, c_{d-1})$ as $(c_0, \ldots, d_i - 1, \ldots, c_{d-1})$. The grid "wraps around" in the ith dimension. A full *torus* is a grid that is periodic in all d dimensions.

The placement of the MPI processes in a grid via the MPI_Cart_create operation by intention carries with it an implied, preferred communication pattern, namely that each process is likely (in the application) to communicate with its immediate neighbors in the grid along the d dimensions. It is implied that a process with coordinate vector $(c_0, \ldots, c_i, \ldots, c_{d-1})$ will most likely communicate (only, in a preferred way) with the $2d$ processes $(c_0, \ldots, c_i \pm 1, \ldots, c_{d-1})$ for each $i, 0 \leq i < d$. If the grid is not periodic in dimension i, then some neighbors might not exist, which is represented by MPI_PROC_NULL, the non-existent process mentioned in Sect. 3.2.7.

The MPI_Cart_create takes a new type of argument, the reorder flag. Setting this flag allows the MPI library to attempt to rerank (reorder) the processes in the output communicator, so as to better reflect the process communication pattern that is implied by the process grid organization. We say that an MPI process has been *reranked* in a new communicator created from an existing one if the ranks of the process in the two communicators are different. The reorder idea is that processes that are expected to communicate by being neighbors in the Cartesian grid are reranked to processes on processor-cores in the physical system that are also close to each other, for instance, by having a direct communication link. Concretely, processes with ranks i and j in cartcomm that are Cartesian neighbors may have different ranks $i' \neq i$ and $j' \neq j$ in comm for processes i' and j' that are "physically close". Whether, how and to what extent an MPI library does such a reranking and what the benefits will be in concrete applications is entirely implementation and situation dependent.

The MPI_Cart_get operations are, again, mostly for library building purposes and can be used to query a communicator created with MPI_Cart_create for information (but, perhaps surprisingly, not for information on whether any reordering was done). Whether the communicator is indeed Cartesian can be checked with the MPI_Topo_test operation which will in that case return the value MPI_CART.

For setting up Cartesian communicators over an existing communicator of size p (that is, with p MPI processes), the MPI_Dims_create function can be helpful for factoring p into d factors that are (as) close to each other (as possible). The factors are returned in non-increasing order in the dims input/output array that must be initialized to non-negative values. Positive entries indicate factors that are already set and fixed, so only

$$\frac{p}{\prod_{\mathtt{dims}[i] > 0} \mathtt{dims}[i]}$$

will be factored over the zero-entries in dims. The denominator must divide p exactly.

```
int MPI_Cart_rank(MPI_Comm cartcomm, const int coords[],
                  int *rank);
int MPI_Cart_coords(MPI_Comm cartcomm, int rank, int maxdims,
                    int coords[]);
int MPI_Cart_shift(MPI_Comm cartcomm,
                   int direction, int disp,
                   int *rank_source, int *rank_dest);
```

The functions MPI_Cart_rank and MPI_Cart_coords are used to translate between ranks and coordinate vectors. Cartesian communicators, in combination with MPI_Comm_split, will be used later to facilitate the implementation of the

SUMMA matrix–matrix multiplication algorithm (see Sect. 3.2.30). The shift operation MPI_Cart_shift can be used to compute the ranks of processes along the ith direction (dimension) by giving an integer, not necessarily non-negative displacement that takes coordinate c_i into $(c_i + \text{disp}) \bmod p$. We will see an example in Sect. 3.2.10.

Here is now a part of an MPI program for setting up (and freeing) Cartesian communicators for all dimensions $d, 0 \leq d < p$ where p is the number of processes in a given communicator comm, and verifying the row-major placement of the MPI processes in each of the created, non-periodic Cartesian grids:

```
MPI_Comm_size(comm,&p);

int reorder = 0; // no reorder attempt
for (d=1; d<=p; d++) {
  int dims[d], periods[d];
  int coords[d];

  MPI_Comm cartcomm;
  int size;
  int r, rr, dd, i;

  for (i=0; i<d; i++) dims[i] = 0;
  MPI_Dims_create(p,d,dims);

  for (i=0; i<d; i++) periods[i] = 0;
  MPI_Cart_create(MPI_COMM_WORLD,d,dims,periods,0,&cartcomm);
  assert(cartcomm!=MPI_COMM_NULL);

  MPI_Comm_size(cartcomm,&size);
  for (r=0; r<size; r++) {
    MPI_Cart_coords(cartcomm,r,d,coords);
    for (i=0; i<d; i++) {
      assert(0<=coords[i]&&coords[i]<dims[i]);
    }
    assert(coords[d-1]==r%dims[d-1]);
    rr = 0; dd = 1;
    for (i=d-1; i>=0; i--) {
      rr += coords[i]*dd;
      dd *= dims[i];
    }
    assert(rr==r);
  }

  MPI_Comm_free(&cartcomm);
}
```

The idea of specifying a likely pattern of most intense communication, based on which the MPI library can attempt to rerank processes, is generalized with the so-called *distributed graph communicators*. Such communicators are created by specifying a communication graph of possibly weighted communication edges between processes. Edge weights could model, for instance, communication volume between two processes, or frequency of communication, or other properties of the application, but the meaning of the weights is not specified in the MPI standard. The specified communication pattern is used for two purposes by the MPI library. First, by setting the reorder flag to true (=1), the MPI library can attempt to rerank the processes (see the above discussion) such that process ranks that are adjacent in the communication graph by heavy communication edges are placed on processes that are "close" to each other in the calling communicator comm. Second, the communication graph defines the so-called *neighborhoods* for a special kind of collective operations, the so-called *neighborhood collectives* that are explained briefly in Sect. 3.2.33. The functionality is a bit tedious at first sight and listed here for completeness, but not treated further in these lectures. To use it, it is necessary to consult the MPI standard [83, Chapter 8, Chapter 6].

```
int MPI_Dist_graph_create(MPI_Comm comm,
                          int n,
                          const int sources[],
                          const int degrees[],
                          const int destinations[],
                          const int weights[],
                          MPI_Info info,
                          int reorder,
                          MPI_Comm *graphcomm);
int MPI_Dist_graph_create_adjacent(MPI_Comm comm,
                          int indegree,
                          const int sources[],
                          const int sourceweights[],
                          int outdegree,
                          const int destinations[],
                          const int destweights[],
                          MPI_Info info,
                          int reorder,
                          MPI_Comm *graphcomm);
int MPI_Dist_graph_neighbors_count(MPI_Comm graphcomm,
                          int *indegree,
                          int *outdegree,
                          int *weighted);
int MPI_Dist_graph_neighbors(MPI_Comm graphcomm,
                          int maxindegree,
                          int sources[],
                          int sourceweights[],
                          int maxoutdegree,
                          int destinations[],
                          int destweights[]);
```

A distributed graph communicator can, like the case was for Cartesian communicators, be queried. The MPI_Topo_test operation will return the value MPI_DIST_GRAPH.

Process ranking (reordering) in MPI (sometimes called *process mapping*) via MPI_Comm_split, MPI_Cart_create, and MPI_Dist_graph_create is always realized in the following way. The MPI processes are bound to processor-cores and compute nodes in the system. Processes are (almost always) statically bound to some part of the system and do not move. Each MPI process belongs to one or more communicators. What can be different from one communicator to another is only the rank that a process may have. So, in MPI, processes are not moved or remapped, but the ranks can change from communicator to communicator. Assume that two processes in the input communicator comm have rank i and rank j and are adjacent (neighbors) in a distributed graph or Cartesian grid. In the resulting, reordered communicator, the ranks i and j may now be the ranks of processes (in comm) that happen to be close to each other in the system, for instance, by residing on the same compute node. Thus, process reordering and process mapping are both misnomers. The MPI mechanisms are merely reordering ranks (reranking).

Since processes themselves do not move, this means that possibly data from the process with rank i in the input communicator comm may have to be transferred to the process that has rank i in the a newly created (Cartesian or graph) communicator. Should such data transfer be necessary, the application programmer must implement it explicitly. Therefore, programs often do the process mapping early in the application before the processes generate or read much data.

To support moving data between communicators where the same process may have different ranks in the communicators, MPI provides mechanisms for translating between ranks in one communicator and another. Some will be described in Sect. 3.2.10. The communicator comparing function MPI_Comm_compare may also be of some use here to detect whether process ranks are different in two communicators.

3.2.9 MPI Concepts: Objects and Handles

The most important MPI object is the *communicator*. A communicator is the concrete representation of an ordered domain of MPI processes that can communicate with each other. It is a *distributed object*, meaning that it can be accessed and used by all the processes that have a reference to the object. MPI objects are referenced via predefined MPI handle types, of which there are quite a few, but not all that many. MPI objects can, like the communicators, be distributed and accessible by a whole set of processes, or be *local objects* that are only accessible by the single process having the handle to the object.

Handles are mostly *opaque* (with one important exception that will be treated next). Their implementation is unspecified in the MPI standard. An object referenced by a handle can be accessed and used only through the functions defined on the corresponding type of handle. The most important MPI objects and corresponding handles are the following:

- MPI_Comm for communicators, distributed (Sect. 3.2.7).
- MPI_Win for communication windows, represent a communication domain and associated pieces of memory, distributed (Sect. 3.2.22).
- MPI_Datatype for so-called datatypes that describe process local layout and structure of data to be communicated, local (Sect. 3.2.15).

- **MPI_Group** for ordered sets of processes as an object that can be manipulated by process local operations, local (Sect. 3.2.10).
- **MPI_Status** for information returned from a (point-to-point) communication operation, local. This is the exception to the opaqueness property of handles (see shortly).
- **MPI_Request** for information about a pending, possibly not yet completed communication operation (mostly point-to-point, but also collective and one-sided), local.
- **MPI_Op** for binary operators for the reduction collectives, local.
- **MPI_Errhandler** for action to be taken on discovery of an error or failure, see remark on error handling in MPI (Sect. 3.2.6), local, and not treated further in these lectures.
- **MPI_Info** for specifying additional information when creating (certain kinds of) objects like distributed graph communicators and communication windows. Local, and not treated in these lectures.

3.2.10 MPI Concept: Process Groups*

Process groups are local objects with handle type `MPI_Group` that represent ordered sets of processes. No communication operations are defined on process groups; the groups are for processes to locally compute other ordered sets of processes. Groups are used as input to a number of other, often collective MPI functions that involve many processes as arguments. The `MPI_Comm_create` operation for partitioning a communicator as specified by process local groups of processes was one example (Sect. 3.2.7).

Initialization of the MPI library does not construct any process groups in the way that `MPI_COMM_WORLD` is constructed. Instead, a local group object can be extracted from a distributed communicator object. The `MPI_Comm_group` call is a local operation that a process can perform on a communicator. It returns the ordered set of processes of the communicator as a local group object. A process can query its rank in a group. If it does not belong to the group, the special value `MPI_UNDEFINED` is returned.

```
int MPI_Comm_group(MPI_Comm comm, MPI_Group *group);

int MPI_Group_rank(MPI_Group group, int *rank);
int MPI_Group_size(MPI_Group group, int *size);
```

Operations on groups are somewhat set-like, but the order plays a role.

```
int MPI_Group_translate_ranks(MPI_Group group1,
                              int n, const int ranks1[],
                              MPI_Group group2,
                              int ranks2[]);

int MPI_Group_union(MPI_Group group1, MPI_Group group2,
                    MPI_Group *newgroup);
```

3.2 Ninth Block (3–4 Lectures)

```
int MPI_Group_intersection(MPI_Group group1,
                           MPI_Group group2,
                           MPI_Group *newgroup);
int MPI_Group_difference(MPI_Group group1,
                         MPI_Group group2,
                         MPI_Group *newgroup);

int MPI_Group_incl(MPI_Group group,
                   int n, const int ranks[],
                   MPI_Group *newgroup);
int MPI_Group_excl(MPI_Group group,
                   int n, const int ranks[],
                   MPI_Group *newgroup);

int MPI_Group_range_incl(MPI_Group group,
                         int n, int ranges[][3],
                         MPI_Group *newgroup);
int MPI_Group_range_excl(MPI_Group group,
                         int n, int ranges[][3],
                         MPI_Group *newgroup);

int MPI_Group_compare(MPI_Group group1,
                      MPI_Group group2,
                      int *result);

int MPI_Group_free(MPI_Group *group);
```

We give three important examples of uses of MPI process groups. The first example shows how to create a communicator that does not contain a certain, specified process. This is helpful and sometimes needed for applications following the *master-worker* pattern (see Sect. 1.3.7) where one master process, determined by its rank, has a special role and should be excluded from communication between the non-masters (worker processes). Such a communicator was also created in the last example of Sect. 3.2.7.

```
MPI_Group group, workers;
MPI_Comm work;

master = ...; // some arbitrary master (rank) in comm
MPI_Comm_rank(comm,&rank);
MPI_Comm_size(comm,&size);

MPI_Comm_group(comm,&group); // get the group
MPI_Group_excl(group,1,&master,&workers); // exclude master
MPI_Comm_create(comm,workers,&work);
if (rank==master) assert(work==MPI_COMM_NULL); // excluded!
else { // relative order of worker processes preserved
  int r;
  MPI_Comm_rank(work,&r);
  if (rank<master) assert(r==rank); else assert(r==rank-1);
}
MPI_Group_free(&group);
```

The group of processes from the given communicator `comm` is extracted, each process computes a group excluding the given master process (given as a process rank between 0 and the size of `comm`), and this group is used as input argument to the `MPI_Comm_create` function. Each process computes the same group. The `master` process, that is not part of the group, is returned the `MPI_COMM_NULL` value, whereas the workers are returned a handle to a new `work` communicator. This communicator can now be used for any kind of communication supported by MPI between the worker processes. In the example in Sect. 3.2.7, we saw the same effect achieved less tediously with the `MPI_Comm_split` collective operation.

The second example computes, for each process in a d-dimensional Cartesian grid (communicator), a group consisting of the $2d+1$ neighboring processes along the d dimensions, including the process itself. It is assumed that the arrays `dims` and `periods` have been correctly and sensibly initialized (see the example in Sect. 3.2.8) prior to the `MPI_Cart_create` call by all processes. All other variables are likewise assumed to have been declared and sensibly initialized.

```
MPI_Cart_create(comm,d,dims,periods,0,&cartcomm);
assert(cartcomm!=MPI_COMM_NULL);

MPI_Comm_group(cartcomm,&group);

MPI_Comm_rank(cartcomm,&r);
k = 0;
neighbors[k++] = r;
for (i=0; i<d; i++) {
  MPI_Cart_shift(cartcomm,i,1,&r1,&r2);
  if (r1!=MPI_PROC_NULL) neighbors[k++] = r1;
  if (r2!=MPI_PROC_NULL&&r1!=r2) neighbors[k++] = r2;
}
assert(k<=2*d+1);
MPI_Group_incl(group,k,neighbors,&neighborgroup);
// neighborgroup now ready for use
```

The `neighborgroup` computed for each process contains the local, implied grid neighborhood. This will be used later for synchronizing one-sided communication operations (Sect. 3.2.22).

The third and last example shows how to translate ranks between two communicators. Assume that a new communicator `comm_new` has been created out of an old one `comm_old` (with `MPI_Comm_split`, `MPI_Cart_create`, `MPI_Dist_graph_create` or other operation), possibly with reranking and possibly fewer processes. What we would need to know is this: For the process with rank i in the old communicator, what is the rank j in the old communicator of the process that has rank i in the new communicator? This information is needed in case data have to be transferred from process i in the old communicator to the process that now has rank i in the new communicator.

```
int i; // misused as array of one element in translation
MPI_Comm_rank(comm_old,&i);
```

```
MPI_Comm_group(comm_old,&group_old);
MPI_Comm_group(comm_new,&group_new);

MPI_Group_translate_ranks(group_new,1,&i,group_old,&j);

MPI_Group_free(&group_old);
MPI_Group_free(&group_new);
```

Since communication in MPI must always be done between processes belonging to the same communicator with that communicator as a handle to the communication operations either the new or the old communicator must be used for the data transfer. After the rank translation call, process i in the old communicator can send its data to process j (also in the old communicator), because process j is the process that has rank i in the new communicator comm_new.

3.2.11 Point-to-Point Communication

Processes that belong to the same communication domain by having a handle to the same communicator can communicate with each other within (or: relative to) that communicator. We first describe the classical MPI message-passing model of point-to-point communication between pairs of processes.

It is important that MPI communication between processes in a communicator has no connectivity restrictions. Any process can communicate with any other process as if the processes would be running on processors in a fully connected network (See Sect. 3.1.1). It is the task of the MPI library and runtime (routing) system to facilitate such communication. Recall that MPI does not provide a cost model for communication between processes and does not provide any performance guarantees on commutation or other operations. In particular, it cannot without further ado be assumed that communication costs are homogeneous, that is, similar for any pair of communicating processes regardless of their ranks in their communicator. Neither can it be assumed that communication costs are independent of the overall traffic in the system, that is, concurrent communication between other MPI processes possibly in other communicators (and even entirely unrelated processes from different programs running concurrently on the system). Communication costs must therefore be measured by appropriate benchmarks from which reasonable assumptions may follow.

It is also important that communication in MPI is always reliable. This means that a transmitted message can *always* be assumed to arrive uncorrupted and in full. In case the Parallel Computing system and communication network on which the MPI program is running are not reliable, it is again the task of the MPI library and runtime system to ensure reliable communication.

Finally, point-to-point communication is *ordered*. This means that a sequence of messages sent from one process to another will (eventually) become available at the receiving process in that order.

In point-to-point communication, two processes are explicitly involved. A sending process belonging to a communication domain (communicator) specifies an amount of data to be sent to an explicitly given *determinate* receiving process which must be prepared to receive at least the sent amount of data. The next two functions are the basic MPI point-to-point communication operations.

```
int MPI_Send(const void *buf,
             int count, MPI_Datatype datatype,
             int dest, int tag, MPI_Comm comm);
int MPI_Recv(void *buf, int count, MPI_Datatype datatype,
             int source, int tag, MPI_Comm comm,
             MPI_Status *status);
```

Data to be sent and received are specified by the first three arguments: A buffer address pointing to the part of memory where data are (to be) located, an element count, and an argument describing the structure of each element (see Sect. 3.2.15). Data buffers are always managed by the user and have to be allocated before use. In C, buffers are represented by a pointer (address) to the start of the buffer, and there are in general no restrictions on where and how buffers to be used with MPI are allocated. Larger buffers should always be allocated dynamically in heap storage with a memory allocator like `malloc()`. It is likewise the users responsibility to `free()` allocated storage after use. Problems with non-allocated, wrongly allocated and prematurely or never freed storage are the most frequent and tiresome for the C programmer.

By posting the `MPI_Send` call, a sending process initiates and completes sending data (a `count` number of elements of given structure specified by the `datatype`) to the receiving process. The sending process returns from the call when the data are safely on their way (in the communication system) and the send buffer can again be used for other data. By posting the `MPI_Recv` call, a receiving process declares itself ready to receive up to the described amount of data (`count` number of elements of given structure specified by the `datatype`) from a sending process. The call completes when the data sent have been received correctly and without loss (see discussion above). Thus, for point-to-point communication to take place, both sending and receiving process are explicitly involved. The receiving process must have allocated enough buffer space (the `buf` argument) to hold the data that are being sent. For communication to take place, the sending and the receiving process must give the same integer *message tag* to the message. The sending process must give the rank of the receiving process and this receiving process must be prepared to receive from that process. However, wildcards for process rank and tag are possible, as will be discussed later. Thus, sending of messages is *determinate*, but receiving is not.

The send-receive functionality illustrates another important MPI principle. All(most all) space for MPI data, notably data buffers but also argument lists etc., is in *user space* and managed by the application programmer. It is important to always have allocated enough buffer space for data that are being sent and received and to later free this space to avoid running out of memory. This is sometimes forgotten, with dire consequences. Memory corruption due to insuffi-

3.2 Ninth Block (3–4 Lectures)   205

cient buffer space is one of the most frequent errors in MPI programs, frustrating and often hard to find, since memory corruption (and program crash!) may become manifest only later in the program execution and not immediately at the function call that caused the memory corruption.

To illustrate point-to-point communication between processes in a communicator, here is an MPI implementation of the broadcast operation described in Definition 12 and discussed intensively later (Sect. 3.2.28). The process with rank root is the process having the data, and count is the number of elements. The elements that are being communicated are simple C integers of type int, which in MPI are described by the MPI datatype MPI_INT. The program is written to work for any number of processes larger than one.

```
#define TAG 1000

MPI_Comm_rank(comm,&rank);
MPI_Comm_size(comm,&size);
assert(size>1);

int *buffer = ...; // allocate and free
if (rank==root) {
  MPI_Send(buffer,count,MPI_INT,(rank+1)%size,TAG,comm);
} else if (rank==(root-1+size)%size) {
  MPI_Status status;

  MPI_Recv(buffer,count,MPI_INT,(rank-1+size)%size,TAG,
          comm,&status);
} else {
  MPI_Status status;

  MPI_Recv(buffer,count,MPI_INT,(rank-1+size)%size,TAG,
          comm,&status);
  MPI_Send(buffer,count,MPI_INT,(rank+1)%size,TAG,comm);
}
```

The processes in this algorithm are organized as a processor ring, and the number of dependent communication steps (rounds) for the algorithm is $p-1$, where p is the number of MPI processes. See Sect. 3.2.13 for more on possible analysis of message-passing algorithms. The ith process in the ring (counting from the root process) needs to wait for process $i-1$ to have received the data, etc. Theorem 14 tells that this algorithm is poor. Here is another implementation of the broadcast operation that is likewise poor, but not equally so and not for the same reasons (why?).

```
#define TAG 1000

MPI_Comm_rank(comm,&rank);
MPI_Comm_size(comm,&size);

int *buffer = ...; // allocate and free
if (rank==root) {
  int i;
```

```
    for (i=0; i<size; i++) {
      if (i==root) continue;
      MPI_Send(buffer,count,MPI_INT,i,TAG,comm);
    }
  } else {
    MPI_Recv(buffer,count,MPI_INT,root,TAG,comm,
             MPI_STATUS_IGNORE);
  }
```

This example shows the use of a special value as status argument, namely MPI_STATUS_IGNORE. This value can be used when the return status of the receive operation is not needed. Otherwise, the status object, referenced through the corresponding handle, contains information on the completion of the receive operation, i.e., whether an error occurred, from which process the data were received, and how much data was received.

Since receive calls like MPI_Recv can specify (in their count argument) more elements than actually sent in a send call, functionality is needed for the receiving process to find out how much data was actually received (and, by implication, sent). This information is available in the process local status object. The following two functions MPI_Get_count and MPI_Get_elements that operate on status objects are defined for this purpose. The datatype is an input argument, which imposes an interpretation of the received data and is needed in order to correctly compute the number (count) of units of such datatype that were received (MPI_Get_count). For complex datatype units that comprise many simple datatype elements, the MPI_Get_elements call instead computes the number of *basic datatype* elements that were received (see Sect. 3.2.15). For basic (simple, non-complex) datatypes like MPI_INT as used in the code examples above, the MPI_Get_count and MPI_Get_elements calls are equivalent and will return the same count.

```
int MPI_Get_count(const MPI_Status *status,
                  MPI_Datatype datatype, int *count);
int MPI_Get_elements(const MPI_Status *status,
                     MPI_Datatype datatype, int *count);
```

The status object/handle is peculiar in MPI. Handles were said to be opaque, but handles of type MPI_Status are only half so. Status objects have three predefined fields, namely MPI_SOURCE, MPI_TAG, and MPI_ERROR. These are important for non-determinate communication as will be explained in Sect. 3.2.12.

Algorithms often desire or even require that a process can both send and receive a message in a single communication round as, for instance, permitted in the one-ported, fully bidirectional send-receive communication model (see Sect. 3.1.2). The example below, where the processes are organized in a ring like in the first broadcast implementation above, will obviously lead to a *communication deadlock*. Each process is waiting to receive data from its predecessor process in the ring, but these data cannot be sent, since this process is also waiting to receive data from its predecessor process, etc.

3.2 Ninth Block (3–4 Lectures)

```
#define TAG 1000

MPI_Status status;
int *a = ...;
int *b = ...;

// receive from predecessor
MPI_Recv(a,count,MPI_INT,(rank-1+size)%size,TAG,
         comm,&status);
// send to successor
MPI_Send(b,count,MPI_INT,(rank+1)%size,TAG,comm);
```

MPI provides an `MPI_Sendrecv` operation to handle such situations. It combines the functionality and parameters of a blocking send and a blocking receive operation. With `MPI_Sendrecv` as declared below, a process can at the same time in the same, single communication operation send and receive data to and from two other processes in the communicator — that could actually be the same process or even the process itself — without the risk of deadlocking. When an `MPI_Sendrecv` call returns, data have left the send buffer (as with `MPI_Send`), which can then be reused for other data and received into the receive buffer (as with `MPI_Recv`). The status of the receive part is recorded in the `status` object.

```
int MPI_Sendrecv(const void *sendbuf,
                 int sendcount, MPI_Datatype sendtype,
                 int dest, int sendtag,
                 void *recvbuf,
                 int recvcount, MPI_Datatype recvtype,
                 int source, int recvtag,
                 MPI_Comm comm, MPI_Status *status);
int MPI_Sendrecv_replace(void *buf,
                 int count, MPI_Datatype datatype,
                 int dest, int sendtag,
                 int source, int recvtag,
                 MPI_Comm comm, MPI_Status *status);
```

Send and receive buffers must not overlap in any way, since this would lead to an indeterminate situation (a race condition): Did the send part take place, in part or in total, before or after the receive part? Which data were actually sent? It is the programmers responsibility to make sure that the non-overlapping buffer property is indeed guaranteed. Neither compiler nor MPI library will or can check this. Such unintentionally overlapping buffers are another common source of often very hard to find errors in MPI programs. In case data should be sent from some buffer and (later) be received into the same buffer, the `MPI_Sendrecv_-replace` operation can be used. It most likely will allocate some intermediate space for the receive part and later copy the received data back, therefore entailing potentially significant extra costs. As always, the MPI standard neither prescribes nor forbids any particular implementation. If one needs to know, only benchmarking and MPI library code inspection, if possible (as it would be for an open source library), can help.

With `MPI_Sendrecv` the deadlock situation from above is resolved:

```
#define TAG 1000

MPI_Status status;
int *a = ...;
int *b = ...;

MPI_Sendrecv(b,count,MPI_INT,(rank+1)%size,TAG,
             a,count,MPI_INT,(rank-1+size)%size,TAG,
             comm,&status);
```

3.2.12 Determinate vs. Non-determinate Communication

A sending process always specifies a determinate, specific receiver by its rank in the communicator. A sending process also gives each message sent a specific tag. In MPI, a *message tag* is just a non-negative integer that is attached as a label to a message (up to a specified upper bound given by MPI_TAG_UB). The message tag can be used by the receiver to distinguish one kind of tagged message from other kinds of tagged messages and to select which message is to be received by an MPI_Recv call in case more than one message has been sent from one or more other processes.

As seen above, a receiving process can explicitly specify the rank of the sending process from which it wants to receive a message with a specific tag. In contrast to sending processes, receiving processes can also receive from a *non-determinate* process. This is done by specifying a wildcard MPI_ANY_SOURCE for the rank argument. This will enable the receiving process to receive the message from any of the processes in the communicator. Likewise, the tag argument can be given a wildcard MPI_ANY_TAG.

Whereas programs using only determinate ranks in the communication operations are communication deterministic, programs using the MPI_ANY_SOURCE wildcard can be non-deterministic. Non-deterministic programs can, not surprisingly, cause problems not encountered with deterministic programs. The following examples illustrate these points.

Point-to-point communication is ordered. If messages are sent by a sequence of MPI_Send (or MPI_Sendrecv) operations with the same tag from the same process, the messages will become ready to be received by the destination process in exactly that order. This is referred to as *ordered communication* in MPI. The program below illustrates the advantages of the ordering constraint. Data from two buffers with different numbers of elements and different element types, the first with 500 integers (MPI_INT) and the second with 100 doubles (MPI_DOUBLE), are sent from process 0 to the last process with rank $p - 1$ (p being the number of processes in communicator comm). It is good SPMD style to always write MPI programs so that they work for any number of processes, which is the case here for any number of processes larger than one, as asserted. An open else instead of the else if (rank==size-1) conditional would lead to a communication deadlock when the number of processes is larger than two.

3.2 Ninth Block (3–4 Lectures)

```
MPI_Comm_rank(comm,&rank);
MPI_Comm_size(comm,&size);
assert(size>1);

if (rank==0) {
  int buf1[500];
  double buf2[100];

  MPI_Send(buf1,500,MPI_INT,size-1,TAG,comm);
  MPI_Send(buf2,100,MPI_DOUBLE,size-1,TAG,comm);
} else if (rank==size-1) {
  MPI_Status status;

  int buf1[1000];
  double buf2[200];
  int cc;

  MPI_Recv(buf1,1000,MPI_INT,0,TAG,comm,&status);
  MPI_Get_elements(&status,MPI_INT,&cc);
  assert(cc<1000);
  assert(cc==500);

  MPI_Recv(buf2,200,MPI_DOUBLE,0,TAG,comm,&status);
  MPI_Get_elements(&status,MPI_DOUBLE,&cc);
  assert(cc<200);
  assert(cc==100);
}
```

Since less data than expected by the receiving process are sent with each message, the exact number of data elements received in each of the messages has to be computed by the `MPI_Get_elements` operation. The assertions assert that the (stack) allocated buffers are not overflowing. For the `MPI_Recv` operation, the `count` argument is an upper bound on the number of elements that can be received. This upper bound should, of course, be no larger than the actual number of elements in the buffer used for reception. Again, the compiler can and will not check this: It is entirely the programmer's responsibility to ensure that buffers are not overwritten. Overwriting will most likely cause segmentation faults (crash!) at some point in the program execution. It is also worth noticing that the message tag has nothing to do with the type of the messages being communicated: the same tag is used for both the `MPI_INT` and the `MPI_DOUBLE` messages. The message tag is a label associated with a message that can be used to ensure that a receive operation indeed receives the message intended for that operation. Stack allocation, especially of variable sized arrays (in C99 terms called variable length arrays) instead of heap allocation with `malloc()`, is an often convenient and sometimes defensible practice in C programs, but it should be used with caution. The stack space is not as large as the heap and can easily be exhausted, especially with recursive algorithms. The compiler or C runtime will not notice: a program crash inevitably ensues.

In the next example, the data to be sent to process $p-1$ come from two different processes. In order to avoid waiting times, the receiving process uses

MPI_ANY_SOURCE to be able to receive the message from whichever source process becomes ready first. Here, both buffers contain C integers, and both sending processes use the same message tag. Since the two sent messages have different numbers of elements, the receiving process must ensure that both receive buffers are large enough to hold the number of elements in the largest message. This is the price of non-determinacy. The MPI_SOURCE field of the MPI_Status object is used to distinguish the messages based on the source of origin.

```
MPI_Comm_rank(comm,&rank);
MPI_Comm_size(comm,&size);
assert(size>2);  // at least three processes

if (rank==0) {
  int buf1[500];

  MPI_Send(buf1,500,MPI_INT,size-1,TAG,comm);
} else if (rank==1) {
  int buf2[100];

  MPI_Send(buf2,100,MPI_INT,size-1,TAG,comm);
} else if (rank==size-1) {
  MPI_Status status;

  int buf1[1000];
  int buf2[1000];
  int cc;

  // receive from "first" sender
  MPI_Recv(buf1,1000,MPI_INT,MPI_ANY_SOURCE,TAG,
           comm,&status);
  MPI_Get_elements(&status,MPI_INT,&cc);
  assert(cc<1000);
  if (status.MPI_SOURCE==0) {
    assert(cc==500);
  } else {
    assert(status.MPI_SOURCE==1);
    assert(cc==100);
  }

  // and then from second sender
  MPI_Recv(buf2,1000,MPI_INT,MPI_ANY_SOURCE,TAG,
           comm,&status);
  MPI_Get_elements(&status,MPI_INT,&cc);
  assert(cc<1000);
  if (status.MPI_SOURCE==0) {
    assert(cc==500);
  } else {
    assert(status.MPI_SOURCE==1);
    assert(cc==100);
  }
}
```

In this program, two processes send messages, and one process receives both messages: all messages sent are indeed received. MPI programs should maintain

3.2 Ninth Block (3–4 Lectures)

this property, and it is good MPI programming discipline to always keep this in mind. Cleanup of sent but unreceived messages will, as far as they have not caused a deadlock, be done by the `MPI_Finalize` call.

Non-determinacy can easily lead to incorrect, possibly crashing programs. In the next, erroneous program, the sending processes send different types and numbers of elements (`MPI_INT` and `MPI_DOUBLE`). For the receiving process, however, it has been forgotten that the two messages may arrive in any order depending on the relative timing of the two sending processes (and possibly other factors).

```
MPI_Comm_rank(comm,&rank);
MPI_Comm_size(comm,&size);
assert(size>2); // at least three processes

if (rank==0) {
  int buf1[500];

  MPI_Send(buf1,500,MPI_INT,size-1,TAG,comm);
} else if (rank==1) {
  double buf2[100];

  MPI_Send(buf2,100,MPI_DOUBLE,size-1,TAG,comm);
} else if (rank==size-1) {
  MPI_Status status;

  int     buf1[1000];
  double  buf2[200]; // buffer space could be insufficient
  int cc;

  // receive "first" message: data type could be wrong
  MPI_Recv(buf1,1000,MPI_INT,MPI_ANY_SOURCE,TAG,
           comm,&status);
  MPI_Get_elements(&status,MPI_INT,&cc);
  assert(cc<1000);
  if (status.MPI_SOURCE==0) {
    assert(cc==500);
  } else {
    assert(status.MPI_SOURCE==1);
    assert(cc==100);
  }

  // receive "second" message: data type could be wrong
  MPI_Recv(buf2,200,MPI_DOUBLE,MPI_ANY_SOURCE,TAG,
           comm,&status);
  MPI_Get_elements(&status,MPI_DOUBLE,&cc);
  assert(cc<1000);
  if (status.MPI_SOURCE==0) {
    assert(cc==500);
  } else {
    assert(status.MPI_SOURCE==1);
    assert(cc==100);
  }
}
```

The program may crash, possibly with an MPI error message that a received message has been truncated or that one of the assertions was violated. Note that data may also be incorrectly received: the datatype argument will not always correspond to the type of the sent data (see later).

The correct order of the received messages can be enforced by using different tags for the two messages. Of course, this sacrifices the potential performance advantage of non-determinacy. The program below is correct. The first `MPI_Recv` operation by process $p - 1$ can only receive a message with tag `TAG0`. Such a message will eventually be sent by process 0. The next to be received message must have tag `TAG1` and, also, such a message will eventually be sent by process 1.

```
#define TAG0 500
#define TAG1 501

if (rank==0) {
  int buf1[500];

  MPI_Send(buf1,500,MPI_INT,size-1,TAG0,comm);
} else if (rank==1) {
  double buf2[100];

  MPI_Send(buf2,100,MPI_DOUBLE,size-1,TAG1,comm);
} else if (rank==size-1) {
  MPI_Status status;

  int     buf1[1000];
  double buf2[200];
  int cc;

  MPI_Recv(buf1,1000,MPI_INT,MPI_ANY_SOURCE,TAG0,
           comm,&status);
  MPI_Get_elements(&status,MPI_INT,&cc);
  assert(cc==500);

  MPI_Recv(buf2,200,MPI_DOUBLE,MPI_ANY_SOURCE,TAG1,
           comm,&status);
  MPI_Get_elements(&status,MPI_DOUBLE,&cc);
  assert(cc==100);
}
```

Message tags are a specialty of point-to-point communication. They can be highly useful when it is necessary to be able to distinguish between different kinds of messages and can provide an extra piece of information "on top of" the messages themselves that would otherwise have to be incorporated manually as part of each message. One-sided communication (Sect. 3.2.22) and collective communication (Sect. 3.2.28) do not provide message or communication tags.

In order to find out whether a message from a determinate or non-determinate source with a given or wildcard tag is ready to be received, MPI provides calls to probe for such possible messages. The following operation sets a status object for an "incoming" message that is ready to be received.

```
int MPI_Probe(int source, int tag, MPI_Comm comm,
              MPI_Status *status);
```

A probe call returns when a message with the specified characteristics (source and tag) is ready to be received but the message is not received by the probe and the data are not accessible. To actually receive the message and make the data available in a receive buffer, an `MPI_Recv` or other point-to-point message receive operation must be executed.

Advanced note: This separation into the probing for a message and the actual reception of the message can cause problems (race conditions) when MPI is used in a multi-threaded program, for instance, with OpenMP or `pthreads` where many threads perform MPI operations and a different thread from the one doing the probe can have received the message. The MPI specification provides means to deal with this situation, but this is beyond these lectures.

3.2.13 Point-to-Point Communication Complexity and Performance

When two MPI processes become ready to communicate at the same time, one process sending m units of data and the other being ready to receive (at least) m units, the time for transmitting the m data units may naively be modeled as $\alpha + \beta m$ for a given, constant start-up *latency* α and a cost per unit β (see Sect. 3.1.3). In a more refined model, α and β might depend on the placement of the two communicating processes in the system, the *mapping* of the communicator to the processors, the total number of processes in the communicator, and on the overall traffic in the communication system during the communication operation.

Alternatively, we can account for this data transmission as one *communication step*, irrespectively of the amount of data being transmitted. Several independent pairs of processes can, if the underlying communication network is strong enough by having enough bisection width, communicate independently, and if all processes communicate (roughly) the same amount of data m, we count such a set of concurrent communication operations between many pairs of processes as one communication step. In the case where many processors are communicating more or less at the same time and performing their respective communication steps concurrently we commonly use the term *communication round* for all these concurrent steps.

The *communication round complexity* for a message-passing algorithm can be accounted for as the number of such *communication rounds* required from the start of all processes until the last process has completed its last communication step. In a bit more general terms, this amounts to finding the longest (weighted) path of communication operations from a first to a last process in a communication DAG (Directed Acyclic Graph) that describes the communication operations of the algorithm. In the communication DAG, there is an edge from process i to process j when process i sends a message that is received by process j, that is, when both processes are *ready* to communicate. The communication round complexity is the length of a longest path of dependent send and receive communication operations in the DAG corresponding to the execution of the algorithm. The DAG way of accounting for the complexity of a message passing (MPI) algorithm can also be used for the case where the amount of data communicated between pairs of processes is not the same in each round in which case the DAG edges are assigned weights reflecting the cost of the corresponding communication step according to some model of pairwise communication costs.

The linear array broadcast implementation from Sect. 3.2.11 was claimed to take $p-1$ communication steps. This can be seen by an inductive argument. If there are only two processes, one communication step is obviously required and suffices. With $p > 2$ processes, the root process first sends the data to the next process in one communication step. This process now behaves like a root for $p-1$ processes. By induction, the broadcast can now be completed in $p-2$ steps, for a total of $(p-2)+1 = p-1$ steps.

Below is a better broadcast algorithm that completes in $\lceil \log_2 p \rceil$ steps, matching the lower bound on the number of communication rounds for broadcast of Theorem 14. The algorithm assumes that broadcast is from the process with rank 0 in the communicator and to ensure this it uses a virtual ranking `virt` of the processes. This trick of reranking by subtracting a `root` rank is often useful and will perform better than creating a new communicator with the desired property.

```
MPI_Comm_rank(comm,&rank);
MPI_Comm_size(comm,&size);

virt = (rank-root+size)%size; // rerank by hand

// buffer of some type from somewhere
if (virt!=0) {
  int d = 1;
  while (virt>=d) {
    dist = d; d <<= 1; // multiply by two
  }
  // rerank back from virt to real rank
  MPI_Recv(buffer,count,datatype,(virt-dist+root+size)%size,
           TAG,comm,MPI_STATUS_IGNORE);
  dist = d;
} else dist = 1;
while (virt+dist<size) {
  MPI_Send(buffer,count,datatype,(virt+dist+root)%size,
           TAG,comm);
  dist <<= 1; // multiply by two
}
```

3.2.14 MPI Concepts: Semantic Terms

The simple send and receive operations, as well as all the other operations discussed in the previous sections, are *blocking*. This is an MPI specific semantic term, which means that the call returns when the operation is complete, locally, from the calling process' point of view. With MPI_Send in particular, this (just) means that data are out of the send-buffer, which can now be reused for other purposes, for instance, as buffer for the next MPI_Send operation. Also, other resources have been given free and can be reused. When, e.g., MPI_Comm_split returns, a new communicator has been created and is ready for use from the calling process' point of view. The *blocking* property does not imply anything about what other processes have done. It simply and only means that an operation has been completed locally by the calling process according to the semantics of the operation. For instance, return from an MPI_Send call does not mean that the data have been received by the receiving process which might not even have posted its MPI_Recv call. Data might not even be anywhere near the receiving process. They could, for instance, have been buffered somewhere by the MPI library, a technique that is often used for small messages and can in some situations have some performance advantages. On the other hand, for a blocking operation to complete, some action by other processes *may* be necessary. For instance, very large data in MPI_Send calls are typically not buffered by MPI libraries. In such cases, point-to-point communication can complete only when sending and receiving processes are both active. Obviously, a blocking MPI_Recv call cannot complete before the data have been sent, so action by the sending process can in this case indeed be inferred.

As counterpart, MPI defines operations that are *nonblocking*. Such operations will return *immediately* (whatever that means, how fast is immediate?) and *always* independently of actions being taken by other processes. They are therefore also called *immediate operations* and prefixed with an I (Capital "I") after MPI_ in MPI.

An MPI operation is said to have *local completion* if it can always complete independently of action by other processes. Trivial examples seen so far are the blocking operations MPI_Comm_rank and MPI_Comm_size. The MPI_Send operation is blocking but does not have local completion since action by the receiving process *may* be required. The same holds for the MPI_Recv operation, which always requires action by a sending process. The collective MPI_Comm_create operation is blocking and does not have local completion: Some action may be and mostly is required by the other processes in order to create the new communicator and make it possible to complete the call. A process calling a blocking collective operation will in most cases not be able to complete and return from the call before at least some of the other processes in the communicator has at least performed the corresponding call.

The counterpart to local completion is *non-local completion*, which means that, in order for an operation to complete, action by other processes *may* be needed. Here, *action* means that other processes perform MPI calls that enable the operation to complete. Per definition, nonblocking MPI calls have local completion.

As discussed for the blocking `MPI_Send` operation, an implementation of MPI that does intermediate buffering may make it possible for an `MPI_Send` call to complete and return even without the receiving process having posted a matching `MPI_Recv` call. But it may also not. Relying on such implementation specific behavior is bad and dangerous practice, since it makes programs non-portable. The program may run on one machine with one MPI library, but it may stop working on the next machine with a different MPI library. Or it may work for small data and problem sizes and then suddenly and mysteriously stop working when problem and data get larger. The practice of (perhaps unbeknownst to the programmer) relying on implementation dependent behavior of MPI operations is called *unsafe programming*.

Concretely, for blocking communication with `MPI_Send` and `MPI_Recv`, one should always write the application under the assumption that completion is indeed non-local and such that there will always, eventually, be a matching `MPI_Recv` call for any `MPI_Send` call executed by some process.

Here is a typical example of unsafe communication with processes communicating in a rank ordered ring pattern. All processes initiate a (blocking) `MPI_Send` call to the next process in the ring, after which they receive data from the previous process in the ring. The `MPI_Send` may — or may *not* — be able to complete, depending on the message count and on implementation details of the MPI library. If it cannot complete, a deadlock ensues. The nasty thing about this kind of code is that it may well work under the right circumstances and then suddenly stop working (deadlock) when conditions change. That is why this style of programming is called *unsafe*. Unsafe programs are in particular not portable.

```
#define TAG 1000

MPI_Status status;
int *a = ...;
int *b = ...;

assert(size%d==0); // even number of processes required
MPI_Send(b,count,MPI_INT,(rank+1)%size,TAG,comm);
MPI_Recv(a,count,MPI_INT,(rank-1+size)%size,TAG,
         comm,&status);
```

In some of the examples above, message tags were used to enforce a certain order on received messages. This usage of tags can easily result in an unsafe program as the example below shows. The first `MPI_Send` call may not be able to complete.

```
#define TAG1 100
#define TAG2 101

if (rank==0) {
  int buf1[500];
```

3.2 Ninth Block (3–4 Lectures)

```
    double buf2[100];

    MPI_Send(buf2,100,MPI_DOUBLE,size-1,TAG2,comm);
    MPI_Send(buf1,500,MPI_INT,size-1,TAG1,comm);
} else if (rank==size-1) {
    // order, buf2 smaller than buf1, but no overflow
    MPI_Status status;

    int buf1[1000];
    double buf2[200];
    int cc;

    MPI_Recv(buf1,1000,MPI_INT,0,TAG1,comm,&status);
    MPI_Get_elements(&status,MPI_INT,&cc);
    assert(cc<=1000);
    assert(cc==500);

    MPI_Recv(buf2,200,MPI_DOUBLE,0,TAG2,comm,&status);
    MPI_Get_elements(&status,MPI_DOUBLE,&cc);
    assert(cc<=200);
       assert(cc==100);
}
```

Care is needed to ensure that a program is not unsafe. Sometimes this can be difficult. For a more interesting example, we return to the stencil pattern as discussed in Sect. 3.2.8 and Sect. 1.3.5. The communication part of a two-dimensional stencil code below shows. Here, the processes have been organized as a two-dimensional Cartesian communicator, see Sect. 3.2.8. For each process, the ranks of the (up to four) neighboring processes, `left`, `right`, `up` and `down`, are computed with the `MPI_Cart_shift` functionality (some of these ranks may be `MPI_PROC_NULL`). Each process has out and in buffers for its four neighboring processes from where it needs to both send and receive data. The stencil update and the four send and receive operations are repeated until some convergence criterion is fulfilled and the `done` flag is set to `true` (=1). This will most likely depend on the not shown computations within each iteration (there could also be a fixed, predetermined number of iterations). A first attempt could look as follows.

```
#define STENTAG 11

int left, right;
int up,   down;

MPI_Cart_shift(cartcomm,1,1,&left,&right);
MPI_Cart_shift(cartcomm,0,1,&up,   &down);

double *out_left, *out_right, *out_up, *out_down;
double *in_left,  *in_right,  *in_up,  *in_down;

... // allocate and set buffers
```

Table 3.1 Some C datatypes and their corresponding MPI_Datatype

C language type	Corresponding MPI datatype
char	MPI_CHAR
short	MPI_SHORT
int	MPI_INT
long	MPI_LONG
float	MPI_FLOAT
double	MPI_DOUBLE

```
int done = 0;
while (!done) { // iterate until convergence
  ...  // the stencil update (computation)

  MPI_Send(out_left , n,MPI_DOUBLE,left , STENTAG,cartcomm);
  MPI_Send(out_right,n,MPI_DOUBLE,right,STENTAG,cartcomm);
  MPI_Send(out_up ,  n,MPI_DOUBLE,up,    STENTAG,cartcomm);
  MPI_Send(out_down , n,MPI_DOUBLE,down , STENTAG,cartcomm);

  MPI_Recv(in_left ,  n,MPI_DOUBLE,left , STENTAG,cartcomm,
           MPI_STATUS_IGNORE);
  MPI_Recv(in_right , n,MPI_DOUBLE,right,STENTAG,cartcomm,
           MPI_STATUS_IGNORE);
  MPI_Recv(in_up,    n,MPI_DOUBLE,up,    STENTAG,cartcomm,
           MPI_STATUS_IGNORE);
  MPI_Recv(in_down , n,MPI_DOUBLE,down , STENTAG,cartcomm,
           MPI_STATUS_IGNORE);

  done = 1; // some termination criterion
}
```

Depending on the completion semantics, the four send operations may not be able to complete before the corresponding receive operations have been initiated, which in that case will not be possible: the program deadlocks. It is a good exercise to reflect on this example and on how the code can be made safe and portable. We will return to this example several times.

3.2.15 MPI Concept: Specifying Data with (Derived) Datatypes

Data to be communicated in MPI are always specified the same way. A block of elements is described by a triple consisting of a starting address (or offset) in memory (buffer), a number of elements (count), and a structure/layout of the elements (datatype). As a mnemonic for the MPI communication operations it is helpful to keep in mind that data are always triples of buffer,count,datatype;

this greatly reduces the number of arguments one has to think of and makes it easy to guess/reconstruct the signature of many MPI operations.

The third argument in the triple, the `MPI_Datatype`, describes the structure or layout of the data elements to be communicated (sent or received) locally, for calling the process. For basic, simple, non-complex objects like the `int`s and `double`s in a C program, there are corresponding, predefined MPI handles like `MPI_INT` and `MPI_DOUBLE` that describe to the MPI library that the bits and bytes in a data buffer represent these kinds of objects.

For the simple, and most common case of elements from a consecutive buffer, for instance, an array of elements of some simple (programming language) type being communicated, the datatype argument just tells the MPI library that the bytes are to be interpreted as the corresponding programming language type is represented in memory. There is, therefore, an MPI datatype for each simple, elementary programming language datatype. Some of the corresponding MPI datatypes for C types are shown in Table 3.1. Fortran has Fortran-like names for the corresponding MPI types.

Correct MPI programs require that data elements (corresponding to some programming language type) that are sent as a sequence of MPI datatypes are received as a sequence of the same MPI datatypes. Observing this requirement ensures that the bits and bytes that are sent and received are interpreted and handled in the intended way, both by the sending and by the receiving process. It is important to understand that the programming language type of objects (buffers) are not and cannot be *a priori* known to the MPI library. Therefore, the library has to be instructed in each communication operation by each involved process. The MPI datatype information is *not* in any way part of the transmitted data. It is entirely the programmer's responsibility to ensure that all communicated data are given the right MPI datatype by both sending and receiving processes. Neither compiler nor MPI library can and will (for performance reasons) check this. For this same reason, MPI does not perform type conversion (as known from, e.g., C). If a data buffer is sent as a sequence of `MPI_INT` objects and received as a sequence of `MPI_FLOAT` objects, no useful outcome can be expected. Most certainly, the `int`s will not be converted to `float`s in a semantically meaningful way!

The next three small examples illustrate this. In the first example, some `long`s are sent correctly as `MPI_LONG`, but wrongly received as `MPI_DOUBLE` and stored in a (large enough, presumably) buffer of `double`s.

```
if (rank==0) {
  long a[n];
  MPI_Send(a,n,MPI_LONG,size-1,TAG,comm);
} else if (rank==size-1) {
  double a[n];
  MPI_Recv(a,n,MPI_DOUBLE,0,TAG,comm,MPI_STATUS_IGNORE);
}
```

In the second example, `double`s sent correctly are received as a sequence of `MPI_BYTE` elements. This may or may not give correct results but is in any case a dangerously incorrect MPI programming style.

```
double a[n];
if (rank==0) {
  MPI_Send(a,n,MPI_DOUBLE,size-1,TAG,comm);
} else if (rank==size-1) {
  MPI_Recv(a,n*sizeof(double),MPI_BYTE,0,TAG,
           comm,MPI_STATUS_IGNORE);
}
```

In the third and last example, the objects are sent and received as streams of uninterpreted bytes. This is not technically wrong, but any type information on how **doubles** are to be handled (e.g., Endianness) is lost.

```
double a[n];
if (rank==0) {
  MPI_Send(a,n*sizeof(double),MPI_BYTE,size-1,TAG,comm);
} else if (rank==size-1) {
  MPI_Recv(a,n*sizeof(double),MPI_BYTE,0,TAG,
           comm,MPI_STATUS_IGNORE);
}
```

The next purpose of MPI datatypes is to be able to describe layouts of complex, often non-consecutive data in process local memory. This gives the MPI library the possibility to read and write data elements from specific locations and not necessarily as a consecutive stream of elements in a simple, linear buffer (array). Natural examples are the columns of two-dimensional matrices, submatrices of larger, d-dimensional matrices, complex C structures with different component types, etc. The MPI concept of a datatype is, thus, somewhat different from the same-named, semantic programming language concept. In MPI, a datatype describes the (local, spatial) structure of data objects to be communicated and carries little, further semantics.

The idea of the MPI *user-defined datatype* or *derived datatype* mechanism is to be able to encapsulate such complex data layouts into a single unit which can then be used as the unit of communication in *all* MPI communication operations. A derived datatype represents an ordered list of simple, basic datatypes (as we have seen: `MPI_INT`, `MPI_DOUBLE`, `MPI_CHAR`, etc.) together with a displacement or relative offset for each simple element. The offset for an element gives the linear position of the element in memory relative to a given base address, e.g., the buffer argument supplied in the MPI communication calls.

An explicit list of basic element datatypes with their displacements is in MPI terms called a *type map*. The type map, which is an opaque construct that is not directly accessible to the programmer, is used locally by communicating MPI processes to access the basic elements in local memory in the order and displacement implied by the list, regardless of whether processes are sending or receiving data. A type map is, thus, a purely process local construct and the type maps of one process are not known to any other processes. Identical type maps for different processes can, of course, be constructed by the programmer, but MPI itself cannot and does not exchange type maps or any other type information. Datatypes and type maps are not *first-class citizens* in MPI.

3.2 Ninth Block (3–4 Lectures)

Communication in MPI can be thought of as a stream of elements described by a corresponding stream of simple, basic datatypes. This stream of simple, basic datatypes can be thought of as the type map stripped of the displacements. Such an ordered sequence of basic datatypes is, in MPI terms, called a *type signature*. When two processes are communicating with point-to-point send and receive operations, the signature of the data that are sent must be a prefix of the signature of the data that the receiving process is prepared to receive (since the receive operation can specify a larger element count than the send operation). Again, the signatures are not part of the data that are being communicated. It is purely the programmers responsibility to guarantee that the signature rule is obeyed. By this choice, it is possible with the help of the programmer to do type safe communication in MPI without the burden (and performance disadvantage) of having to communicate additional type meta information.

In MPI, neither type maps nor type signatures are represented explicitly by lists of basic datatypes and displacements. Instead, MPI provides a number of constructors for compactly describing more and more irregular layouts of data in memory. Layouts described by this mechanism are called *derived* or *user-defined datatypes*. In Sect. 3.2.20 the MPI datatype constructors will be briefly explained. A derived datatype can be used in any MPI communication operation and in all operations that take `MPI_Datatype` arguments.

A type map, as represented by a derived datatype, is a complex object encompassing possibly many different, basic datatypes together with their displacements. The *size* of a derived datatype is the number of bytes required (locally, for the process) to represent all the basic datatypes in the derived datatype. The *extent* of a derived datatype is a quantity in bytes associated with a derived datatype which is necessary to define what happens when a derived datatype is used in communication operations with an element count larger than one. The signature of the derived datatype is the unit of communication. A count $c, c > 1$ tells that more than one such unit is to be communicated. The ith unit, $0 \leq i < c$, is taken from relative offset $i \cdot$ `extent` from the given communication buffer address, where `extent` is the extent of the datatype. The following MPI calls return the size and extent of both simple, basic, predefined datatypes and user-defined datatypes.

```
int MPI_Type_size(MPI_Datatype datatype, int *size);
int MPI_Type_extent(MPI_Datatype datatype, MPI_Aint *extent);

int MPI_Type_get_extent(MPI_Datatype datatype,
                        MPI_Aint *lb, MPI_Aint *extent);
int MPI_Type_get_true_extent(MPI_Datatype datatype,
                             MPI_Aint *true_lb,
                             MPI_Aint *true_extent);
```

Often, but not always, the extent of a derived datatype corresponds to the "footprint" in memory of the type layout described by that datatype. This is the linear difference between the simple element (basic datatype) with the smallest displacement and the simple element with the largest displacement plus the size of that element. The datatype constructors all have associated rules for how the extent of the resulting derived datatype is computed. There are, however, special type constructors for creating datatypes with a different (arbitrary) extent, a feature that is extremely powerful for advanced usage of MPI. Therefore, the extent is not simply the "memory footprint" of the layout. However, the memory footprint is needed when new memory for some complex layout needs to be allocated. For this, the special call MPI_Type_get_true_extent is defined. Unfortunately, even this is not always sufficient for computing the right amount of memory space. Memory allocation and derived datatypes in MPI need care.

The calls returning an extent have arguments of type pointer to MPI_Aint. This argument type is not an MPI handle, but the type of an object that can represent an *address-sized integer*. In many cases (compilers, systems), an MPI_Aint is indeed different from a C int (64 versus 32 bits). The MPI_Aint type is used for many MPI operations where it is important that an argument is a process local address; but is not used very consistently in the MPI standard.

3.2.16 MPI Concept: Matching Communication Operations

In order for point-to-point communication between two processes to be successful, the MPI_Send and MPI_Recv operations must *match*. First of all, the two processes must make their calls on the same communicator: In MPI, communication on one communicator can *never* interfere with communication on another communicator. In particular, communication with an MPI_Send on one communicator and an otherwise correct MPI_Recv operation on another communicator will never take place and will result in a deadlock. The destination rank given by the sending process must match the rank given by the receiving process. Either the receiving process gives the rank of the sending process explicitly or the MPI_ANY_SOURCE wildcard. Likewise, the message tags must be the same or the receiving operation must use the MPI_ANY_TAG wildcard. As mentioned, it is perfectly legal for a process to communicate with itself. However, with blocking operations only, it is not possible to do this in a safe way; at least one of either the send or receive operations has to be nonblocking. Alternatively, the MPI_Sendrecv operation can be used. Also, in such cases care has to be taken that receive and send buffer (which are on the same process) do not overlap in anyway. Otherwise, the result would depend on the exact order in which data elements are received and sent and put into the respective buffers: a kind of race condition that is not allowed in correct MPI programs.

Second, the amount of data sent in the MPI_Send operation must be at most the amount of data that the MPI_Recv operation is prepared to receive, as specified by its count and datatype arguments. The MPI types of the sent and received elements must correspond. Technically, this means that the signature of the sent data must be a prefix of the signature of the data specified in the receive call. As discussed, MPI cannot (without considerable overhead) and does not check for this.

3.2 Ninth Block (3–4 Lectures)

When an `MPI_Send` and an `MPI_Recv` call match, communication can take place and the MPI implementation guarantees that data are eventually correctly received. There is no need for low-level consistency or correctness checks on behalf of the user code.

Communication with the special `MPI_PROC_NULL` process always matches but has no effect, neither in an `MPI_Send` nor in an `MPI_Recv` operation.

3.2.17 Nonblocking Point-to-Point Communication

MPI defines nonblocking point-to-point communication counterparts for the simple `MPI_Send`, `MPI_Recv`, and `MPI_Probe` operations.

```
int MPI_Isend(const void *buf,
              int count, MPI_Datatype datatype,
              int dest, int tag,
              MPI_Comm comm, MPI_Request *request);
int MPI_Irecv(void *buf, int count, MPI_Datatype datatype,
              int source, int tag,
              MPI_Comm comm, MPI_Request *request);
int MPI_Iprobe(int source, int tag, MPI_Comm comm, int *flag,
               MPI_Status *status);
```

With the MPI 4.0 version of the MPI standard, also nonblocking versions of the `MPI_Sendrecv` and `MPI_Sendrecv_replace` operations have been introduced.

```
int MPI_Isendrecv(const void *sendbuf,
                  int sendcount, MPI_Datatype sendtype,
                  int dest, int sendtag,
                  void *recvbuf,
                  int recvcount, MPI_Datatype recvtype,
                  int source, int recvtag,
                  MPI_Comm comm,
                  MPI_Request *request);
int MPI_Isendrecv_replace(void *buf,
                          int count, MPI_Datatype datatype,
                          int dest, int sendtag,
                          int source, int recvtag,
                          MPI_Comm comm,
                          MPI_Request *request);
```

All these operations return "immediately": What exactly this means and how fast "immediate" is, is, by the nature of the MPI standard specification, not defined. The important point is that the operations have entirely local completion semantics and return independently of any MPI actions taken by any other processes. Nonblocking point-to-point operations can, therefore, be used to avoid situations that might otherwise lead to a deadlock (unsafe code) with blocking communication.

These nonblocking send and receive operations take the same input parameters as their blocking counterparts, but have a new output argument, the MPI_- Request object (handle). The MPI_Request object can be used to query the completion status of the corresponding operation and to enforce completion. A nonblocking MPI_Isend call with ensuing, enforced completion has the same effect (semantics) as a blocking MPI_Send call. That is, enforced completion means only that the send operation has been completed from the process' point of view and does not imply anything about the receiving process, not even that is has reached the matching receive call. The nonblocking counterpart of the probe operation, the MPI_Iprobe, does, curiously, not return an MPI_Request object. Instead, the completion of the probe for a matching, incoming message is indicated in the flag return argument (pointer).

There is a whole repertoire of operations for checking and enforcing completion of immediate, pending MPI_Isend and MPI_Irecv communication operations. These calls can either test whether an operation, referred to by an MPI_Request object, is complete, which is signalled in a flag return argument, or enforce (by waiting) the completion of an operation. There are calls that operate on a list of request objects, rather than a single object. They can test for or enforce completion of either some single (arbitrary) operation, some, or all operations in the set of requests (given as an input array). For complete operations, their status is returned in corresponding MPI_Status objects, just as was the case for the blocking MPI_Send and MPI_Recv calls.

```
int MPI_Wait(MPI_Request *request, MPI_Status *status);
int MPI_Test(MPI_Request *request,
             int *flag, MPI_Status *status);

int MPI_Waitany(int count, MPI_Request requests[], int *indx,
                MPI_Status *status);
int MPI_Testany(int count, MPI_Request requests[], int *indx,
                int *flag, MPI_Status *status);
int MPI_Waitall(int count, MPI_Request requests[],
                MPI_Status statuses[]);
int MPI_Testall(int count, MPI_Request requests[],
                int *flag, MPI_Status statuses[]);
int MPI_Waitsome(int incount, MPI_Request requests[],
                 int *outcount,
                 int indices[], MPI_Status statuses[]);
int MPI_Testsome(int incount, MPI_Request requests[],
                 int *outcount,
                 int indices[], MPI_Status statuses[]);

int MPI_Request_free(MPI_Request *request);
```

The nonblocking communication operations separate the initialization and the completion of an operation and can be most convenient for writing safe programs that cannot deadlock in any possible situation. For instance, the MPI_Sendrecv operation is equivalent to either

3.2 Ninth Block (3–4 Lectures)

```
MPI_Request  request;
MPI_Status   status;
MPI_Isend(sendbuf,sendcount,sendtype,dest,sendtag,
          comm,&request);
MPI_Recv(recvbuf,recvcount,recvtype,source,recvtag,
         comm,&status);
MPI_Wait(&request,MPI_STATUS_IGNORE);
```

or

```
MPI_Request  request;
MPI_Status   status;
MPI_Irecv(recvbuf,recvcount,recvtype,source,recvtag,
          comm,&request);
MPI_Send(sendbuf,sendcount,sendtype,dest,sendtag,comm);
MPI_Wait(&request,&status);
```

or even

```
MPI_Request  request[2];
MPI_Status   status[2];
MPI_Irecv(recvbuf,recvcount,recvtype,source,recvtag,
          comm,&request[0]);
MPI_Isend(sendbuf,sendcount,sendtype,dest,sendtag,
          comm,&request[1]);
MPI_Waitall(2,request,status);
```

where, for the last code snippet, the status of the receive operation is in `status[0]`.

The unsafe, two-dimensional stencil code built from blocking `MPI_Send` and `MPI_Recv` operations (Sect. 3.2.14) can now be made safe and deadlock-free by simply using nonblocking send and receive operations. This may have the additional performance advantage that communication can take place in the order in which processes become ready and not in the fixed order given by the sequence of blocking `MPI_Send` and `MPI_Recv` operations.

```
MPI_Request  request[8];

int done = 0;
while (!done) { // iterate until convergence
   ... // the stencil update (computation)

   int k = 0;
   MPI_Isend(out_left, n,MPI_DOUBLE,left, STENTAG,cartcomm,
             &request[k++]);
   MPI_Isend(out_right,n,MPI_DOUBLE,right,STENTAG,cartcomm,
             &request[k++]);
   MPI_Isend(out_up,   n,MPI_DOUBLE,up,    STENTAG,cartcomm,
             &request[k++]);
   MPI_Isend(out_down, n,MPI_DOUBLE,down,  STENTAG,cartcomm,
             &request[k++]);

   MPI_Irecv(in_left,  n,MPI_DOUBLE,left,  STENTAG,cartcomm,
             &request[k++]);
```

```
    MPI_Irecv(in_right, n,MPI_DOUBLE,right,STENTAG,cartcomm,
              &request[k++]);
    MPI_Irecv(in_up,    n,MPI_DOUBLE,up,   STENTAG,cartcomm,
              &request[k++]);
    MPI_Irecv(in_down,  n,MPI_DOUBLE,down, STENTAG,cartcomm,
              &request[k++]);

    MPI_Waitall(k,request,MPI_STATUSES_IGNORE);

    done = 1; // some termination criterion
}
```

The special value `MPI_STATUSES_IGNORE` argument indicates that all statuses should be ignored and no output status array is given.

The stencil computation can also be made safe and deadlock-free with the combined `MPI_Sendrecv` operation. The trick is to communicate along the dimensions of the process grid, one after the other, in a ring-like fashion, receiving from the left process and sending to the right process etc. Since all processes perform their communication operations in a symmetric fashion, no deadlocks can occur. For a possible performance advantage, `MPI_Isendrecv` could have been used as in the above code.

```
int done = 0;
while (!done) { // iterate until convergence
    ... // the stencil update (computation)

    MPI_Sendrecv(out_left,  n,MPI_DOUBLE,left, STENTAG,
                 in_right,  n,MPI_DOUBLE,right,STENTAG,cartcomm,
                 MPI_STATUS_IGNORE);
    MPI_Sendrecv(out_right, n,MPI_DOUBLE,right,STENTAG,
                 in_left,   n,MPI_DOUBLE,left, STENTAG,cartcomm,
                 MPI_STATUS_IGNORE);
    MPI_Sendrecv(out_up,    n,MPI_DOUBLE,up,   STENTAG,
                 in_down,   n,MPI_DOUBLE,down, STENTAG,cartcomm,
                 MPI_STATUS_IGNORE);
    MPI_Sendrecv(out_down,  n,MPI_DOUBLE,down, STENTAG,
                 in_up,     n,MPI_DOUBLE,up,   STENTAG,cartcomm,
                 MPI_STATUS_IGNORE);

    done = 1; // some termination criterion
}
```

3.2.18 Exotic Send Operations*

MPI provides a few more send operations with additional semantic content. These operations come in both blocking and nonblocking variants. There is a *synchronous send* operation, where local completion implies that the receiving process has indeed started reception of the message by a matching receive operation. There is a *buffered send* operation, where data are explicitly stored in

3.2 Ninth Block (3–4 Lectures)

a local buffer in order to provide local completion semantics. The local buffer is allocated in user space which, for this use, needs to be explicitly attached to the MPI library. Finally, there is a *ready send* operation, which can be used provided that a matching receive operation has already been posted before the buffered send. Send-receive communication can possibly be implemented more efficiently under this precondition. The ready send operation was included in MPI to enable such implementations. Using it correctly requires additional explicit or implicit synchronization and is rather left to experts.

These more exotic send functions are listed below, in order of exoticness, but are not covered further in these lectures.

```
int MPI_Ssend(const void *buf,
              int count, MPI_Datatype datatype,
              int dest, int tag, MPI_Comm comm);
int MPI_Bsend(const void *buf,
              int count, MPI_Datatype datatype,
              int dest, int tag, MPI_Comm comm);
int MPI_Rsend(const void *buf,
              int count, MPI_Datatype datatype,
              int dest, int tag, MPI_Comm comm);

int MPI_Buffer_attach(void *buffer, int size);
int MPI_Buffer_detach(void *buffer_addr, int *size);
```

The nonblocking counterparts are listed below.

```
int MPI_Issend(const void *buf,
               int count, MPI_Datatype datatype,
               int dest, int tag,
               MPI_Comm comm, MPI_Request *request);
int MPI_Ibsend(const void *buf,
               int count, MPI_Datatype datatype,
               int dest, int tag,
               MPI_Comm comm, MPI_Request *request);
int MPI_Irsend(const void *buf,
               int count, MPI_Datatype datatype,
               int dest, int tag,
               MPI_Comm comm, MPI_Request *request);
```

Any type of send operation can match with any type of receive operation, whether blocking or nonblocking. There is only one kind of receive operation in MPI and completion of a receive operation signifies that data have been received correctly from a matching, sending process.

For completeness, we mention that it is/should be technically possible to cancel a message. However, the semantics and guarantees of this operation are not clear and relying on this functionality is never recommended in MPI programs.

```
int MPI_Cancel(MPI_Request *request);
int MPI_Test_cancelled(const MPI_Status *status, int *flag);
```

3.2.19 MPI Concept: Persistence*

An additional, recently extended MPI concept, which we do not cover in these lectures, is *persistent (point-to-point) communication*. The idea is to be able to separate the initialization of a communication operation (argument parsing, reservation of memory and communication resources, algorithmic preprocessing) from the operation itself to make it possible to execute the operation many times with the same arguments. Persistent operations aim to make it possible to *amortize* possibly expensive set-up costs over many uses of the same operation.

Concretely, MPI reuses `MPI_Request` handles as objects to store the precomputed information for a persistent communication operation. The MPI standard defines a persistent counterpart for all the different types of send operations as well as for the receive operation. New operations are introduced to (re)start any single or a whole set of persistent communication operations.

```
int MPI_Send_init(const void *buf,
                  int count, MPI_Datatype datatype,
                  int dest, int tag,
                  MPI_Comm comm, MPI_Request *request);
int MPI_Ssend_init(const void *buf,
                  int count, MPI_Datatype datatype,
                  int dest, int tag,
                  MPI_Comm comm, MPI_Request *request);
int MPI_Bsend_init(const void *buf,
                  int count, MPI_Datatype datatype,
                  int dest, int tag,
                  MPI_Comm comm, MPI_Request *request);
int MPI_Rsend_init(const void *buf,
                  int count, MPI_Datatype datatype,
                  int dest, int tag,
                  MPI_Comm comm, MPI_Request *request);
int MPI_Recv_init(void *buf,
                  int count, MPI_Datatype datatype,
                  int source, int tag,
                  MPI_Comm comm, MPI_Request *request);

int MPI_Start(MPI_Request *request);
int MPI_Startall(int count, MPI_Request requests[]);
```

Both of the start calls are local and nonblocking, although the init calls may take a non-trivial amount of time depending on the amount of preprocessing that can be or is done (MPI libraries may vary). Thus, the persistent communication operations behave like the corresponding nonblocking operations. Completion can be checked or enforced with the same operations on the `MPI_Request` object as explained in Sect. 3.2.17.

3.2.20 More on User-Defined, Derived Datatypes*

The datatype argument of the communication operations seen so far describes the process local unit of communication and the count argument the number of such units. The units we have seen in the small examples hitherto corresponded to the basic C datatypes like ints to the MPI_Datatype MPI_INT, etc. (Table 3.1). A process local communication unit can be more complex, though, and describe a whole sequence of basic datatypes together with their relative displacements in memory. Such a description is called the *type map* of a memory layout.

Type maps are represented by MPI_Datatype objects (handles) in a more concise form than by explicit, possibly very long lists of basic datatypes and their displacements. MPI provides a set of constructors for constructing new, more complex datatypes out of already existing ones: MPI objects cannot be changed, only new objects can be created from existing ones. Datatype objects are called *derived datatypes* and are means to describe the structure (layout in memory) of complex data in the local memory of a process.

A set of fundamental constructors are listed below in order of increasing generality. That is, the structures that can be described by one constructor can also be described by the following ones. These, on the other hand, can describe layouts that cannot be described by a previous one.

```
int MPI_Type_contiguous(int count, MPI_Datatype oldtype,
                        MPI_Datatype *newtype);
int MPI_Type_vector(int count, int blocklength, int stride,
                    MPI_Datatype oldtype,
                    MPI_Datatype *newtype);
int MPI_Type_create_indexed_block(int count, int blocklength,
                                  const int displacements[],
                                  MPI_Datatype oldtype,
                                  MPI_Datatype *newtype);
int MPI_Type_indexed(int count, const int blocklengths[],
                     const int displacements[],
                     MPI_Datatype oldtype,
                     MPI_Datatype *newtype);
int MPI_Type_create_struct(int count,
                           const int blocklengths[],
                           const MPI_Aint displacements[],
                           const MPI_Datatype types[],
                           MPI_Datatype *newtype);
```

Note that the naming of these type creating functions is somewhat inconsistent. This has historical reasons, and the MPI archeologist can mine out which.

Before a derived datatype can be used in communication operations, it must be *committed* to the MPI library. The MPI_Type_commit operation is a designated point in the program execution where the MPI library can perform optimizations on the type map description. Such optimizations that can be costly can hopefully be amortized over many uses of the same, derived datatype. As with other MPI objects, derived datatypes should be freed after use: They may take up (rarely, but sometimes considerable) resources.

```
int MPI_Type_commit(MPI_Datatype *datatype);

int MPI_Type_free(MPI_Datatype *datatype);
```

As with all user-created MPI objects, derived datatypes explicitly created in the application must be freed. The predefined datatypes `MPI_INT`, `MPI_DOUBLE`, etc., cannot be freed.

The constructors listed above describe the following sorts of data layouts. As can be seen from the interface listings, all constructors take (various kinds of) repetition counts, lists of displacements, and previously defined units of communication described as derived datatypes.

1. A *contiguous type* describes a contiguous repetition of an already described unit, where one unit follows immediately after the previous one. More formally, with c units of extent e, the ith unit has displacement $ie, 0 \leq i < c$.
2. A *vector type* describes a regularly strided (spaced) repetition of blocks of an already described unit. More formally, with c blocks of extent e and stride s, the ith unit has displacement $ise, 0 \leq i < c$.
3. A *block index type* describes a sequence of contiguous blocks of previously described units, each with a specific, relative displacement. All blocks have the same size in number of units. More formally, with c blocks of b units with extent e and displacement d_i for the ith block, the elements have displacements $d_i e + je, 0 \leq i < c, 0 \leq j < b$.
4. An *index type* describes a sequence of blocks of previously described units, each with a specific, relative displacement; blocks may have different sizes in number of units. More formally, with c blocks of b_i units with extent e and displacement d_i for the ith block, the elements have displacements $d_i e + je, 0 \leq i < c$ and $0 \leq j < b_i$.
5. A *structured type* describes a sequence of blocks of previously described units, each with a specific, relative displacement, blocks may have different sizes in number of units, and the units of the blocks may be different, previously described units. In contrast to the previous constructors, displacements d_i are in bytes. Since the units may be different, they have possibly different extents e_i. More formally, with c blocks of b_i units with extent e_i and displacement d_i for the ith block, the elements have displacements $d_i + je_i, 0 \leq i < c$ and $0 \leq j < b_i$.

The elements in contiguous blocks of units are spaced from each other by the *extent* of the unit, see Sect. 3.2.15. Likewise, all relative displacements are multiples of the extent of the unit. Only for structured types, the displacements are given in bytes: A single extent multiplier in the displacement makes no sense since the different blocks can have different units. The extent of a newly constructed derived datatype (unit) is the linear distance from the beginning of the first block to the end of the last block in the unit.

It is worth noticing that with the types of constructors described above, it is indeed possible to construct type maps where some data elements have the same displacement. Such type maps are not *per se* illegal or disallowed. A type map with this property is said to have *overlapping entries*. The rules for matching communication are intended to enforce that the outcome of a communication operation is determinate. Thus, in particular datatypes used as arguments for receive buffers in receive operations must not have overlapping entries. For datatypes used as send arguments, this is not a problem and allowed; whether this is good programming practice is a different matter. Such usages should be carefully deliberated.

A first example illustrates the probably most common, often convenient and efficient use of the vector datatype. For this, we elaborate on the stencil example introduced in Sect. 3.2.14, where the placement of data and communication buffers was left open until now. In the distributed stencil computation, a large matrix $M[m, n]$ is subdivided into p smaller submatrices each with roughly the same number of elements. The stencil computation updates each matrix element $M[i, j]$ by a function (for instance, an average) over the neighboring elements. A common stencil is the 5-point stencil, where $M[i, j]$ is updated by a function of the five elements $M[i, j], M[i, j+1], M[i, j-1], M[i+1, j], M[i-1, j]$.

Now, for each MPI process, let `matrix` be the submatrix for the process. We implement a weakly scaling version of the stencil computation, in which the size of the local matrix is kept constant. We let `m` and `n` be the number of local rows and local columns, respectively. Thereby, the total size of the matrix M over all the p processes is $p(\texttt{m} \times \texttt{n})$. It is convenient to actually think of the matrix (and the submatrices) as having two additional rows and two additional columns, thus being of size $M[m+2][n+2]$, and such that row and column elements $M[-1, j], M[m, j], M[i, -1]$ and $M[i, n]$, for $-1 \leq j < n+1$ and $-1 \leq i < m+1$, can be addressed in the stencil computation. These extra rows and columns are called the (sub)matrix *halo* (see also the similar code in Sect. 1.3.5).

In C, each submatrix with its halo can be allocated dynamically by declaring a pointer to rows of size $n+2$ and then allocating space for $m+2$ such rows. This is shown in the code below, which also shows how the address of matrix element [0][0] is shifted, such that the halo rows and columns can be addressed by indices -1 and m and n, respectively. Be careful when later freeing this dynamically allocated memory.

```
m = ...;
n = ...; // small weak scaling example

double (*matrix)[n+2];
matrix = (double(*)[n+2])malloc((m+2)*(n+2)*sizeof(double));
// shift matrix; be careful with free
matrix =
  (double(*)[n+2])((char*)matrix+(n+2+1)*sizeof(double));

// initialize matrix including halo
for (i=-1; i<m+1; i++) {
  for (j=-1; j<n+1; j++) {
    matrix[i][j] = ...;
  }
}
```

We have already seen how the MPI processes can be organized into a two-dimensional mesh with the MPI_Cart_create operation (Sect. 3.2.8), such that each process has neighboring processes in the left, right, up and down directions (some of which are possibly MPI_PROC_NULL). The halos of the process local submatrices represent rows and columns of the full matrix that are present at two processes. In that sense, the process submatrices have overlapping rows and columns and this overlap has to be kept consistent. Thus, the local stencil updates can be performed for all matrix entries $M[i,j], 0 \leq i < m, 0 \leq j < n$, provided that the halo rows and columns have been filled in advance with the corresponding elements from the submatrices at the neighboring processes. The halo column $M[i,-1]$ must be filled with elements from the rightmost column of the left neighbor process, the halo row $M[-1,j]$ analogously with elements from the bottom row of the top neighbor, and so on. Since the matrices in C are in row major order, the rows for the up and down neighbors are consecutive, one-dimensional arrays in memory and can readily be sent and received. The columns, however, are not consecutive but consist of the first element of each row. With the row length being $n+2$ elements, this layout of data in memory can be described as an MPI vector type with element blocks of one element that are strided $n+2$ elements apart. A corresponding datatype for communication of such layouts is created by the MPI_Type_vector constructor and committed for use with MPI_Type_commit. The addresses of the communication buffers of the rows and columns to be sent to neighboring processes are now the addresses of the matrix element $M[0,0]$ (for left and up neighbor) and $M[0, n-1]$ (for the right neighbor) and $M[m-1, 0]$ (for the down neighbor). The addresses of the rows and columns to be received and stored in the halo rows and columns are $M[0,-1]$ (left), $M[0,n]$ (right), $M[-1,0]$ (up) and $M[m,0]$ (down).

```
int left, right;
int up,   down;

MPI_Cart_shift(cartcomm,1,1,&left,&right);
MPI_Cart_shift(cartcomm,0,1,&up,   &down);

MPI_Datatype column;

MPI_Type_vector(m,1,n+2,MPI_DOUBLE,&column);
MPI_Type_commit(&column);

double *out_left, *out_right, *out_up, *out_down;
double *in_left,  *in_right,  *in_up,  *in_down;

out_left  = &matrix[0][0];
out_right = &matrix[0][n-1];
out_up    = &matrix[0][0];
out_down  = &matrix[m-1][0];

in_left   = &matrix[0][-1];
in_right  = &matrix[0][n];
in_up     = &matrix[-1][0];
in_down   = &matrix[m][0];
```

3.2 Ninth Block (3–4 Lectures) 233

```
MPI_Request request[8];

int done = 0;
while (!done) { // iterate until convergence
    ... // the stencil update (computation)

  int k = 0;
  MPI_Isend(out_left,  1,column,     left, STENTAG,cartcomm,
            &request[k++]);
  MPI_Isend(out_right,1,column,     right,STENTAG,cartcomm,
            &request[k++]);
  MPI_Isend(out_up,    n,MPI_DOUBLE,up,   STENTAG,cartcomm,
            &request[k++]);
  MPI_Isend(out_down,  n,MPI_DOUBLE,down, STENTAG,cartcomm,
            &request[k++]);

  MPI_Irecv(in_left,   1,column,     left, STENTAG,cartcomm,
            &request[k++]);
  MPI_Irecv(in_right,  1,column,     right,STENTAG,cartcomm,
            &request[k++]);
  MPI_Irecv(in_up,     n,MPI_DOUBLE,up,   STENTAG,cartcomm,
            &request[k++]);
  MPI_Irecv(in_down,   n,MPI_DOUBLE,down, STENTAG,cartcomm,
            &request[k++]);

  MPI_Waitall(k,request,MPI_STATUSES_IGNORE);

  done = 1; // some termination criterion
}

MPI_Type_free(&column);
```

Alternatively to the vector type, a resized double datatype with an extent of $n+2$ doubles could have been used (see the discussion below). It is an instructive exercise to work this out in detail and to compare it against the solution just described.

In the next example the MPI_Type_vector constructor is used to describe an n column submatrix of an $m \times (np)$ matrix with m rows and np columns, where p is the number of MPI processes. In the program, all processes have a matrix of this size and send their first n columns to the process with rank 0. This process reconstructs a full $m \times (np)$ matrix by putting the p column submatrices together one after the other. Matrices are maintained per hand in row-major order. The elements corresponding to n consecutive columns are, thus, blocks of n elements starting at each multiple inp of np for $i, 0 \leq i < m$. The resulting, full $m \times (np)$ matrix is stored at process 0 in a separately allocated, new matrix. Thus, it cannot happen that a process sends and receives data from overlapping memory regions.

```
int m, n; int i, j;

m = ...;
```

```
n = ...;

double *matrix;
matrix = (double*)malloc(m*size*n*sizeof(double));

MPI_Datatype cols;
MPI_Type_vector(m,n,n*size,MPI_DOUBLE,&cols);
MPI_Type_commit(&cols);

MPI_Request request;
MPI_Isend(matrix,1,cols,0,MATTAG,comm,&request);
if (rank==0) {
  double *newmatrix;
  newmatrix = (double*)malloc(m*size*n*sizeof(double));

  for (i=0; i<size; i++) {
    MPI_Recv(newmatrix+i*n,1,cols,i,MATTAG,
             comm,MPI_STATUS_IGNORE);
  }
}
MPI_Wait(&request,MPI_STATUS_IGNORE);

MPI_Type_free(&cols);
```

In this example where communication is of the individual $m \times n$ submatrices and the receiving process gives the displacement for each received submatrix explicitly, the extent of the vector datatype does not play a role. This is not always so. Sometimes, the default extent of a derived datatype is not what is effectively needed in order to access and store the data correctly in the right locations. An important type creating function for controlling the extent of a datatype, outside the scope of these lectures, is the resizing function. It allows to create a new datatype with arbitrary extent from an existing, derived datatype.

```
int MPI_Type_create_resized(MPI_Datatype oldtype,
                            MPI_Aint lb, MPI_Aint extent,
                            MPI_Datatype *newtype);
```

Should displacements in multiples of the extent of the MPI_Datatype old unit not be sufficient(ly expressive), also constructors where all strides and displacements are given in bytes are provided.

```
int MPI_Type_hvector(int count,
                     int blocklength, MPI_Aint stride,
                     MPI_Datatype oldtype,
                     MPI_Datatype *newtype);
int MPI_Type_hindexed(int count, const int blocklengths[],
                      const MPI_Aint displacements[],
                      MPI_Datatype oldtype,
                      MPI_Datatype *newtype);
int MPI_Type_create_hvector(int count,
                            int blocklength, MPI_Aint stride,
                            MPI_Datatype oldtype,
                            MPI_Datatype *newtype);
```

3.2 Ninth Block (3–4 Lectures)

```
int MPI_Type_create_hindexed_block(int count, int blocklength,
                   const MPI_Aint displacements[],
                   MPI_Datatype oldtype,
                   MPI_Datatype *newtype);
int MPI_Type_create_hindexed(int count,
                   const int blocklengths[],
                   const MPI_Aint displacements[],
                   MPI_Datatype oldtype,
                   MPI_Datatype *newtype);
```

Complex, composite layouts corresponding to distributed arrays and subarrays can be described with the following two composite derived datatype constructors that are also well beyond the scope of these lectures (but sometimes seen in applications).

```
int MPI_Type_create_darray(int size, int rank, int ndims,
                   const int gsizes[],
                   const int distribs[],
                   const int dargs[],
                   const int psizes[],
                   int order, MPI_Datatype oldtype,
                   MPI_Datatype *newtype);
int MPI_Type_create_subarray(int ndims,
                   const int sizes[],
                   const int subsizes[],
                   const int starts[],
                   int order, MPI_Datatype oldtype,
                   MPI_Datatype *newtype);
```

We finally mention that MPI provides a special datatype for opaque, compact storage of data described by derived datatypes. The datatype for such data is MPI_PACKED. Three functions make it possible to pack and unpack data into this format. This functionality should ideally never be needed.

```
int MPI_Pack(const void *inbuf,
          int incount, MPI_Datatype datatype,
          void *outbuf, int outsize, int *position,
          MPI_Comm comm);
int MPI_Unpack(const void *inbuf, int insize, int *position,
          void *outbuf,
          int outcount, MPI_Datatype datatype,
          MPI_Comm comm);
int MPI_Pack_size(int incount, MPI_Datatype datatype,
          MPI_Comm comm, int *size);
```

3.2.21 MPI Concept: Progress

When is (point-to-point) communication that is eventually to happen, for instance, by a pair of correctly matching send and receive operations, actually happening? What are the guarantees in MPI that an application will *progress* and eventually complete? The naïve and expected (non-)answer is "as fast and efficiently as possible for the underlying communication network, but possibly depending on the overall load of the system".

The MPI standard does not prescribe how the communication system (hardware and software) is to be implemented. It only loosely states the *progress rule* that correct communication that could happen eventually should happen, at the very latest when `MPI_Finalize` or some other MPI operation is invoked. This gives a lot of freedom to MPI library implementers, and implementers indeed exploit this freedom. There are three basic implementation alternatives to ensure progress in MPI. Progress can be enforced by either:

1. hardware, meaning communication network and network processor or network interface card,
2. separate progress thread in the MPI runtime system or
3. MPI library calls which interact with the runtime system and advance not yet completed communication operations.

Since MPI library implementations rely, to different extents, on all three mechanisms, it is usually good advice and good practice to make MPI calls regularly in the application to ensure that the communication in the application is progressing.

3.2.22 One-Sided Communication

With the two-sided, point-to-point communication model seen so far, the two communicating processes are both explicitly involved in the communication to take place, one specifying where data to be sent are located in process local memory and how they are structured, the other specifying where the data to be received are to go in that process' local memory and how they are structured. Communication can take place when both processes have posted their respective calls. The processes complete according to the semantics described so far for the communication operations and modes that are being used.

In contrast, with MPI *one-sided communication*, one process alone explicitly initiates the communication. It, therefore, has to specify what is happening at both sides. MPI provides one-sided communication operations for retrieving data (`MPI_Get`) from another process, for transferring data to another process (`MPI_-Put`), for transferring data to and performing an operation (an `MPI_Op`, see later) at another process (`MPI_Accumulate`), as well as a number of special, atomic operations on data at other processes. These communication initiating operations are all nonblocking. The process that initiates the communication operation is, in MPI terms, referred to as the *origin process* and the process to or from which data are transferred or retrieved as the *target process*. In order to ensure that a data transfer has taken place and is completed, whether at origin or at target process, an explicit synchronization must be performed. This can involve both

3.2 Ninth Block (3–4 Lectures)

origin and target processes. With one-sided communication, synchronization is, thus, explicit and decoupled from the communication operation. This was different for point-to-point communication. There, synchronization and completion is coupled to the communication operation, regardless of whether it is blocking or nonblocking. In contrast to point-to-point communication, all one-sided communication calls are *nonblocking* in the MPI sense.

In the distributed memory programming model, processes do not share address spaces in any way. An address (pointer) at one process has no meaning at another process. Thus, means are needed to make it possible for an origin process to address data at a target process. In MPI, processes participating in one-sided communication expose parts of their memory in a special, distributed data structure called a *communication window*. Communication windows are represented by a new kind of handle of type MPI_Win. Data at target processes are referenced by non-negative, relative displacements and translated into actual addresses in the exposed memory. MPI provides the collective MPI_Win_create operation for creating communication windows. In the call, each process supplies the process local address of the memory to expose, the size (in bytes) of the memory it will expose, and a process local displacement unit to be used when translating displacements into addresses. The MPI operations for managing windows and memory are shown below.

```
int MPI_Win_create(void *base, MPI_Aint size, int disp_unit,
                   MPI_Info info, MPI_Comm comm,
                   MPI_Win *win);
int MPI_Win_free(MPI_Win *win);

int MPI_Win_get_group(MPI_Win win, MPI_Group *group);

int MPI_Win_allocate(MPI_Aint size, int disp_unit,
                     MPI_Info info,
                     MPI_Comm comm, void *baseptr,
                     MPI_Win *win);
int MPI_Win_allocate_shared(MPI_Aint size, int disp_unit,
                            MPI_Info info, MPI_Comm comm,
                            void *baseptr, MPI_Win *win);
int MPI_Win_shared_query(MPI_Win win,
                         int rank,
                         MPI_Aint *size, int *disp_unit,
                         void *baseptr);
int MPI_Win_create_dynamic(MPI_Info info, MPI_Comm comm,
                           MPI_Win *win);
int MPI_Win_attach(MPI_Win win, void *base, MPI_Aint size);
int MPI_Win_detach(MPI_Win win, const void *base);
```

Window creation is a collective operation for the processes in the communicator used in the call. This means that all processes in the communicator must eventually call MPI_Win_create. Memory per process that is to be exposed to other processes must have been allocated in advance, either with a C standard memory allocator like malloc() or with a special, dedicated memory-allocator that is defined in the MPI specification and implemented by the library. Using stack allocated data in a communication window is dangerous practice since this memory can disappear before the window is freed: a very subtle source of memory bugs. The rationale for having special allocators is that a HPC system may have special regions of memory that are particularly well suited to one-sided communication, e.g., can be read and written by other processors with special instructions or can be efficiently shared between some MPI processes (e.g., processes on the same shared memory compute node). The special allocator (with its special free operation) makes it possible to enforce the use of such memory regions in a portable way. Window objects should, as always, be freed when no longer used in the application which is done by the collective `MPI_Win_free` call. However, allocated and exposed memory must be freed explicitly; freeing memory is not taken care of by `MPI_Win_free` for windows created with `MPI_Win_create`.

The `MPI_Info` object makes it possible to provide additional information on the use of the communication window to the MPI library. The special `MPI_INFO_NULL` value is always a valid argument and is the only type of MPI "info" that we will consider in these lectures.

```
int MPI_Alloc_mem(MPI_Aint size, MPI_Info info,
                  void *baseptr);
int MPI_Free_mem(void *base);
```

```
int MPI_Win_get_info(MPI_Win win, MPI_Info *info_used);
int MPI_Win_set_info(MPI_Win win, MPI_Info info);
```

The one-sided communication operations are listed below.

```
int MPI_Get(void *origin_addr,
            int origin_count, MPI_Datatype origin_datatype,
            int target_rank, MPI_Aint target_disp,
            int target_count, MPI_Datatype target_datatype,
            MPI_Win win);
int MPI_Put(const void *origin_addr,
            int origin_count, MPI_Datatype origin_datatype,
            int target_rank, MPI_Aint target_disp,
            int target_count, MPI_Datatype target_datatype,
            MPI_Win win);

int MPI_Accumulate(const void *origin_addr,
                   int origin_count,
                   MPI_Datatype origin_datatype,
                   int target_rank, MPI_Aint target_disp,
                   int target_count,
                   MPI_Datatype target_datatype,
                   MPI_Op op, MPI_Win win);
int MPI_Get_accumulate(const void *origin_addr,
                       int origin_count,
                       MPI_Datatype origin_datatype,
                       void *result_addr,
                       int result_count,
                       MPI_Datatype result_datatype,
```

3.2 Ninth Block (3–4 Lectures)

```
                    int target_rank,
                    MPI_Aint target_disp,
                    int target_count,
                    MPI_Datatype target_datatype,
                    MPI_Op op, MPI_Win win);

int MPI_Fetch_and_op(const void *origin_addr,
                     void *result_addr,
                     MPI_Datatype datatype,
                     int target_rank,
                     MPI_Aint target_disp,
                     MPI_Op op, MPI_Win win);
int MPI_Compare_and_swap(const void *origin_addr,
                         const void *compare_addr,
                         void *result_addr,
                         MPI_Datatype datatype,
                         int target_rank,
                         MPI_Aint target_disp,
                         MPI_Win win);
```

The `MPI_Get` and `MPI_Put` calls are the two basic one-sided communication calls. Each specifies data for the operation at the calling *origin process* in the usual form of a base address, an element count, and a datatype that describes the kind and structure of the elements (Sect. 3.2.15 and Sect. 3.2.20). What is to happen at the *target process* is likewise specified with the operation in the form of a relative displacement, an element count, and a datatype. Data at both origin and target processes can be arbitrarily structured, and any predefined or committed user-defined derived datatype can be used for both `origin_datatype` and `target_datatype`. The two datatypes can even be different. However, for a one-sided communication call to be correct, the signature of the data to be transmitted must be a prefix of the signature of the data to be received. Thus, for `MPI_Get`, the sequence of data elements described by `target_count` and `target_datatype` must be a prefix of the data elements described by `origin_count` and `origin_datatype`. For `MPI_Put`, it is the other way around. This is the same as the rule for point-to-point communication (Sect. 3.2.15). As with point-to-point communication, also `MPI_PROC_NULL` can be used as rank for the target process: no communication will take place.

The one-sided communication calls are like the nonblocking point-to-point operations: They only indicate that communication eventually is to take place. When this exactly happens depends on the synchronization mechanisms that will be used and to a very large extent on the MPI library implementation. In order to be able to write provably correct programs, MPI poses strict conditions on which data elements can be written where. These rules, in effect, state that no data element may possibly be (over)written by more than one one-sided communication operation before synchronization has taken place; programs that violate this rule are simply erroneous. As with so many other things in MPI, it is solely the programmer's responsibility to ensure that this cannot happen. Thus, two or more `MPI_Put` operations are not allowed to put any data to the same target address. Two or more `MPI_Get` operations are not allowed to retrieve data to the same origin address. Concurrent `MPI_Get` and `MPI_Put` operations that reference the same address are also not allowed; this situation is a classical *data race*. Different one-sided communication operations cannot be kept separate from each other by means of message tags as was the case for point-to-point communication.

A one-sided communication operation that accesses data at a target process with some displacement `disp`, will access the address

$$\texttt{base} + \texttt{disp} \cdot \texttt{disp\_unit}$$

where both `base` and `disp_unit` are the values provided by the target process in the `MPI_Win_create` call. In standard uses of one-sided communication, all processes give the same `disp_unit`.

The `MPI_Accumulate` call is like an `MPI_Put` operation, but will apply the supplied MPI binary `MPI_Op` operator (see later) on the supplied origin and the stored target elements. The `MPI_Accumulate` operation is an exception to the stated rules: several concurrent operations can update the same elements. Such concurrent updates are performed like atomic operations but are atomic only per element. The `MPI_Get_accumulate` retrieves the old element values from the exposed target memory before doing the accumulation. Only the predefined `MPI_Op` operators and no user-defined operators can be used (think about why this is the case).

The atomic *Fetch-And-Operate (FAO)* and *Compare-And-Swap (CAS)* operations provide atomic operation functionality to MPI and can be used (only) on single elements of a predefined datatype. An efficient MPI library implementation may be able to execute these calls by native, atomic operations, at least under some circumstances.

3.2.23 One-Sided Completion and Synchronization

A one-sided communication operation by itself is nonblocking and neither determines when data are transferred between origin and target processes nor when data will be available at either of the processes. This must be enforced by explicit synchronization operations.

In order to understand, work with, and reason about one-sided communication, MPI employs a so-called *communication epoch* model. From each process' point of view, one-sided communication takes place in disjoint epochs. Epochs are opened and closed by synchronization operations. A process that wants to access the window memory of some other process must open a next epoch for *access* to that process (*access epoch*). A process whose window memory may be accessed by another process must open an epoch for *exposure* to that process (*exposure epoch*).

The MPI one-sided communication model provides two kinds of synchronization operations for opening epochs: With *active synchronization*, both origin and

3.2 Ninth Block (3–4 Lectures)

target processes actively open their respective access and exposure epochs. With *passive synchronization*, the origin process alone will open an epoch for access (at the origin process) and exposure (at the target process). Epochs must be explicitly closed. When an origin process closes its access epoch, all one-sided communication operations will be completed from the origin process' point-of-view. In particular, all data elements retrieved by `MPI_Get` or `MPI_Get_accumulate` operations will be available for use. When a target process closes its exposure epoch, all one-sided communication operations on that target will be complete at the target. In particular, data transferred with `MPI_Put` will be available for use. Operations for closing epochs are thus blocking.

MPI provides two kinds of operations for active synchronization. The `MPI_Win_fence` is a collective operation over all processes belonging to the window. An `MPI_Win_fence` will close a preceding epoch and open an access epoch with access to all other processes and an exposure epoch giving exposure to all other processes for each of the processes. The `MPI_Win_fence` operation has non-local completion semantics and may thus have to wait for other processes to perform their corresponding `MPI_Win_fence` call.

Dedicated, more specific control over access and exposure is provided by the `MPI_Win_start` and `MPI_Win_post` operations. The first provides access to a group of processes (represented as `MPI_Group` objects, see Sect. 3.2.10), the second one grants exposure to a group of processes. Access and exposure epochs are explicitly closed with `MPI_Win_complete` and `MPI_Win_wait`, respectively. The `MPI_Win_test` operation is a nonblocking version of `MPI_Win_wait`. The `MPI_Win_start` operation has non-local completion semantics and may thus have to wait for the processes that are to be accessed to perform their `MPI_Win_post` call. The `MPI_Win_post` operation has local completion semantics. Therefore, in the frequent case where a process both seeks access and grants access to other processes, the `MPI_Win_post` call should be performed before the `MPI_Win_start` call. The other order is *unsafe* and the program may deadlock.

```
int MPI_Win_fence(int assert, MPI_Win win);

int MPI_Win_post(MPI_Group group, int assert,
                 MPI_Win win); // for exposure
int MPI_Win_start(MPI_Group group, int assert,
                  MPI_Win win); // for access
int MPI_Win_complete(MPI_Win win);
int MPI_Win_wait(MPI_Win win);
int MPI_Win_test(MPI_Win win, int *flag);

int MPI_Win_lock(int lock_type, int rank, int assert,
                 MPI_Win win);
int MPI_Win_unlock(int rank, MPI_Win win);

int MPI_Win_lock_all(int assert, MPI_Win win);
int MPI_Win_unlock_all(MPI_Win win);
```

The `MPI_Win_lock` and `MPI_Win_unlock` operations passively open a target exposure epoch and an origin access epoch. A target can be opened for exclusive access by the locking origin process alone by providing the `MPI_LOCK_EXCLUSIVE` lock type. A target can be opened for shared, concurrent access by more than one MPI process by providing the `MPI_LOCK_SHARED` lock type.

These operations have nothing to do with locks (mutexes) in the sense seen so far (see Sect. 2.2.6): They do not and cannot provide mutual exclusion. When a target process is "locked" exclusively, data can indeed be accessed by the one-sided communication operations but since the `MPI_Put` and `MPI_Get` operations are nonblocking, nothing can be done with this data. The exception is, of course, the `MPI_Accumulate` operations. In order to use the data from `MPI_Get`, access and exposure epochs have to be closed by the `MPI_Win_unlock` call. When this happens, another process may come between and "lock" the target and change the data. Read (get), compute and update (put) under mutual exclusion is, thus, not provided.

3.2.24 Example: One-Sided Stencil Updates

As an example we now implement the stencil update that we saw before with blocking and nonblocking point-to-point communication using one-sided communication instead. For this, a window is created from the Cartesian communicator that was created for defining the neighborhoods, see Sect. 3.2.14. An advantage of this implementation over the point-to-point implementations is flexibility: It could be that not all four neighbors have to be updated in some iteration. With the one-sided implementation, the corresponding `MPI_Get` calls could simply be dropped. We here give a full-fledged implementation also using a vector datatype (see Sect. 3.2.20).

First, we implement the stencil update with active, collective `MPI_Win_fence` synchronization for opening access and exposure epoch on all processes, for all processes.

```
int left, right;
int up,   down;

MPI_Cart_shift(cartcomm,1,1,&left,&right);
MPI_Cart_shift(cartcomm,0,1,&up,  &down);

MPI_Datatype column;

MPI_Type_vector(m,1,n+2,MPI_DOUBLE,&column);
MPI_Type_commit(&column);

double *out_left, *out_right, *out_up, *out_down;
double *in_left, *in_right, *in_up, *in_down;

out_left  = &matrix[0][0];
out_right = &matrix[0][n-1];
out_up    = &matrix[0][0];
out_down  = &matrix[m-1][0];
```

3.2 Ninth Block (3–4 Lectures)

```
in_left   = &matrix[0][-1];
in_right  = &matrix[0][n];
in_up     = &matrix[-1][0];
in_down   = &matrix[m][0];

MPI_Win win;

MPI_Win_create((double*)matrix-(n+2+1),
               (m+2)*(n+2)*sizeof(double),sizeof(double),
               MPI_INFO_NULL,cartcomm,&win);

int disp_left, disp_right, disp_up, disp_down;

disp_left  = (n+2)+n;
disp_right = (n+2)+1;
disp_up    = m*(n+2)+1;
disp_down  = (n+2)+1;

int done = 0;
while (!done) { // iterate until convergence
    ... // the stencil update (computation)

  MPI_Win_fence(MPI_MODE_NOPRECEDE,win);

  MPI_Get(in_left,    1,column,left,
          disp_left,  1,column,win);
  MPI_Get(in_right,   1,column,right,
          disp_right,1,column,win);
  MPI_Get(in_up,      n,MPI_DOUBLE,up,
          disp_up,    n,MPI_DOUBLE,win);
  MPI_Get(in_down,    n,MPI_DOUBLE,down,
          disp_down,  n,MPI_DOUBLE,win);

  MPI_Win_fence(MPI_MODE_NOSUCCEED,win);
  // data available

  done = 1; // some termination criterion
}

MPI_Win_free(&win);
MPI_Type_free(&column);
```

Because of the collective nature of the MPI_Win_fence operations, the processes are "more synchronized" than needed. Each process needs to access window memory of its at most four neighboring processes and likewise provide exposure to these processes. For such situations, the dedicated synchronization mechanism could be more efficient, providing a looser form of synchronization.

```
int neighbors[4];

MPI_Group group;
MPI_Group accessexposure;
```

```
MPI_Comm_group(cartcomm,&group);
int k = 0;
if (left !=MPI_PROC_NULL)   neighbors[k++] = left;
if (right!=MPI_PROC_NULL)   neighbors[k++] = right;
if (up   !=MPI_PROC_NULL)   neighbors[k++] = up;
if (down !=MPI_PROC_NULL)   neighbors[k++] = down;
MPI_Group_incl(group,k,neighbors,&accessexposure);

int done = 0;
while (!done) { // iterate until convergence
    ... // the stencil update (computation)

  MPI_Win_post(accessexposure,0,win);
  MPI_Win_start(accessexposure,0,win);

  MPI_Put(out_left,   1,column,left,
          disp_left,  1,column,win);
  MPI_Put(out_right,  1,column,right,
          disp_right,1,column,win);
  MPI_Put(out_up,     n,MPI_DOUBLE,up,
          disp_up,    n,MPI_DOUBLE,win);
  MPI_Put(out_down,   n,MPI_DOUBLE,down,
          disp_down,  n,MPI_DOUBLE,win);

  MPI_Win_complete(win);
  MPI_Win_wait(win);

  done = 1; // some termination criterion
}
```

3.2.25 Example: Distributed Memory Binary Search

The following binary search example illustrates a situation where one-sided communication is a more suitable model than two-sided point-to-point communication with MPI_Send and MPI_Recv. The situation here is that a process needs data from some other process, which is, however, not aware of that need. One-sided communication makes it possible for the process that knows to alone do the communication!

Let a be a distributed array with local blocks, all of the same size n. Assume that the distributed array is ordered: Within each process local block, the elements are ordered, and the elements of the block of some process are smaller than or equal to the elements of the local block of the next (higher ranked) process. The total number of array elements over all processes is n*p where p is the number of processes in our communication window. We want to do binary search in such an array. Each process should be allowed to initiate a search for some element x. The result shall be a global index i, such that $a[i] \leq x < a[i+1]$. From the global index, the process where the element x was found and the relative index in the block of that process can easily be computed.

3.2 Ninth Block (3–4 Lectures)

In the code, we assume that the window `win` has already been created and that the local `a` arrays are of C type `float`.

```
int l, u, m;
int target, locali;

float ma;

l = -1; u = n*p; // total size of distributed array

do {
  m =(l+u)/2;

  target = m/n; // locate middle element
  locali = m%n;

  MPI_Win_lock(MPI_LOCK_SHARED,target,0,win);
  MPI_Get(&ma,1,MPI_FLOAT,
          target,locali,1,MPI_FLOAT,win); // get middle
  MPI_Win_unlock(target,win);

  if (x<ma) u = m; else l = m;
} while (l+1<u);
```

Binary search takes $O(\log n)$ iterations in each of which the searching process passively synchronizes (with `MPI_Win_lock`) with a target process, which is determined by dividing the index m to be accessed with the block size. The displacement to be accessed is the index modulo the block size. Since the target process only reads elements, `MPI_LOCK_SHARED` exposure at the target is sufficient and can allow other MPI processes to search concurrently.

Merging by co-ranking can be implemented by similar considerations. It is a good exercise to do this.

3.2.26 Additional One-Sided Communication Operations*

The one-sided communication model provides communication operations that return an `MPI_Request` object that can be used for individually testing or enforcing completion of that operation, similar to the nonblocking point-to-point communication operations. They are listed here for completeness.

```
int MPI_Rput(const void *origin_addr,
             int origin_count, MPI_Datatype origin_datatype,
             int target_rank, MPI_Aint target_disp,
             int target_count, MPI_Datatype target_datatype,
             MPI_Win win, MPI_Request *request);
int MPI_Rget(void *origin_addr,
             int origin_count, MPI_Datatype origin_datatype,
             int target_rank, MPI_Aint target_disp,
             int target_count, MPI_Datatype target_datatype,
             MPI_Win win, MPI_Request *request);
```

```
int MPI_Raccumulate(const void *origin_addr,
                    int origin_count,
                    MPI_Datatype origin_datatype,
                    int target_rank, MPI_Aint target_disp,
                    int target_count,
                    MPI_Datatype target_datatype,
                    MPI_Op op,
                    MPI_Win win,MPI_Request *request);
int MPI_Rget_accumulate(const void *origin_addr,
                        int origin_count,
                        MPI_Datatype origin_datatype,
                        void *result_addr,
                        int result_count,
                        MPI_Datatype result_datatype,
                        int target_rank,
                        MPI_Aint target_disp,
                        int target_count,
                        MPI_Datatype target_datatype,
                        MPI_Op op,
                        MPI_Win win, MPI_Request *request);
```

3.2.27 MPI Concept: Collective Semantics

So far, we have seen many examples of MPI operations that are collective in the sense that they have to be called by all processes belonging to the input communicator. More concretely, if a collective operation C on a communicator comm is called by some process in comm, then all other processes in comm must also eventually call C and no other collective operation C' before C on comm. By this rule, for each communicator the application programmer must ensure that all collective calls are done in the same order by all processes in the communicator. As with other calls and operations in MPI, disregarding this rule and doing something else is plain wrong and the outcome undefined. Concretely, this means that any behavior is possible: deadlock, memory corruption, immediate program crash, and even successful completion with apparently sensible results. The latter is the most misleading and dangerous behavior!

Collective operations like C are always invoked *symmetrically*. That is, the same function C is called by all processes, but the processes can give different parameters, and the arguments can have a different meaning on the different processes (see shortly). For all collectives, arguments must be given *consistently* over the calling processes. This means different things for different collectives. For instance, for the MPI_Comm_create collective operation (see Sect. 3.2.7), there are rules for the input group arguments, namely that all processes that belong to a group given as input by some process must call with an equivalent group argument. Recall that groups are process local objects; in the collective call, all processes in the group must have created a group for the same set of processes in the same order. Disregarding such rules on consistent arguments is erroneous. There is no guarantee on how an MPI library may react (deadlock, crash, weird results, ...).

Here are two further examples illustrating the consistency rules, anticipating the collective operations to be discussed in the next section. The MPI_Bcast operation broadcasts a buffer of some number of elements from a root process to all other processes in the communicator. It is a consistency requirement that all processes specify the same root process and exactly the same number of elements (adhering to the type signature rules). In the first example (below), the non-root processes inadvertently give a larger element count than the root process. The program may well run with some MPI libraries, but the outcome will sooner or later prove fatal: the last, fourth element in the dims array has never been received by the non-root processes. The dims[3] element could be anything.

```
MPI_Comm_rank(comm,&rank);
MPI_Comm_size(comm,&size);

if (rank==root)
  MPI_Bcast(&dims[0],3,MPI_INT,root,comm);
} else {
  MPI_Bcast(&dims[0],4,MPI_INT,root,comm);
}
```

In the second example, the non-roots give the fixed root value 0 for the fourth argument of the MPI_Bcast call. The consistency requirement for MPI_Bcast is, however, that all processes must give the same value for the root argument. The program will most likely hang with most MPI libraries when root is *not* process 0 in the communicator (deadlock!).

```
MPI_Comm_rank(comm,&rank);
MPI_Comm_size(comm,&size);

if (rank==root)
  MPI_Bcast(&dims[0],4,MPI_INT,root,comm);
} else {
  MPI_Bcast(&dims[0],4,MPI_INT,0,comm);
}
```

Note that the broadcast collective operation is invoked symmetrically by all processes making their MPI_Bcast call. The different, asymmetric outcomes for the different processes (root and non-roots) are determined by the supplied input arguments. In contrast, point-to-point communication (and also one-sided communication) is *non-symmetric*: There are distinct send operations, like MPI_-Send, MPI_Isend, ..., and different, distinct receive operations, like MPI_Recv and MPI_Irecv. A process determines its role (sender or receiver) in the communication operation by the appropriate, different calls. One-sided communication is the extreme case: Only one side makes a communication call at all. Also some of the synchronization operations are asymmetric. A sometimes seen beginner's mistake is to try to perform a broadcast by letting the root process call MPI_-

Bcast while non-roots try to get the data by calling MPI_Recv. Such programs (almost) never work (if they do, by luck).

Collective operations seen so far, and also those that will be introduced in the next section, are all *blocking* in the MPI sense. When a process returns from a collective call C, the operation has been completed from that process' point of view. All resources needed for the call have been given free by the call and can be reused. In collective operations for exchanging information between processes, this in particular means that data are out of the process' send buffers, and have been delivered in its receive buffers. Send buffers can again be used freely to store new data for the following communication operations, and values in receive buffers can be used for computation by the process.

Like for point-to-point communication, also some nonblocking collective operations have been defined in MPI. The semantic rules are slightly different than those for nonblocking point-to-point communication. Nonblocking collective operations are beyond these lectures (some will be mentioned for completeness, though, see Sect. 3.2.32).

Blocking collective operations have *non-local completion*. This means (as for point-to-point communication) that for a process to complete a collective call, it may require, and in most cases does require (!), that the other processes in the communicator actively engage in the operation. The rules for correct usage of collective operations exactly ensure that for any collective call C made by some process eventually all processes in the communicator will have made the collective call to C. At the latest at that point, C can be completed for the processes.

On the other hand, collective operations are and should indeed be thought of as *non-synchronizing* by the application programmer. A process returning from its blocking collective call C cannot make any inference about what any of the other processes have done or not done. Some processes may not even have reached the point in their code where they perform the C call! There is one conspicuous, obvious exception to this rule (think ahead).

A program using collective operations that relies on synchronizing behavior or makes any such assumptions is called *unsafe*. We stress again: Unsafe programming is a pernicious practice. An unsafe program may well run under some circumstances (MPI library, system, number of compute-nodes, problem size, ...) and then suddenly not run anymore (or produce wrong results) when circumstances change. Unsafe programs are non-portable programs!

3.2.28 Collective Communication and Reduction Operations

Collective communication in MPI, the third important communication model (for thoughtful amusement, see [46]), more specifically refers to the small set of 17 functions or patterns for data exchange and reductions over all processes in a communicator (see Sect. 1.3.12 and Sect. 1.3.14). These 17 collective operations are what is commonly meant by the term (MPI) *collectives*.

3.2 Ninth Block (3–4 Lectures) 249

The MPI collectives are broadly of the following types (see Sect. 1.3.12):

- A *barrier operation* ensures that all processes have reached a certain point in their execution.
- A *broadcast operation* transfers the same data from one designated process to all other processes.
- A *gather operation* collects data from all processes on one designated process.
- A *scatter operation* transfers different, individual data from one designated process to each of the other processes.
- An *allgather operation*, also known as all-to-all broadcast, gathers the same data to all processes, or, equivalently, broadcasts data from each process to all other processes.
- An *all-to-all operation*, also known as *personalized exchange* or *transpose*, transfers different, individual data from each process to each of the other processes.
- A *reduction operation* applies a binary, associative operator in order to data contributed by all processes and makes the result available to one or all processes in total or in part.
- A *scan operation* performs a prefix sums computation in rank order on data contributed by the processes.

The designated process for the broadcast, gather and scatter operations is called the *root process* or just *root*. The operations exist in different variants according to the amount of data that are supplied and collected by the processes. Variants where each process either receives or sends the same amount of data to other processes are called *regular*. Variants where different pairs of processes may send and/or receive amounts of data that are different from other processes' amounts are called *irregular*. For historical reasons, the irregular variants of the MPI collective operations are sometimes (but not always) called "vector" (and "vee") variants. Data are always specified as blocks of elements, each block by a count and (derived) datatype argument. It is sometimes helpful, especially for the reduction and scan operations, to think of input and output as mathematical vectors of elements, most often of the same, basic datatype like MPI_INT, MPI_FLOAT, etc.

The usage of the terms is not always consistent and different people sometimes mean different things or use different words. It is maybe helpful as a mnemonic to classify the collectives based on the regularity of data exchanged and whether some process has a special role: Regular vs. irregular ("vector") and rooted (asymmetric) vs. non-rooted (symmetric). See Table 3.2 for such a classification using the names given to the collective operations by MPI.

The performance and concrete implementation of the collectives are, as for everything else in MPI, *not* specified by the MPI standard. In order to say something about what can be expected, in particular, to make performance predictions, assumptions have to be imposed from the outside.

Time complexities of the regular collectives in a simple, homogeneous, linear-cost transmission model (see Sect. 3.1.3) on fully connected networks with one-ported communication capabilities with p processors and total data m are as

Table 3.2 Classification of the MPI collectives along the dimensions of pairwise data regularity and rootedness (symmetry)

	Regular MPI_Barrier	Irregular (vector)
Rooted (asymmetric)	MPI_Bcast MPI_Gather MPI_Scatter MPI_Reduce	MPI_Gatherv MPI_Scatterv
Non-rooted (symmetric)	MPI_Allgather MPI_Alltoall MPI_Allreduce MPI_Reduce_scatter_block MPI_Reduce_scatter MPI_Scan MPI_Exscan	MPI_Allgatherv MPI_Alltoallv MPI_Alltoallw

Table 3.3 Time complexity of the MPI collective operations in the linear-cost communication model under fully-connected network (and one-ported) communication assumptions. The total problem size is m and the number of processes p

Collective	Time complexity $T_p(m)$
MPI_Barrier	$O(\log p)$
MPI_Bcast	$O(m + \log p)$
MPI_Gather	$O(m + \log p)$
MPI_Scatter	$O(m + \log p)$
MPI_Allgather	$O(m + \log p)$
MPI_Alltoall	Between $O(m + p)$ and $O(m \log p)$
MPI_Reduce	$O(m + \log p)$
MPI_Allreduce	$O(m + \log p)$
MPI_Reduce_scatter_block	$O(m + \log p)$
MPI_Scan	$O(m + \log p)$
MPI_Exscan	$O(m + \log p)$

stated in Table 3.3. On networks that are not fully connected, having diameter larger than one (see Sect. 3.1.1), the time complexities are as stated in Table 3.4. Finding the algorithms that achieve these bounds is not at all trivial. A good starting point for the interested reader is [26] and [24] with interesting trade-offs for all-to-all communication. For collective algorithms, it is important that the dominating terms in the upper bound, which often correspond to the number of communication rounds or critical path length, have small constants. Analyzing (and improving) these constant terms is important.

The interface specifications for the regular communication/data exchange collectives are listed below. The MPI_Barrier operation is special: It does not communicate any data but has the sole effect of logically synchronizing the processes.

3.2 Ninth Block (3–4 Lectures)

Table 3.4 Time complexity of the MPI collective operations in the linear-cost communication model under non-fully connected network assumptions. The total problem size is m and the number of processes p and the network diameter d

Collective	Time complexity $T_p(m)$
MPI_Barrier	$O(d)$
MPI_Bcast	$O(m+d)$
MPI_Gather	$O(m+d)$
MPI_Scatter	$O(m+d)$
MPI_Allgather	$O(m+d)$
MPI_Alltoall	$O(m+pd)$
MPI_Reduce	$O(m+d)$
MPI_Allreduce	$O(m+d)$
MPI_Reduce_scatter_block	$O(m+d)$
MPI_Scan	$O(m+d)$
MPI_Exscan	$O(m+d)$

All processes in the communicator must eventually call the barrier operation, and no process is allowed to return from this blocking call before all other processes have made their call to MPI_Barrier. This is the only collective with synchronizing behavior where a process that returns from its call can infer and rely on (all) other processes also having made the call. For all other blocking collectives, the return from a call by a process means only that the operation has been completed from that process' point of view. It is not possible to infer anything about the other processes in general and some may not even have made the corresponding call. Relying on synchronizing behavior of collectives is *unsafe programming* and can lead to unpleasant surprises with errors that can be very hard to debug.

We now give the signatures for and discuss the individual MPI collectives.

```
int MPI_Barrier(MPI_Comm comm);

int MPI_Bcast(void *buffer, int count, MPI_Datatype datatype,
              int root, MPI_Comm comm);

int MPI_Gather(const void *sendbuf,
               int sendcount, MPI_Datatype sendtype,
               void *recvbuf,
               int recvcount, MPI_Datatype recvtype,
               int root, MPI_Comm comm);
int MPI_Scatter(const void *sendbuf,
                int sendcount, MPI_Datatype sendtype,
                void *recvbuf,
                int recvcount, MPI_Datatype recvtype,
                int root, MPI_Comm comm);
int MPI_Allgather(const void *sendbuf,
                  int sendcount, MPI_Datatype sendtype,
```

```
                    void *recvbuf,
                    int recvcount, MPI_Datatype recvtype,
                    MPI_Comm comm);
int MPI_Alltoall(const void *sendbuf,
                    int sendcount, MPI_Datatype sendtype,
                    void *recvbuf,
                    int recvcount, MPI_Datatype recvtype,
                    MPI_Comm comm);
```

For the `MPI_Bcast` operation, the designated *root process* (the process with rank equal to `root`) transfers the data stored at the address `buffer` to the other processes in the communicator used in the call. Data consists of `count` elements of type and structure described by the `datatype` argument. The processes can give different datatype and count arguments, but all processes must specify *exactly* the same type signature: the same lists of elements of a basic datatype. The collective rule is, thus, stricter than the signature rules for point-to-point and one-sided communication. Also, all processes must give the same value for the `root` argument; if they do not, a deadlock is likely to occur. Whether this actually happens depends on the concrete MPI library implementation and on the context of the call.

For the other collectives, similar rules apply. Data leaving a process are specified in the send buffer arguments and data to be received by a process in the receive buffer arguments. Again, signatures between processes where a data transfer is to take place must be *identical*. For the rooted collectives, all processes must give the same root argument. In all the collective operations, count and datatype arguments together describe one *block* of data. As we will see, some collectives send and/or receive p blocks of data on communicators with p processes. It is solely the programmers responsibility to ensure that in such cases, enough memory space has been allocated. Forgetting this is a common mistake, the consequence of which is almost always memory corruption and program crash.

The `MPI_Gather` operation collects different data from all p processes to the designated root process. The data to be stored at the root process are stored starting at the `recvbuf` address. The data from each process will consist of `recvcount` elements, all of the type and structure described by the `recvtype` (derived) datatype. The data from the processes are stored in *rank order*, with the data from process i at the address

$$\texttt{recvbuf} + i \cdot \texttt{recvcount} \cdot \texttt{extent}$$

where `extent` is the extent in bytes of the `recvtype` datatype, as can be found by the `MPI_Type_get_extent` call (explained Sect. 3.2.15). Thus, p blocks with the same structure are received, offset from the address given by `recvbuf`. The `recvbuf` must have been allocated large enough for these p blocks. The data that a process contributes are stored starting at the `sendbuf` address. Each process contributes `sendcount` elements of type and structure given by `sendtype`. Each process' send signature must be identical to the signature of the received data. For all non-root processes, the receive buffer arguments are not significant. All processes contribute data to the root, including the root itself! The data from the root to the root are stored at the address `recvbuf + root · recvcount · extent`. This may incur a memory copy operation at the root process. Such a perhaps costly (perhaps not) memory copy can be avoided by letting the root process give the special address argument `MPI_IN_PLACE` for the `sendbuf` argument. This means that the root process does not care about data from itself and nothing will be copied into the `recvbuf` from the root. Many other collective operations have the same "problem", and the `MPI_IN_PLACE` argument can be applied in many cases.

The `MPI_Scatter` operation is the counterpart (some say "dual") of the `MPI_Gather` operation. Data blocks stored at the root process in rank order in `sendbuf` are transmitted to the other processes from this buffer. The data for process i are stored at the address

$$\texttt{sendbuf} + i \cdot \texttt{sendcount} \cdot \texttt{extent}$$

where `extent` is the extent (in bytes) of the `sendtype` (derived) datatype. Same rules and considerations as for `MPI_Gather` apply, including the caveat on sufficient buffer space. Also here, the `MPI_IN_PLACE` argument can be given as the `recvbuf` argument at the root to prevent that data are copied from the send buffer to the receive buffer at the root.

Here is an example illustrating the use of the `MPI_Gather` collective together with derived datatypes. An $m \times (np)$ matrix is to be put together from column submatrices of n columns (out of np columns in total) at the root process. This is done by gathering the column submatrices at the root. It is a good exercise to recap the extent rules for `MPI_Gather` and figure out why it is necessary to modify the extent of the receive datatype (by creating a new datatype with the `MPI_Type_create_resized` operation, see Sect. 3.2.20).

```
double (*matrix)[n];
matrix = (double(*)[n])malloc(m*n*size*sizeof(double));

MPI_Datatype vec, cols;
MPI_Type_vector(m,n,n*size,MPI_DOUBLE,&vec);
MPI_Type_create_resized(vec,0,n*sizeof(double),&cols);
MPI_Type_commit(&cols);

double (*fullmatrix)[size*n];
if (rank==root) {
  fullmatrix =
    (double(*)[size*n])malloc(m*n*size*sizeof(double));
}

MPI_Gather(matrix,m*n,MPI_DOUBLE,fullmatrix,1,cols,root,
           comm);

MPI_Type_free(&vec);
MPI_Type_free(&cols);

free(matrix);
if (rank==root) free(fullmatrix);
```

The `MPI_Allgather` operation has the same effect as if each process would be the root in an `MPI_Gather` operation and would send the same data in each of these `MPI_Gather` operations; that is, the same effect as p (the number of MPI processes in the communicator) `MPI_Gather` operations with root arguments $i = 0, \ldots, p - 1$ and the same other arguments. Equivalently, `MPI_Allgather` has the same effect as if each process $i, 0 \leq i < p$ would copy its data from its `sendbuf` to the address `recvbuf` $+ \, i \cdot$ `recvcount` $\cdot$ `extent` and perform a broadcast operation from this buffer of `recvcount` elements described by the `recvtype` datatype. In other words, data from all processes are gathered in rank order by all processes. The `MPI_IN_PLACE` argument can be used to indicate that the data from a process are already in the correct position in that process' `recvbuf`. Thus, the copy operation above could be saved. The MPI rules for `MPI_IN_PLACE` for `MPI_Allgather` are strict, though, and require that if some process give `MPI_IN_PLACE` as `sendbuf` argument, then all processes must do so.

Finally, in the `MPI_Alltoall` operation, each process has individual ("personalized") data to transmit to each other process. The data for process $i, 0 \leq i < p$ are stored starting from address

$$\texttt{sendbuf} + i \cdot \texttt{sendcount} \cdot \texttt{sendextent}$$

and the data from process j are received and stored starting at address

$$\texttt{recvbuf} + j \cdot \texttt{recvcount} \cdot \texttt{recvextent} \quad .$$

The data sent to each process consist of `sendcount` elements of type and structure described by `sendtype`, and the data received of `recvcount` elements are as described by `recvtype`. As can be seen, the `MPI_Alltoall` operation has the same effect as p `MPI_Scatter` operations with roots $i = 0, \ldots, p - 1$ or as p `MPI_Gather` operations with roots $i = 0, \ldots, p - 1$. For completeness, we mention that the `MPI_IN_PLACE` argument can also be used with `MPI_Alltoall`, but with a quite different meaning and flavor: The `MPI_IN_PLACE` argument can be given for the `sendbuf` argument in which cases data are sent from and received (replaced) in the same `recvbuf` address (in rank order). If used, all processes must call with the `MPI_IN_PLACE` argument.

For the gather, scatter, allgather, and all-to-all operations, also so-called irregular or "vector" variants are defined in MPI. The interface specifications for these irregular communication/data exchange collectives are listed below.

```
int MPI_Gatherv(const void *sendbuf,
                int sendcount, MPI_Datatype sendtype,
                void *recvbuf,
                const int recvcounts[],
                const int recvdispls[],
                MPI_Datatype recvtype, int root,
                MPI_Comm comm);
int MPI_Scatterv(const void *sendbuf,
                const int sendcounts[],
                const int senddispls[],
```

3.2 Ninth Block (3–4 Lectures)

```
                 MPI_Datatype sendtype,
                 void *recvbuf,
                 int recvcount, MPI_Datatype recvtype,
                 int root,
                 MPI_Comm comm);

int MPI_Allgatherv(const void *sendbuf,
                 int sendcount, MPI_Datatype sendtype,
                 void *recvbuf,
                 const int recvcounts[],
                 const int recvdispls[],
                 MPI_Datatype recvtype,
                 MPI_Comm comm);

int MPI_Alltoallv(const void *sendbuf,
                 const int sendcounts[],
                 const int senddispls[],
                 MPI_Datatype sendtype,
                 void *recvbuf,
                 const int recvcounts[],
                 const int recvdispls[],
                 MPI_Datatype recvtype,
                 MPI_Comm comm);
int MPI_Alltoallw(const void *sendbuf,
                 const int sendcounts[],
                 const int senddispls[],
                 const MPI_Datatype sendtypes[],
                 void *recvbuf,
                 const int recvcounts[],
                 const int recvdispls[],
                 const MPI_Datatype recvtypes[],
                 MPI_Comm comm);
```

Each of these operations perform the same kind of communication/data exchange operations as their regular counterpart, but the amount of data contributed can vary between processes. For instance, the MPI_Gatherv operation transfers data from all processes to a given root process. Data to be transferred are specified by the send buffer argument triple (sendbuf, sendcount, and sendtype) and the processes may, in contrast to the MPI_Gather operation, specify different numbers of elements to be transferred. The root process has a vector (hence the "vector" suffix v to these operations) of counts where recvcounts[i] specifies the count of elements (of type recvtype) from process i. The send signature of process i specified by the sendcount and sendtype arguments must be identical to the signature at the root process given by recvcounts[i] and recvtype. At the root the data are gathered starting at memory address recvbuf. More precisely, the data from process i are stored starting at address

$$\text{recvbuf} + \text{recvdispls}[i] \cdot \text{extent}$$

where `extent` is the extent (in bytes) of the `recvtype` derived datatype. Thus, the displacement vector `recvdispls` is the relative offset or displacement of the data from each process in units of the extent of the receive type.

The `MPI_Scatterv`, `MPI_Allgatherv`, and `MPI_Alltoallv` operations are similar. Where several blocks of data are to be transferred to other processes, there are `sendcounts` and send `senddispls` vectors in the argument lists, and where several data blocks are to be transferred from other processes there are `recvcounts` and receive `recvdispls` vectors in the argument lists. There are a single send and a single receive datatype argument, `sendtype` and `recvtype`, respectively, describing the type and structure of all data sent or received. The `MPI_Alltoallw` operation is different in this respect. This special collective has a separate datatype argument for each data block to and from each of the other processes.

Using irregular collectives can be tedious. Assume a root process has to gather different amounts of data from the other processes, like the column vector `MPI_-Gather` application above, but now with possibly different numbers of columns for each process. The root may, however, not know in advance how much data it is going to receive from each of the other processes. Since the `MPI_Gatherv` collective needs the `recvcounts` and `recvdispls` vectors to be set up correctly, the element counts must first be collected from all processes. For this, the regular `MPI_Gather` operation can be used. So, first the element counts are gathered at the root and stored in the `recvcounts` vector, based on which appropriate displacements are computed (in the example, data are stored consecutively, but this must not necessarily always be so). Finally, the data can be correctly collected with the `MPI_Gatherv` operation.

```
// gather counts from all processes
MPI_Gather(&sendcount,1,MPI_INT,recvcounts,1,MPI_INT,root,
          comm);
if (rank==root) {
  // compute displacements, on root only
  recvdispls[0] = 0;
  // data to be received consecutively (prefix sums)
  for (i=1; i<size; i++) {
    recvdispls[i] = recvdispls[i-1]+recvcounts[i-1];
  }
}
// gather the possibly different amounts of data
MPI_Gatherv(sendbuf,sendcount,sendtype,recvbuf,
            recvcounts,recvdispls,recvtype,root,comm);
```

The `MPI_IN_PLACE` argument can also be used for the irregular communication/data exchange collectives. Sometimes, this is convenient, and it can sometimes even give a performance benefit.

The *reduction collectives* additionally perform computation on the data supplied by the processes making the collective call. Here, it is convenient to think of the processes as supplying vectors of some count number of elements of a basic datatype (like `MPI_INT`, `MPI_FLOAT`, `MPI_LONG`, `MPI_DOUBLE`, etc.), although derived datatypes can be used in some circumstances. These vectors are reduced

3.2 Ninth Block (3–4 Lectures)

element by element using a binary operator supplied in the call and result in a result vector with the same number of elements. The interface specifications for the reduction type collectives are listed below.

```
int MPI_Reduce(const void *sendbuf, void *recvbuf,
               int count, MPI_Datatype datatype,
               MPI_Op op, int root, MPI_Comm comm);
int MPI_Allreduce(const void *sendbuf, void *recvbuf,
                  int count, MPI_Datatype datatype,
                  MPI_Op op, MPI_Comm comm);
int MPI_Reduce_scatter_block(const void *sendbuf,
                             void *recvbuf,
                             int recvcount,
                             MPI_Datatype datatype,
                             MPI_Op op, MPI_Comm comm);
int MPI_Reduce_scatter(const void *sendbuf, void *recvbuf,
                       const int recvcounts[],
                       MPI_Datatype datatype,
                       MPI_Op op, MPI_Comm comm);
int MPI_Scan(const void *sendbuf, void *recvbuf,
             int count, MPI_Datatype datatype,
             MPI_Op op, MPI_Comm comm);
int MPI_Exscan(const void *sendbuf, void *recvbuf,
               int count, MPI_Datatype datatype,
               MPI_Op op, MPI_Comm comm);
```

Let $\oplus$ be an associative, binary operator operating elementwise on vectors x and y with the same number of elements c. The reduction collective operations perform a reduction like

$$z = x_0 \oplus x_1 \oplus \cdots \oplus x_{p-1}$$

where x_i is the vector supplied by MPI process i and p the number of processes. Brackets can be left away due to associativity; $x \oplus (y \oplus z) = (x \oplus y) \oplus z$. Operators are not assumed to be commutative but many commonly used operators are commutative $(+, \max, \ldots)$. If operator $\oplus$ is commutative, then

$$z = x_0 \oplus x_1 \oplus \cdots \oplus x_{p-1}$$
$$= x_{\pi(0)} \oplus x_{\pi(1)} \oplus \cdots \oplus x_{\pi(p-1)}$$

for any permutation $\pi : \{0, \ldots, p-1\} \to \{0, \ldots p-1\}$. This can possibly be exploited by the reduction algorithms underlying an MPI library implementation and sometimes is. However, reductions are preferred to be performed in *rank order* and MPI libraries normally try to respect this as far as possible. The special MPI query operation `MPI_Op_commutative` can be used to find out whether a given (user-defined) operator is commutative.

MPI provides a number of predefined operators working on vectors of basic datatypes stored consecutively in send and receive buffers with a count of ele-

Table 3.5 Binary operators for collective reduction operations

Operator	MPI
Sum	MPI_SUM
Product	MPI_PROD
Minimum	MPI_MIN
Maximum	MPI_MAX
Logical (wordwise) and, or, exclusive or	MPI_LAND, MPI_LOR, MPI_LXOR
Bitwise and, or, exclusive or	MPI_BAND, MPI_BOR, MPI_BXOR
Minimum with location	MPI_MINLOC
Maximum with location	MPI_MAXLOC

ments. Operators are identified by the MPI_Op handle. It is also possible for the application programmer to define own operators by attaching a function with a predefined signature to an operator handle, but this is beyond the scope of these lectures. The standard MPI operators are listed in Table 3.5. All these operators are (mathematically) commutative and associative.

In the reduction and scan collectives, all processes must give the same MPI_Op argument, otherwise the results are undefined (as can be imagined). All processes must give input vectors with the same number of elements (of the same basic datatype).

Elementwise binary reduction by some operator $\oplus$ on two input vectors of c elements means, for instance, that

$$\begin{pmatrix} x_{c-1} \\ \vdots \\ x_1 \\ x_0 \end{pmatrix} + \begin{pmatrix} y_{c-1} \\ \vdots \\ y_1 \\ y_0 \end{pmatrix} = \begin{pmatrix} x_{c-1} + y_{c-1} \\ \vdots \\ x_1 + y_1 \\ x_0 + y_0 \end{pmatrix}$$

for when $\oplus$ is the $+$ operator MPI_SUM, and

$$\min\left\{\begin{pmatrix} x_{c-1} \\ \vdots \\ x_1 \\ x_0 \end{pmatrix}, \begin{pmatrix} y_{c-1} \\ \vdots \\ y_1 \\ y_0 \end{pmatrix}\right\} = \begin{pmatrix} \min\{x_{c-1}, y_{c-1}\} \\ \vdots \\ \min\{x_1, y_1\} \\ \min\{x_0, y_0\} \end{pmatrix}$$

when $\oplus$ is the minimum operator MPI_MIN.

The reduction collectives differ in the way the output vector is stored. For the MPI_Reduce operation, which takes a root argument, the computed c-element result vector z is stored in the receive buffer at the root. The recvbuf argument is significant only for the root process. For the MPI_Allreduce operation, all processes receive the computed result z in their respective receive buffers. With the MPI_Reduce_scatter_block and MPI_Reduce_scatter operations, the c-element result vector z is split into subvectors $z^0, z^1, \ldots z^{p-1}$ of $c_0, c_1, \ldots c_{p-1}$

3.2 Ninth Block (3–4 Lectures)

elements, respectively, with $c = \sum_{i=0}^{p-1} c_i$, and the vector z_i stored in the receive buffer at process i. For MPI_Reduce_scatter_block, all c_i are equal and so subvectors have the same number of elements given by a single count argument, whereas for MPI_Reduce_scatter the c_i counts are stored in the input vector recvcounts with recvcounts[i] = c_i. All processes must give the same recvcounts vector as input. The MPI_Reduce_scatter operation is the irregular ("vector" variant) and MPI_Reduce_scatter_block the regular variant of this collective operation (see Table 3.2). The MPI_IN_PLACE argument can be given as sendbuf argument in some cases. For MPI_Reduce, the root process (only) can specify that the input vector is to be taken from the recvbuf address where the result of the reduction is also stored by giving MPI_IN_PLACE as sendbuf argument. For MPI_Allreduce, MPI_Reduce_scatter_block, and MPI_Reduce_scatter, if one process gives the MPI_IN_PLACE argument, then all processes must give the MPI_IN_PLACE argument.

A simple, but very common application of collective reduction operations is checking for agreement on some Boolean outcome. Say, all processes need to agree on some convergence criterion which follows from all processes having locally satisfied some criterion. Agreement can be checked by performing a reduction with a Boolean (logical) "and" operation and then making sure that all processes receive the result. The case could occur in a stencil computation, which is iterated until convergence by all processes is reached (see Sect. 1.3.5). It could be implemented with an MPI_Allreduce operation with the logical "and" operation MPI_LAND; the MPI_IN_PLACE argument is convenient here.

```
while (!done) {
  ... // the stencil update (computation)

  int k = 0;
  MPI_Isend(out_left,c,MPI_DOUBLE,left,TAG,cartcomm,
            &request[k++]);
  MPI_Isend(out_right,c,MPI_DOUBLE,right,TAG,cartcomm,
            &request[k++]);
  MPI_Isend(out_up,c,MPI_DOUBLE,up,TAG,cartcomm,
            &request[k++]);
  MPI_Isend(out_down,c,MPI_DOUBLE,down,TAG,cartcomm,
            &request[k++]);

  MPI_Irecv(in_left,c,MPI_DOUBLE,right,TAG,cartcomm,
            &request[k++]);
  MPI_Irecv(in_right,c,MPI_DOUBLE,left,TAG,cartcomm,
            &request[k++]);
  MPI_Irecv(in_up,c,MPI_DOUBLE,down,TAG,cartcomm,
            &request[k++]);
  MPI_Irecv(in_down,c,MPI_DOUBLE,up,TAG,cartcomm,
            &request[k++]);

  MPI_Waitall(k,request,MPI_STATUSES_IGNORE);

  done = 1; // some local convergence criterion
  MPI_Allreduce(MPI_IN_PLACE,&done,1,MPI_INT,MPI_LAND,
```

```
                cartcomm);
  // global agreement, same number of iterations
}
```

The two scan collective operations MPI_Scan and MPI_Exscan implement the *inclusive prefix sums* and *exclusive prefix sums* operations (elementwise, on c-element vectors), respectively, see Sect. 1.4.5. The ith elementwise inclusive or exclusive prefix sum is stored at process i. Processes can use the MPI_IN_PLACE argument to indicate that input is to be taken from the recvbuf address (where the result is also placed).

An important, later addition to MPI, is the capability to locally apply a binary operator on two input vectors. The operator can be any of the predefined MPI_Op operators (or even a user-defined operator). This local operation is shown below; the second argument is both the second input and the address where the result is stored. This is sometimes convenient and sometimes not; there is (unfortunately) no three-argument version $a = b + c$ of this local operation in MPI [117].

```
int MPI_Reduce_local(const void *inbuf, void *inoutbuf,
                     int count, MPI_Datatype datatype,
                     MPI_Op op);
int MPI_Op_commutative(MPI_Op op, int *commute);
```

Below is an implementation of a $p - 1$ communication round algorithm for MPI_Scan, which illustrates the use of MPI_Reduce_local. A copy of the input in the receive buffer to the send buffer is needed and implemented by an MPI_Sendrecv operation, where each process sends the input data to itself [117]. Here, this operation is done on the special MPI_COMM_SELF communicator which is a predefined singleton communicator that consists of the process itself only. This copy would be unnecessary if the MPI_IN_PLACE argument had been given to the MPI_Scan operation.

```
MPI_Sendrecv(sendbuf,c,MPI_FLOAT,0,SCANTAG,
             recvbuf,c,MPI_FLOAT,0,SCANTAG,MPI_COMM_SELF,
             MPI_STATUS_IGNORE);
if (rank>0) {
  MPI_Recv(tempbuf,c,MPI_FLOAT,rank-1,SCANTAG,comm,
           MPI_STATUS_IGNORE);
  MPI_Reduce_local(tempbuf,recvbuf,c,MPI_FLOAT,MPI_SUM);
}
if (rank<size-1) {
  MPI_Send(recvbuf,c,MPI_FLOAT,rank+1,SCANTAG,comm);
}
```

The algorithm is linear in the number of MPI processes and not fast. It is a good exercise to consider in which ways the algorithm is inefficient (cost) and how it can be improved.

For MPI_COMM_SELF, the following holds.

```
int rank, size;

MPI_Comm_rank(MPI_COMM_SELF,&rank);
MPI_Comm_size(MPI_COMM_SELF,&size);
```

```
assert(size==1);
assert(rank==0);
```

As mentioned, it is possible for the application programmer to define and register own, binary functions as `MPI_Op` operations. The functionality for this is listed below.

```
int MPI_Op_create(MPI_User_function *user_fn, int commute,
                  MPI_Op *op);
int MPI_Op_free(MPI_Op *op);
```

3.2.29 Complexity and Performance of Applications with Collective Operations

Many message-passing algorithms can be expressed entirely or almost entirely in terms of collective operations [46]. This often holds for algorithms following a loosely Bulk Synchronous Parallel pattern described in Sect. 1.3.9: A sequence of steps where processes perform local computations followed by collective operations that summarize and redistribute data over the processes. Level-wise Breadth-First Search (BFS) would follow this pattern. At each level, processes locally explore new vertices reachable from vertices from the current level and then either exchange vertices (by one of the `MPI_Alltoall` collectives) or computes the set of vertices for the next level (by `MPI_Allreduce` or other reduction collective).

Such algorithms could be analyzed in terms of the local computations and the collective operations performed. Assume than an input graph $G = (V, E)$ with n vertices and m edges is given, with depth K from the given start vertex $s \in V$. An implementation maintaining a set of new vertices for each level, represented as a bit-map, would take K iterations and should do at most m vertex updates in total over the K iterations. By a good distribution of the edges E over the processes this can presumably be parallelized to take $O(m/p)$ time steps in total with the p processes. At the end of each iteration, an `MPI_Allreduce` operation is done, for a total time of

$$O(m/p) + K\ T_{\texttt{MPI\_Allreduce}(p)}(n)$$

where $T_{\texttt{MPI\_Allreduce}(p)}(n)$ denotes the time of an `MPI_Allreduce` operation on n-element input sets on a communicator with p processes. Collective operations are expensive so the analysis could therefore focus on the exact number of these operations. The actual runtime and complexity of the collectives, for instance as stated in Table 3.3 and Table 3.4, would not need to be known. Since MPI does not make any performance guarantees or prescribes specific algorithms for the collective operations, the actual complexity for some MPI library cannot be known a priori. Note that there is no claim made that the outlined BFS implementation is in any way best possible or even a (very) good one.

3.2.30 Examples: Elementary Linear Algebra

Matrix–vector multiplication and matrix–matrix multiplication are two elementary operations in linear algebra. The collective operations we have seen in the preceding sections are convenient for solving these problems in parallel without relying on shared memory access to the input and output matrices and vectors.

In such operations, the input matrices and vectors are distributed in some way over the available processes. The output is likewise distributed over the processes in some (possibly other) way. The distribution of input and output should be considered part of the *problem specification* and an algorithm/implementation for solving any such problem must respect the prescribed distribution. If the distribution is different, either another algorithm must be developed, or the distribution must be changed (by some algorithm). Distributions are most often balanced, meaning that with p processes, each process will posses $1/p$ of the total input and compute $1/p$ of the total output. Often, somewhat complex (block cyclic, see 1.3.10) distributions have to be used to achieve a good load balance between the processes. It is obvious that no efficient, parallel algorithm can be allowed to gather the full input or the full output (Amdahl's Law).

We first give two implementations of algorithms for performing matrix–vector multiplication for two different input and output distributions. The full input is a real-valued (double) $m \times n$ matrix M and a real-valued n element vector x. The output is a real-valued m element vector y with $y = Mx$. For simplicity, we assume that p, the number of processes, divides both m and n. It is, of course, a good exercise to generalize the implementations to arbitrary input sizes m and n.

In the first example, the input matrix is distributed row-wise, meaning that each process has m/p full, consecutive rows of the matrix M. Process 0 the first such m/p rows, process 1 the next m/p rows, and so on. The input vector x is likewise distributed in pieces of n/p consecutive elements. The output vector y is to be distributed in the same manner with m/p consecutive elements per process.

Let M_i be the $(m/p) \times n$ part of the matrix of process i. The part of the output for process i can be computed as $y_i = M_i x$. In order to do this computation, the full x vector must be available at all processes which can be accomplished with an `MPI_Allgather` operation. The rest is easy.

```
MPI_Comm_rank(comm,&rank);
MPI_Comm_size(comm,&size);

assert(m%size==0); // regular only
assert(n%size==0);

double *fullvector;
fullvector = (double*)malloc(n*sizeof(double));

MPI_Allgather(vector,n/size,MPI_DOUBLE,
              fullvector,n/size,MPI_DOUBLE,comm);
for (i=0; i<m/size; i++) {
```

3.2 Ninth Block (3–4 Lectures)

```
    result[i] = matrix[i][0]*fullvector[0];
    for (j=1; j<n; j++) {
      result[i] += matrix[i][j]*fullvector[j];
    }
  }
  free(fullvector);
```

The run time complexity of this first algorithm can easily be analyzed as follows. As stated in Table 3.3, the allgather operation can be done in

$$T_{\texttt{MPI\_Allgather}(p)}(n) = O(n + \log p)$$

time. The process local matrix–vector product computation takes $O((m/p)n)$ time, for a total of $O((m/p)n + n + \log p)$ time steps. This is cost-optimal with p in $O(m)$ processors if we assume that $n > \log p$ since sequential matrix–vector multiplication takes $O(mn)$ time steps.

In the second example, the input matrix is distributed column-wise, meaning that each process has n/p consecutive columns with m rows of the matrix M. Process 0 the first such n/p columns, process 1 the next n/p columns, and so on. The input vector x is likewise distributed in pieces of n/p consecutive elements. The output vector y is to be distributed in the same manner with m/p consecutive elements per process.

Let M'_i be the $m \times (n/p)$ part of the matrix of process i. The full output vector y can be computed as $y = \sum_{i=0}^{p-1} M'_i x_i$ and then be distributed into the parts y_i of m/p consecutive elements per process. The summation and subsequent distribution of the parts can be accomplished by an MPI_Reduce_scatter_block operation.

```
MPI_Comm_rank(comm,&rank);
MPI_Comm_size(comm,&size);

assert(m%size==0); // regular only
assert(n%size==0);

double *partial;
partial = (double*)malloc(m*sizeof(double));

for (i=0; i<m; i++) {
  partial[i] = matrix[i][0]*vector[0];
  for (j=1; j<n/size; j++) {
    partial[i] += matrix[i][j]*vector[j];
  }
}

MPI_Reduce_scatter_block(partial,result,m/size,MPI_DOUBLE,
                         MPI_SUM,comm);
free(partial);
```

The run time complexity of the second algorithm can easily be analyzed as follows: The process local work for the initial matrix–vector multiplication is in $O(m(n/p))$. According to Table 3.3, the reduce-scatter operation can be done in

$$T_{\texttt{MPI\_Reduce\_scatter}(p)}(m) = O(m + \log p)$$

time, for a total of $O(m(n/p) + m + \log p)$ time steps. This is cost-optimal for p in $O(n)$ processors if we assume that $m > \log p$.

Summarizing, we have found the following.

Theorem 16. *Matrix–vector multiplication of an $m \times n$ matrix with an n element vector can be done work-optimally on a p processor distributed memory system with message-passing communication in $O(mn/p + \min(m,n) + \log p)$ time steps.*

Which of the two algorithms performs better in practice depends on the actual quality of the implementation of the `MPI_Allgather` and `MPI_Reduce_scatter_block` operations, and on the magnitude of m and n. Keep in mind that the two algorithms assume different distributions of the input matrix! A more scalable algorithm, one for which more processors can be employed with linear speed-up, can be given by combining the two ideas (with a different distribution of the input). It is a good exercise to extend the two algorithms to work also for the case where p divides neither m nor n. The irregular collectives `MPI_Allgatherv` and `MPI_Reduce_scatter` will be of help and actually do most of the (conceptual) work.

The more challenging operation to perform without having the matrices stored in shared memory and being accessible to every thread (process) is matrix–matrix multiplication. Given an $m \times l$ input matrix A, an $l \times n$ input matrix B, compute the $m \times n$ output matrix C as $C = AB$. For simplicity, we assume that the number of processes p is a square (which is not entirely without loss of generality), that is $p = \sqrt{p}\sqrt{p}$ for an integer $\sqrt{p}$, and that $\sqrt{p}$ divides all of m, l, n. The input distribution is balanced such that each process has input submatrices of $(m/\sqrt{p}) \times (l/\sqrt{p}) = ml/p$ and $(l/\sqrt{p}) \times (n/\sqrt{p}) = ln/p$ elements, respectively. The algorithm produces an output submatrix of $(m/\sqrt{p}) \times (n/\sqrt{p}) = mn/p$ elements for each of the p processes.

We organize the processes in a quadratic, 2-dimensional mesh and give each processor a coordinate (i, j), for instance, by creating a Cartesian communicator with `MPI_Cart_create` as shown in Sect. 3.2.8. The submatrices for process i, j are denoted by A_{ij}, B_{ij} and C_{ij}, respectively. Each output submatrix C_{ij} is computed straight ahead by

$$C_{ij} = \sum_{k=0}^{\sqrt{p}-1} A_{ik} B_{kj}$$

We observe that on each row of processes, the same A_{ik} submatrices and on each column of processes, the same B_{kj} submatrices are needed by all processes. This can be accomplished by $\sqrt{p}$ broadcast operations on the rows and on the columns of processes. To implement this conveniently with MPI, communicators for the processes in same the rows and the same columns are needed. Fortunately, creating communicators for processes with the same row coordinate and

3.2 Ninth Block (3-4 Lectures)

processes with the same column coordinate can be done with the proper `MPI_Comm_split` operations. Naturally, this potentially expensive communicator creation should be done once and for all (and reused over many matrix–matrix multiplications). The initial communicator (with a square number of processes) is `comm`.

```
MPI_Comm_size(comm,&size);

int rc[2];       // row-column factorization
int period[2];
int coords[2];   // coordinates of process
int reorder;

rc[0] = 0; rc[1] = 0;
MPI_Dims_create(size,2,rc);
assert(rc[0]==rc[1]);  // number of processes must be square

period[0] = 0;
period[1] = 0;
reorder = 0;

MPI_Cart_create(comm,2,rc,period,reorder,&cartcomm);
MPI_Cart_coords(cartcomm,rank,2,coords);

MPI_Comm_split(cartcomm,coords[0],0,&rowcomm);
MPI_Comm_split(cartcomm,coords[1],0,&colcomm);

int rowrank, colrank;
MPI_Comm_rank(rowcomm,&rowrank);
MPI_Comm_rank(colcomm,&colrank);
assert(rowrank==coords[1]);
assert(colrank==coords[0]);
```

The matrix–matrix multiplication can now easily be implemented as shown in the code below. The row and column communicators are `rowcomm` and `colcomm`. The multiplication and summation of submatrices is done by an efficient, sequential implementation which we have black-box encoded in the fused-matrix–multiply-add procedure `fmma()`. We assume that the matrices are represented by a pointer to an array of the elements in row-major order.

```
int rowsize, colsize;
MPI_Comm_size(rowcomm,&rowsize);
MPI_Comm_size(colcomm,&colsize);
assert(rowsize==colsize);  // size is square

double *Atmp, *Btmp;
// allocate space for temporary matrices

int i;
for (i=0; i<rowsize; i++) {
  double *AA, *BB;

  AA = (i==rowrank) ? A : Atmp;
```

```
MPI_Bcast(AA,m/rowsize*l/rowsize,MPI_DOUBLE,i,rowcomm);

BB = (i==colrank) ? B : Btmp;
MPI_Bcast(BB,l/rowsize*n/rowsize,MPI_DOUBLE,i,colcomm);

fmma(C,AA,BB,m/rowsize,l/rowsize,n/rowsize);
}
```

The parallel running time of the matrix–matrix multiplication implementation can be analyzed as follows: As building block, a sequential matrix–matrix multiplication algorithm is used. We assume it takes $M(m,l,n)$ operations to multiply an $m \times l$ matrix with an $l \times n$ matrix. The cost of adding two matrices is asymptotically much smaller. The algorithm performs $2\sqrt{p}$ MPI_Bcast operations of matrices with $(ml)/p$ and $(ln)/p$ elements, respectively. According to Table 3.3, this can be done in

$$O(\sqrt{p}\frac{ml+ln}{p} + \log\sqrt{p}) = O(\frac{l(m+n)}{\sqrt{p}} + \log p)$$

time steps. The number of process local matrix–matrix multiplications is $\sqrt{p}$, each of which takes $M(m/\sqrt{p}, l/\sqrt{p}, n/\sqrt{p})$ time steps. The sequential matrix–matrix multiplication algorithm we have seen takes $M(m,l,n) = O(mln)$ steps. Using this algorithm gives

$$\sqrt{p}\, O((m/\sqrt{p})(l/\sqrt{p})(n/\sqrt{p})) = O(\frac{mln}{p})$$

with linear speed-up for the multiplication work.

Summarizing, with the standard sequential matrix–matrix multiplication algorithm as plug-in, we have the following:

Theorem 17. *Matrix–matrix multiplication can be done in $O(mln/p + l(m+n)/\sqrt{p} + \log p)$ time steps on a p processor system with message-passing communication which is cost-optimal compared to a sequential $M(m,l,n) = O(mln)$ matrix–matrix multiplication algorithm.*

Speed-up is linear as long as p is in $O((\frac{mn}{m+n})^2)$, assuming that both the first and second term dominate the last $\log p$ term.

This algorithm for matrix–matrix multiplication doing broadcast operations on rows and columns of processes (and improvements thereof) is called SUMMA (Scalable Universal Matrix Multiplication Algorithm) [45].

3.2.31 Examples: Sorting Algorithms

The Quicksort algorithm idea lends itself well also to parallel implementation by point-to-point and collective communication. There are two natural variants. As in the preceding lectures, we assume that good pivots can be found by some means, which is of course crucial for both the theoretical and practical performance. We, however, ignore this aspect here and leave it to others, for instance [9, 10, 92, 94].

For a distributed memory implementation, we assume that the input data (elements from some totally ordered set, like integers, floating point numbers, objects, etc.) have been evenly distributed over the available processes. For input of n elements in total, each process will, thus, have (approximately) n/p elements. The elements are to be globally sorted in such a way that, preferably, each process will have approximately n/p elements of the output. The output must fulfill that, for each process, the elements in the process' part of the output are sorted, and that the elements of process i are all larger than or equal to the elements of process $i-1$ (for $i>0$) and smaller than or equal to the elements of process $i+1$ (for $i<p-1$).

For the parallel Quicksort, we assume that the number of processes p is a power of two, $p=2^k$ for some $k, k\geq 0$. We formulate the algorithm recursively, but recurse on the number of processes. Each recursive call will split (exactly) its number of processes into two halves until one process is left with some array of elements to sort sequentially. An implementation for $p, p>1$ MPI processes in a communicator comm would go as follows.

1. Select a global pivot for the n elements and distribute this pivot to all p processes.
2. Processes locally partition their set of elements into elements smaller than or equal to the global pivot and elements larger than or equal to the global pivot.
3. The processes pairwise exchange elements, such that half the processes will have elements smaller than or equal to the global pivot and the other half of processes will have element larger than or equal to the global pivot. Concretely, this will be done such that processes with rank $i, i<p/2$ will have the smaller elements, and processes $i, i\geq p/2$ will have the larger elements.
4. The communicator comm with the p processes is split into two communicators with processes smaller than $p/2$ and processes larger than or equal to $p/2$, respectively.
5. Each process recursively calls Quicksort on the new communicator of $p/2$ processes to which it belongs.

With only one process, $p=1$, a sequential Quicksort is used to sort the process' $n/p=n$ elements. With such an implementation and a best known implementation of sequential Quicksort, absolute and relative speed-up of the implementation will coincide.

Step 1 will most likely involve one or more collective operations, e.g., MPI_-Bcast. For the local computation in Step 2, a best known sequential implementation for partitioning (in-place) should be used. See, for instance [97, 98, 100]. We note that the global pivot for the processes may actually not be in the set of input elements for any one process. For Step 3, point-to-point communication is used, for example like this (for elements of C type double):

```
double *a;   // input elements
double *b;   // temporary array for communication

int n;       // size of local array a
int nn;      // local pivot index from partition function
int nl, ns;  // number of larger and smaller elements

int half = size/2;
if (rank<half) {
  // will receive elements smaller than pivot
  nl = n-nn;
  MPI_Sendrecv(&nl,1,MPI_INT,rank+half,QTAG,
               &ns,1,MPI_INT,rank+half,QTAG,
               comm,MPI_STATUS_IGNORE);
  n = nn+ns;
  b = (double*)malloc(n*sizeof(double));
  assert(n==0||b!=NULL);

  MPI_Sendrecv(a+nn,nl,MPI_DOUBLE,rank+half,QTAG,
               b+nn,ns,MPI_DOUBLE,rank+half,QTAG,
               comm,MPI_STATUS_IGNORE);
  memcpy(b,a,nn*sizeof(double));
} else {
  // will receive elements larger than pivot
  ns = nn;
  MPI_Sendrecv(&ns,1,MPI_INT,rank-half,QTAG,
               &nl,1,MPI_INT,rank-half,QTAG,
               comm,MPI_STATUS_IGNORE);
  n = n-nn+nl;
  b = (double*)malloc(n*sizeof(double));
  assert(n==0||b!=NULL);

  MPI_Sendrecv(a,ns,MPI_DOUBLE,rank-half,QTAG,
               b,nl,MPI_DOUBLE,rank-half,QTAG,
               comm,MPI_STATUS_IGNORE);
  memcpy(b+nl,a+ns,(n-nl)*sizeof(double));
}
// split communicator and recurse
// free(b); when done
```

In Step 2, the partitioning function has locally partitioned the a array and computed the pivot index nn that separates larger and smaller elements. The processes exchange elements pairwise. The processes with ranks smaller than $p/2$ are to receive the smaller elements, while the higher ranked processes are to receive the larger elements. The first MPI_Sendrecv operation exchanges the number of small and large elements needed for this, based on which the temporary communication array b can be allocated. The element exchange itself is now done by the second MPI_Sendrecv operation. The elements for each process for the recursive call are in the newly allocated b array. Some care has to be taken to make sure such intermediate arrays are properly freed.

3.2 Ninth Block (3–4 Lectures)

Step 4 is again a typical case for the `MPI_Comm_split` operation. This may introduce overhead that can affect overall performance, and it may be worthwhile to consider whether explicit communicator splitting can be avoided.

Assuming that pivots are selected perfectly and lead to even partitions at all levels of the recursions, the running time can be asymptotically estimated with the following recurrence relation. The $O(\log p)$ term is for the collective operations for pivot selection and the $O(n/p)$ term for the element exchange.

$$T(n, p) = O(\log p) + O(n/p) + T(n/2, p/2)$$
$$T(n, 1) = O(n \log n)$$

Since $(n/2)/(p/2) = n/p$, each level of the recursion will contribute the $O(n/p)$ term, and since $\log_2 p$ recursive calls are needed (p is a power of two), the solution is

$$T(n, p) = O(\log^2 p) + (\log_2 p)O(n/p) + O(n/p \log(n/p))$$
$$= O(\log^2 p) + O(\frac{n \log p}{p} + \frac{n \log n - n \log p}{p}))$$
$$= O(\log^2 p) + O((n/p) \log n)$$

with linear speed-up when n is sufficiently large compared to $\log_2^2 p$.

For well-behaved inputs and pivot selection, this implementation can work well in practice, but it does not guarantee that the output is balanced as blocks of n/p elements per process. It is a good exercise to consider how bad the algorithm can behave, and how worst-case inputs may look, also under different assumptions on the pivot selection.

Another common parallel Quicksort implementation variant, which is sometimes referred to as *HyperQuicksort* [121], is to let the processes first sort their n/p elements; this makes perfect pivot selection per process trivial. It possibly also makes it easier to find a good overall pivot. Local arrays are kept sorted through the recursive calls, and in order to maintain sorted order, a merge step is needed after the element exchange. These variants, and others that rely solely on collective communication operations for exchanging data are discussed further and implemented in [115]. A drawback of Quicksort as implemented here is that the number of processes must be a power of two. This is quite a restriction, and it is worthwhile thinking about whether this can be alleviated.

A completely different idea for sorting (non-negative) integers is *counting sort* (or *bucket sort*) which can also be given a parallel, distributed memory implementation. Stable counting sort is a building block in *radix sort*. Given input of n elements (with integer keys), the idea is to count the number of occurrences of each key by using the keys as indices into a counting array. After counting, the counting array can be used to reserve space for buckets in the right (increasing) order of the right sizes for each of the occurring keys. Finally, the elements are put into their corresponding buckets. This can all be done in time proportional to the key range and the number of elements n. When the key range is no larger than $O(n)$ this is linear in $O(n)$.

In a distributed memory setting, each process will have n/p of the elements available. Processes locally compute the sizes of the buckets. For each bucket, the processes must all know the total number of elements for that bucket. This can be computed by an allreduce operation over the `bucketsize` vectors. Each process must also know, for each bucket, how many elements on smaller ranked processes will go into that bucket. This is a natural application of an exclusive prefix sums computation, again over the computed bucket sizes. Here is a part of such a counting sort (bucket sort) implementation.

```
int n = ...;  // key range, number of buckets
int bucketsize[n];
int allsize[n];  // global size of buckets
int presize[n];  // bucketsizes in smaller ranked processes

// local counting
for (i=0; i<n; i++) bucketsize[i] = 0;
for (i=0; i<n; i++) bucketsize[key[i]]++;

MPI_Allreduce(bucketsize,allsize,n,MPI_INT,MPI_SUM,comm);
MPI_Exscan    (bucketsize,presize,n,MPI_INT,MPI_SUM,comm);
```

The counts in the `presize` and `allsize` vectors can now be used to compute which elements are to be sent to other processes and how many elements each process has to receive from other processes. The final element exchange can be done with `MPI_Alltoall` and `MPI_Alltoallv` operations. To complete, local sorting or reordering is needed. It is a good exercise to try to implement this idea in detail.

3.2.32 Nonblocking and Persistent Collective Operations*

The 17 standard collectives explained in the last section are all blocking in the MPI semantic sense. Recent additions to MPI are a whole set of corresponding, nonblocking and also persistent collective operations. Nonblocking collectives are not part of the material of these lectures, but the operations are listed here for completeness. The operations complete "immediately", irrespective of any action taken by the other processes in the communicator (which is what nonblocking means). They return an `MPI_Request` object that can be used to query for and enforce completion of any given operation, just as was the case with the nonblocking point-to-point communication operations (Sect. 3.2.17).

A highly important difference to nonblocking point-to-point communication is that blocking and nonblocking collectives cannot be combined in the sense that some processes invoke the blocking variant and other processes the nonblocking variant and expect a sensible outcome. The reason for this is that blocking and nonblocking implementations may use (completely) different algorithms.

3.2 Ninth Block (3–4 Lectures)

Therefore, the steps taken by a process doing a broadcast with MPI_Ibcast may not match the steps taken by another process doing the broadcast with MPI_Bcast.

The nonblocking, regular exchange operations are the following.

```
int MPI_Ibarrier(MPI_Comm comm, MPI_Request *request);

int MPI_Ibcast(void *buffer,
               int count, MPI_Datatype datatype,
               int root,
               MPI_Comm comm, MPI_Request *request);
int MPI_Igather(const void *sendbuf,
                int sendcount, MPI_Datatype sendtype,
                void *recvbuf,
                int recvcount, MPI_Datatype recvtype,
                int root,
                MPI_Comm comm, MPI_Request *request);
int MPI_Iscatter(const void *sendbuf,
                 int sendcount, MPI_Datatype sendtype,
                 void *recvbuf,
                 int recvcount, MPI_Datatype recvtype,
                 int root,
                 MPI_Comm comm, MPI_Request *request);
int MPI_Iallgather(const void *sendbuf,
                   int sendcount, MPI_Datatype sendtype,
                   void *recvbuf,
                   int recvcount, MPI_Datatype recvtype,
                   MPI_Comm comm, MPI_Request *request);
int MPI_Ialltoall(const void *sendbuf,
                  int sendcount, MPI_Datatype sendtype,
                  void *recvbuf,
                  int recvcount, MPI_Datatype recvtype,
                  MPI_Comm comm, MPI_Request *request);
```

The nonblocking, regular reduction collectives are the following.

```
int MPI_Ireduce(const void *sendbuf, void *recvbuf,
                int count, MPI_Datatype datatype,
                MPI_Op op, int root, MPI_Comm comm,
                MPI_Request *request);
int MPI_Iallreduce(const void *sendbuf, void *recvbuf,
                   int count, MPI_Datatype datatype,
                   MPI_Op op, MPI_Comm comm,
                   MPI_Request *request);
int MPI_Ireduce_scatter_block(const void *sendbuf,
                              void *recvbuf,
                              int recvcount,
                              MPI_Datatype datatype,
                              MPI_Op op, MPI_Comm comm,
                              MPI_Request *request);

int MPI_Iscan(const void *sendbuf, void *recvbuf,
              int count, MPI_Datatype datatype, MPI_Op op,
              MPI_Comm comm, MPI_Request *request);
int MPI_Iexscan(const void *sendbuf, void *recvbuf,
```

```
                 int count, MPI_Datatype datatype, MPI_Op op,
                 MPI_Comm comm, MPI_Request *request);
```

The irregular, nonblocking data exchange operations are the following.

```
int MPI_Igatherv(const void *sendbuf,
                 int sendcount, MPI_Datatype sendtype,
                 void *recvbuf,
                 const int recvcounts[],
                 const int recvdispls[],
                 MPI_Datatype recvtype,
                 int root, MPI_Comm comm,
                 MPI_Request *request);
int MPI_Iscatterv(const void *sendbuf,
                 const int sendcounts[],
                 const int senddispls[],
                 MPI_Datatype sendtype,
                 void *recvbuf,
                 int recvcount, MPI_Datatype recvtype,
                 int root, MPI_Comm comm,
                 MPI_Request *request);
int MPI_Iallgatherv(const void *sendbuf,
                 int sendcount, MPI_Datatype sendtype,
                 void *recvbuf,
                 const int recvcounts[],
                 const int senddispls[],
                 MPI_Datatype recvtype,
                 MPI_Comm comm, MPI_Request *request);
int MPI_Ialltoallv(const void *sendbuf,
                 const int sendcounts[],
                 const int senddispls[],
                 MPI_Datatype sendtype,
                 void *recvbuf,
                 const int recvcounts[],
                 const int recvdispls[],
                 MPI_Datatype recvtype,
                 MPI_Comm comm, MPI_Request *request);
int MPI_Ialltoallw(const void *sendbuf,
                 const int sendcounts[],
                 const int senddispls[],
                 const MPI_Datatype sendtypes[],
                 void *recvbuf,
                 const int recvcounts[],
                 const int recvdispls[],
                 const MPI_Datatype recvtypes[],
                 MPI_Comm comm, MPI_Request *request);
```

Finally, there is the single, irregular nonblocking reduce-scatter operation.

```
int MPI_Ireduce_scatter(const void *sendbuf, void *recvbuf,
                    const int recvcounts[],
                    MPI_Datatype datatype, MPI_Op op,
                    MPI_Comm comm, MPI_Request *request);
```

A nonblocking communicator duplicate operation is also included in MPI.

3.2 Ninth Block (3-4 Lectures)

```
int MPI_Comm_idup(MPI_Comm comm,
                  MPI_Comm *newcomm, MPI_Request *request);
```

The repertoire of nonblocking collective operations in MPI may grow with time. A most recent addition was for instance a complete set of 17 persistent collective operations (see Sect. 3.2.19). These operations have (almost) the same interfaces as the corresponding nonblocking operations and binds all input parameters in a request object that can be (re)used as many times as desired. We do not list the operation interfaces here.

3.2.33 Sparse Collective Communication: Neighborhood Collectives*

A recent addition to MPI is a number of collective communication operations that perform data exchanges not over all processes but only among subsets of the processes. These so-called *neighborhood collectives* are not treated in these lecture notes, but the functionality is mentioned here for completeness.

The idea of sparse (in contrast to dense), neighborhood collective communication is that each process can perform a data exchange operation with a small set of neighboring processes. What a neighboring process is, is defined by defining the set of neighborhoods, collectively, for all processes. In Sect. 3.2.8, two ways of defining neighborhoods by creating new communicators with associated neighborhoods were discussed, in detail `MPI_Cart_create` and briefly touched upon `MPI_Dist_graph_create`.

The collective operations on sparse neighborhoods are of the allgather and all-to-all type and come in both regular and irregular variants, as well as in blocking and nonblocking (and persistent) variants. All neighborhood collectives are strictly collective, that is they have to be called by all processes in the communicators, and no synchronization behavior is implied.

Note that the signatures of these operations are identical to those of the standard collective operations. This can be helpful for remembering how these functions look and what they do [116].

The regular, blocking and nonblocking variants are listed below.

```
int MPI_Neighbor_allgather(const void *sendbuf,
                           int sendcount,
                           MPI_Datatype sendtype,
                           void *recvbuf,
                           int recvcount,
                           MPI_Datatype recvtype,
                           MPI_Comm comm);
int MPI_Neighbor_alltoall(const void *sendbuf,
                          int sendcount,
                          MPI_Datatype sendtype,
                          void *recvbuf,
                          int recvcount,
```

```
                        MPI_Datatype recvtype,
                        MPI_Comm comm);

int MPI_Ineighbor_allgather(const void *sendbuf,
                        int sendcount,
                        MPI_Datatype sendtype,
                        void *recvbuf,
                        int recvcount,
                        MPI_Datatype recvtype,
                        MPI_Comm comm,
                        MPI_Request *request);
int MPI_Ineighbor_alltoall(const void *sendbuf,
                        int sendcount,
                        MPI_Datatype sendtype,
                        void *recvbuf,
                        int recvcount,
                        MPI_Datatype recvtype,
                        MPI_Comm comm,
                        MPI_Request *request);
```

The irregular ("vector"), blocking and nonblocking variants are listed below.

```
int MPI_Neighbor_allgatherv(const void *sendbuf,
                        int sendcount,
                        MPI_Datatype sendtype,
                        void *recvbuf,
                        const int recvcounts[],
                        const int recvdispls[],
                        MPI_Datatype recvtype,
                        MPI_Comm comm);
int MPI_Neighbor_alltoallv(const void *sendbuf,
                        const int sendcounts[],
                        const int senddispls[],
                        MPI_Datatype sendtype,
                        void *recvbuf,
                        const int recvcounts[],
                        const int recvdispls[],
                        MPI_Datatype recvtype,
                        MPI_Comm comm);
int MPI_Neighbor_alltoallw(const void *sendbuf,
                        const int sendcounts[],
                        const MPI_Aint senddispls[],
                        const MPI_Datatype sendtypes[],
                        void *recvbuf,
                        const int recvcounts[],
                        const MPI_Aint recvdispls[],
                        const MPI_Datatype recvtypes[],
                        MPI_Comm comm);

int MPI_Ineighbor_allgatherv(const void *sendbuf,
                        int sendcount,
                        MPI_Datatype sendtype,
                        void *recvbuf,
                        const int recvcounts[],
                        const int recvdispls[],
```

3.2 Ninth Block (3–4 Lectures)

```
                           MPI_Datatype recvtype,
                           MPI_Comm comm,
                           MPI_Request *request);
int MPI_Ineighbor_alltoallv(const void *sendbuf,
                            const int sendcounts[],
                            const int senddispls[],
                            MPI_Datatype sendtype,
                            void *recvbuf,
                            const int recvcounts[],
                            const int recvdispls[],
                            MPI_Datatype recvtype,
                            MPI_Comm comm,
                            MPI_Request *request);
int MPI_Ineighbor_alltoallw(const void *sendbuf,
                            const int sendcounts[],
                            const MPI_Aint senddispls[],
                            const MPI_Datatype sendtypes[],
                            void *recvbuf,
                            const int recvcounts[],
                            const MPI_Aint recvdispls[],
                            const MPI_Datatype recvtypes[],
                            MPI_Comm comm,
                            MPI_Request *request);
```

3.2.34 MPI and Threads*

MPI can be and often is used together with thread interfaces like OpenMP or pthreads. The idea is, for systems with shared memory multi-core nodes that are interconnected by a communication network, to let cores on the shared memory node compute as threads and let only a single or a few MPI processes on the shared memory node perform communication with processes on other nodes using MPI. This is a two-level, heterogeneous, hierarchical, programming model. A limited number of processes per shared memory node communicate with other processes using MPI and threads inside the processes use a thread model to compute in parallel. The threads are the active entities inside the processes. Therefore, such a two-level model raises the question which threads can or are allowed to perform MPI operations (in order to avoid race conditions, deadlocks or other deadly issues)?

MPI answers the question by defining the level of thread support that an MPI library implementation can provide. There are four defined levels of thread support. With MPI_THREAD_SINGLE, only a single thread is allowed to execute which essentially means that thread parallel programming cannot be used! With MPI_THREAD_FUNNELED threads can be used, but only a designated, single *main* or *master* thread can perform MPI calls. With MPI_THREAD_SERIALIZED all threads are allowed to perform MPI calls, but only one at a time. It is the programmer's responsibility to ensure that this is the case, for instance, by using critical sections and other mechanisms provided by the thread model that is used. With `MPI_THREAD_MULTIPLE`, all threads can perform MPI calls and may do so concurrently, in parallel. The levels of thread support are ordered as `MPI_THREAD_SINGLE` < `MPI_THREAD_FUNNELED` < `MPI_THREAD_SERIALIZED` < `MPI_THREAD_MULTIPLE`, meaning that a program that assumes a higher level of thread support, e.g., `MPI_THREAD_SERIALIZED`, may not run correctly if the MPI library supports only a lower level.

Threads levels are controlled and queried by a special initialization function to be used instead of `MPI_Init`. With `MPI_Init_thread`, the user gives a required thread level, and the function returns a thread level that can be supported. If the required thread level cannot be supported, the provided level is the highest thread level of the MPI library implementation. If the required thread level can be supported, the provided level returned is larger than or equal to the required level.

```
int MPI_Init_thread(int *argc, char ***argv,
                    int required, int *provided);
int MPI_Is_thread_main(int *flag);
int MPI_Query_thread(int *provided);
```

3.2.35 MPI Outlook

A number of important aspects and parts of the huge MPI standard were deliberately not treated in these bachelor lecture notes. These include a whole model for input-output and communication with the external file system (MPI-IO), dynamic process management (spawning new MPI processes from an application, connecting running MPI processes), so-called *inter-communicators* (that are important for process management), MPI attributes (a very useful mechanism for library building by which information can be attached to MPI objects [117]), the profiling and tools interfaces (important for library and performance analysis tool building), partitioned point-to-point communication and a few other things. The treatment stayed within the so-called "world model", in which externally started processes are grouped together within the `MPI_COMM_WORLD` communicator. We did not at all cover the alternative "sessions model", in which this is not the case and processes initially have to create the communicator they want to belong to.

The most recent, at the time of writing, version of the MPI standard is MPI 4.1 (November 2nd, 2023). The MPI forum is actively preparing a next version with further additions and corrections to the standard. Some of the important recent additions were and are persistent collective operations (see Sect. 3.2.19), the sessions model, so-called partitioned (point-to-point) communication, additional support for portably adapting applications to specifics of system topologies (`MPI_Comm_split_type` is one function of this kind), and further provisioning for fault tolerant MPI programming.

3.3 Exercises

1. Give a polynomial-time algorithm to compute the diameter diam(G) of a given, directed or undirected graph $G = (V, E)$. Hint: Reapply BFS.
2. Devise an algorithm for the broadcast problem for d-dimensional hypercubes with $p = 2^d$ processors. What is the number of communication rounds taken by your algorithm? How does that relate to the diameter lower bound for the broadcast problem? Is your algorithm optimal?
3. Argue why the communication round complexity for a semantic barrier operation in fully connected, one-ported p-processor communication systems is in $\Theta(\log p)$.
4. Consider a high-performance computing system consisting of a (large) number of shared memory multi-core processor nodes interconnected with a complex communication network. Assume that some processor i is sending a (large) number of message packets $b = b_0, b_1, b_2, \ldots$ one after the other to some other, different processor j in the system. What might be reasons that packets are not necessarily delivered in the sent order to processor j? What would an MPI library need to do in order to guarantee that messages that are sent in sequence are indeed received (seen by the receiving process) in the same order? What if individual packets (that could be parts of larger messages) are lost or corrupted? What would an MPI library implementation have to do?
5. On your favorite system, run the communicator creation example from Sect. 3.2.7 instrumented with print-statements to show the process ranks in old and new communicators. Develop assertions to express the relations between old and new ranks in all the communicators. Extend the example with a partition of the comm communicator duplicate of MPI_COMM_WORLD into two communicators consisting of the processes with rank smaller than some given rank split and the processes with rank larger than or equal to the split process. Create the same communicators by using the process group functionality of Sect. 3.2.10. Verify by assertions and use of MPI_Comm_compare and MPI_Group_compare that the created communicators are indeed equivalent.
6. The following program has the intention of collecting information at a given root process from all processes, somewhat like the MPI_Gather operation can do. The code has numerous safety issues and obviously does not work. Pinpoint the problems and repair the code; there are several possibilities for a "correct" solution, since the outcome has not been explicitly specified.

```
MPI_Comm_rank(comm,&rank);
MPI_Comm_size(comm,&size);

if (rank==root) {
  int all[size];
  MPI_Status status;

  MPI_Send(&rank,1,MPI_INT,root,1000,comm);
  for (i=0; i<size; i++) {
    MPI_Recv(&all[i],1,MPI_INT,MPI_ANY_SOURCE,1000,
```

```
              comm,&status);
  }
  for (i=0; i<size; i++) assert(all[i]==i);
} else {
  MPI_Send(&rank,1,MPI_INT,root,1000,comm);
}
```

7. A root process identified by a `root` rank that is not necessarily 0 is given for a communicator. For some application, a new communicator with the same processes is needed where the root process has rank 0, the process with the next rank has rank 1, the process with the next to next rank has rank 2 and so on. Use `MPI_Comm_split` to create a new communicator where processes have been reranked towards root 0. How costly is explicit communicator creation compared to manual reranking using a virtual rank `virt = (rank-root+size)%size` in the application?
8. Implement the unsafe ring and the unsafe stencil communication patterns from Sect. 3.2.14 using blocking `MPI_Send` and `MPI_Recv` operations. Use a simple, 5-point average element stencil update rule. Devise an experiment to determine at which buffer sizes deadlocks occur (on the system and MPI library available to you). Are these sizes different in the two cases?
9. Implement (incorrect!) programs as in Sect. 3.2.15 where a process sends data as a sequence of `MPI_LONG` to another process that receives the data as a sequence of `MPI_DOUBLE`, and vice versa, and examine the outcome. Are there interesting differences between the two cases? Is the outcome of such communication meaning- or useful?
10. Implement the two-dimensional stencil computation safely and correctly with `MPI_Sendrecv` as described at the end of Sect. 3.2.17; use any non-trivial 5-point stencil update rule (for instance, a simple average). For the communication of submatrix columns, copy the elements of leftmost and rightmost columns into intermediate, consecutive buffers. Likewise, receive the column elements in intermediate, consecutive buffers and copy these into their desired positions in the matrix columns after the communication. Verify correctness by comparison to a sequential implementation. Iterate the stencil computation a number of times (for instance, an input parameter). Repeat and time the whole computation using `MPI_Wtime` over a number of repetitions and compute the average and best completion time (best time for the slowest MPI process over the repetitions). Experiment with different matrix sizes and different numbers of MPI processes in different configurations. Present strong scaling results where the total matrix size is kept independent of the number of processes and weak scaling results where the submatrix size per process is kept constant.
11. Repeat Exercise 10 using instead nonblocking `MPI_Isend` and `MPI_Irecv` communication as explained in Sect. 3.2.17.
12. Repeat now Exercise 11 using instead of nonblocking the persistent `MPI_Send_init` and `MPI_Recv_init` operations that were briefly explained in Sect. 3.2.19. How does the performance of the nonblocking and persistent implementations differ? Use a sufficiently large number of stencil iterations.

13. Given an $m \times n$ matrix in process-local memory for some process. Implement a process local matrix transposition into an $n \times m$ matrix using the MPI_-Type_vector datatype to describe columns of either input or output matrix. The exercise illustrates the problem of doing process-local, MPI type correct data reorganization and the power of MPI datatypes for effecting this. For the process-local copy, use communication on MPI_COMM_SELF and use either MPI_Sendrecv communication or a collective operation like MPI_Allgather (see the end of Sect. 3.2.28).
14. Repeat Exercise 10 eliminating the intermediate, consecutive buffers by using instead an MPI_Type_vector datatype to describe the strided layout of a submatrix column. Benchmark and compare the results to your results from Exercise 10, and discuss (notable) differences. Instead of using an MPI_Type_vector, try MPI_Type_create_resized to create a special **double** datatype with the extent of a full row. What might the advantages of this solution be compared to the (less flexible) MPI_Type_vector solution?
15. The stencil implementations suggested in Sect. 3.2.14 etc. use a decomposition of the $n \times n$ matrix into smaller $n_r \times n_c$ submatrices where r and c are the numbers of row and columns of the MPI process grid. Why is this distribution beneficial compared to, say, a row-wise or a column-wise distribution as used for the matrix–vector multiplication algorithms? For your answers, consider the ratio of communication volume to computation done per MPI process.
16. Implement a two-dimensional, 9-point stencil computation where the update rule (say, average; sometimes used in image processing applications) for a matrix-element $M[i,j]$ depends on 9 neighboring elements (including the element itself), $M[i, j-1], M[i+1, j-1], M[i+1, j], M[i+1, j+1], M[i, j+1], M[i-1, j+1], M[i-1, j], M[i-1, j-1]$ and $M[i,j]$ itself. Some of these elements may be undefined, instead their values are given by border (boundary) conditions. Partition the full matrix into roughly square matrices over a MPI process grid. Implement the communication (horizontal, vertical, diagonal) with MPI_Sendrecv. How can you avoid deadlocks? What is the communication volume as a function of the input matrix size nm? What is the amount of local computation (element updates) per stencil iteration? What is the ratio of computation steps to communication volume? What is the parallel running time per stencil iteration? To how many processors will the implementation scale? A well-known trick can reduce the number of communication operations per process from 8 to 4. What is the idea? Does such an optimization make a difference in performance?
17. Repeat Exercise 10 using instead one-sided MPI_Get or MPI_Put communication as explained in Sect. 3.2.24. Try with both MPI_Win_fence and with MPI_Win_post-MPI_Win_start-MPI_Win_complete-MPI_Win_wait synchronization. Compare the performance of the one-sided implementations against each other, and compare to either of the solutions with point-to-point communication. You may or may not use MPI_Type_vector to ease communication with left and right neighbors.

18. Complete the implementation of the binary search operation with one-sided communication outlined in Sect. 3.2.25. Each process contributes an ordered array (of, say, floats) of n elements for the window. Each process can perform binary search in the array by calling a search function on the window. Return values should be the rank of the process where the element belongs and the relative index (displacement) in the window of that process. Benchmark your implementation with one process and with all processes performing search operations. Consider worst and best cases.

19. Give a full, distributed memory implementation of merging by co-ranking as described in Sect. 1.4.3. Use one-sided communication with MPI_Win_lock and MPI_Win_unlock for implementing a corank() function (see the previous exercise). You can use either one or two windows for storing the ordered, distributed input arrays A and B of n and m elements per process, respectively. The output should be, for each process, an ordered array C of size $n + m$ elements such that all elements at some process $i, 0 < i$ are equal to or larger than all elements at process $i - 1$. Benchmark (weak scaling) your implementation for larger and larger n and m and different numbers of processes p (not only powers-of-two).

20. As in the previous exercise, give a full, distributed memory implementation of merging by co-ranking (Sect. 1.4.3) where now the input arrays A and B of n and m elements in total are distributed as follows. Assume for simplicity that both n and m are divisible by p, the number of MPI processes. Divide the total input into blocks of size roughly $(n+m)/p$. The A array is divided into $\frac{np}{n+m}$ blocks which are assigned to the first MPI processes $0, 1, \ldots \frac{np}{n+m} - 1$. The B array is divided into $\frac{mp}{n+m}$ blocks and assigned to the remaining processes $\frac{np}{n+m}, \ldots, p-1$. Assume here that n, m, p are chosen such that all fractions are nice numbers. This way, each process has a part of the input of (roughly) the same size as all other processes. The total input can be kept in an MPI_Win window. The size of the output per MPI process in the C array should be of the same size as the input for each process, ad each process local C array should be ordered and the C elements at any process i be smaller than or equal to the C element of the next process $i + 1$. Implement a corank() function using passive MPI_Win_lock and MPI_Win_unlock synchronization that can work with this input distribution. Benchmark your implementation as in the previous exercise. This corank() function could be a building block of an MPI mergesort implementation.

21. In the following, typical MPI benchmarking loop for benchmarking a single problem instance for some algorithm use MPI_Reduce to find the parallel time(s) for each problem size for each repetition (recall: defined as the time spent by the slowest process) and store the result at root process 0. Use as few MPI_Reduce calls as possible. Extend your implementation to also compute the load imbalance (defined as the difference between the slowest and the fastest process), again with as few MPI_Reduce calls as possible.

3.3 Exercises

```
double start, stop, spent;
for (r=0; r<REPETITIONS; r++) { // do some repetitions
  MPI_Barrier(comm);
  MPI_Barrier(comm);

  start = MPI_Wtime();

  ... // operation on comm to be benchmarked

  stop = MPI_Wtime();
  spent = stop-start;
  ... // store spent time for this repetition somewhere
}
... // post-process: minimum, average?
... // collect results at process 0
```

Post-processing can either be done on the processes or, centrally, at process 0 which then does the statistics. Consider both options. What might the pros and cons be?

22. Write a series of small programs that illustrate the semantics of the collective operations. Each program should allocate proper send and receive buffers of, say, MPI_INT type, at all processes, either of a small constant number of elements or proportional to p, the number of processes in the communicator. Initialize all buffers with values that make it easy to verify that a) values are exchanged (and reduced) properly with the right results in the receive buffers and b) no send buffers have been modified. Instrument the program first with print statements, and verify by inspection with $p = 1, 2, 4, 5, 7, \ldots$ MPI processes. Then formulate assertions that make it possible to verify exhaustively at larger scale that the collective operations do as claimed.

Start with the simple, regular collectives MPI_Bcast, MPI_Gather, MPI_Scatter, MPI_Allgather, MPI_Alltoall. Proceed to the regular reduction collectives MPI_Reduce, MPI_Allreduce, MPI_Reduce_scatter_block, MPI_Scan, and MPI_Exscan. Time and interest permitting, extend your analysis to the irregular counterparts of these collective operations.

Here is an example:

```
int rank, size;
MPI_Comm_rank(comm,&rank);
MPI_Comm_size(comm,&size);

int n = 2;
int buffer[n+1];

int root = size-1;
if (rank==root) {
  buffer[0] = size;
  buffer[1] = 0;
  buffer[2] = -rank-1;
} else {
  buffer[0] = -rank-1;
  buffer[1] = -rank-1;
```

```
    buffer[2] = -rank-1;
  }
  MPI_Bcast(buffer,n,MPI_INT,root,comm);

  assert(buffer[0]==size);
  assert(buffer[1]==0);
  assert(buffer[2]==-rank-1);

  if (rank==0) {
    printf("Rank %d: buffer=[%d,%d,%d]\n",
           rank,buffer[0],buffer[1],buffer[2]);
  }
```

23. Implement an own vector-scan operation with the same interface and semantics as `MPI_Scan` using the (blocked) Hillis–Steele algorithm of Sect. 1.4.10 (do not call your interface `MPI_Scan`!). Make sure that the implementation is **safe** by using the proper point-to-point communication operations. What is the number of communication rounds? What is the run time complexity of the implementation as a function of the number of processes p and the number of vector elements n per process? Note that no barrier synchronization (like `MPI_Barrier`) is needed.

24. The following two collective MPI calls are supposed to implement a barrier operation in the same way as `MPI_Barrier` does.

```
MPI_Gather(NULL,0,MPI_INT,NULL,0,MPI_INT,0,comm);
MPI_Scatter(NULL,0,MPI_INT,NULL,0,MPI_INT,0,comm);
```

Explain why this is not necessarily semantically equivalent to an `MPI_Barrier`, and why using this implementation as a barrier substitute will be unsafe. Can you come up with code demonstrating that the implementation can go wrong? Give a simple fix, still in terms of `MPI_Gather` and `MPI_Scatter`. Is your fix efficient in the volume of data communicated?

25. In Sect. 3.2.28, it was described how to locally copy data on an MPI process from a send buffer with a send datatype to a receive buffer with a receive datatype using `MPI_Sendrecv` on the special `MPI_COMM_SELF` communicator. Show how to implement the same local copy operation with a collective operation on `MPI_COMM_SELF`. Hint: consider `MPI_Allgather` or `MPI_Alltoall`.

26. Implement `MPI_Allgather` by a series of `MPI_Bcast` operations. Implement `MPI_Allgather` by a series of `MPI_Gather` operations. Repeat the exercise for `MPI_Allgatherv`. Time permitting, time the new implementations in comparison to `MPI_Allgather` for different MPI process configurations and input block sizes (counts).

27. Implement `MPI_Alltoall` by a series of `MPI_Gather` operations. Implement `MPI_Alltoall` by a series of `MPI_Scatter` operations. Repeat the exercise for `MPI_Alltoallv` and `MPI_Alltoallw`. Time permitting, time the new implementations in comparison to `MPI_Alltoall` and variants for different MPI process configurations and input block sizes (counts).

28. Implement `MPI_Reduce_scatter_block` by a series of `MPI_Reduce` operations. Implement `MPI_Reduce_scatter_block` by an `MPI_Reduce` operation

followed by an `MPI_Scatter` operation. Repeat the exercise for `MPI_Reduce_-scatter`. Time permitting, time the new implementations in comparison to `MPI_Reduce_scatter_block` for different MPI process configurations and input block sizes (counts).

29. Devise an MPI program using collective operations for computing the scalar (dot) product of two distributed n-element vectors a and b, i.e., the sum $\sum_{i=0}^{n-1}$ a$[i]$b$[i]$. The vectors are represented as disjoint blocks of consecutive elements of roughly n/p elements, and each process has two such blocks of a and b elements, respectively. Give two variants of the program, one that stores the result (dot product) at a designated root process and one that stores the result at all processes. The programs should work correctly regardless of whether p divides n, p being the number of available MPI processes, preferably also for the case where $n < p$.

30. Finite sets can be represented by bitmaps of n bits where n is the maximum cardinality of such a set: An element is in the set if and only if the corresponding bit is set. Union and intersection of such sets can then easily be computed by "bitwise or" and "bitwise and" operations. Now, let some maximum cardinality n be given, and let sets be represented by m-element arrays of `MPI_LONG` integers with $n = 64m$ (assuming that `sizeof(long)` $= 64$). Give collective calls for computing, for all p processes in a communicator comm, first the union and second the intersection of p such sets, with the resulting set stored at all p processes. Assume now instead that the resulting set from a union or intersection operation is to be stored in a distributed fashion, with roughly n/p bits per process. Give also for this case collective calls for computing the union and intersection of p such sets with the resulting set stored in a distributed fashion. Each of the p input sets, one for each process, is a full set of n bits. Assume first that p divides m. Give also a solution where m is not necessarily divisible by p.

31. Many of the MPI collectives can relatively easily and conveniently be expressed and implemented in terms of other MPI collectives without any further ado like copying data and doing local computations (see previous exercises). Why do you think MPI offers as many collectives as it does? Which ones would you think of as redundant? Are there collectives that are not easily or at all reducible to other collectives?

32. Which, if any, of the following three MPI programs are correct? They are all assumed to broadcast a value from the last process in the communicator and do a barrier. Explain your answers and explain possible outcomes.

   a.
   ```
   int rank, size;
   MPI_Comm_rank(comm,&rank);
   MPI_Comm_size(comm,&size);

   int i, j;
   if (rank%2==0) {
     MPI_Bcast(&i,1,MPI_INT,size-1,comm);
     MPI_Barrier(comm);
   } else {
   ```

```
    MPI_Bcast(&j,1,MPI_INT,size-1,comm);
    MPI_Barrier(comm);
}
```

b.
```
int rank, size;
MPI_Comm_rank(comm,&rank);
MPI_Comm_size(comm,&size);

int i, j;
if (rank%2==0) {
    MPI_Bcast(&i,1,MPI_INT,size-1,comm);
    MPI_Barrier(comm);
} else {
    MPI_Barrier(comm);
    MPI_Bcast(&j,1,MPI_INT,size-1,comm);
}
```

c.
```
int rank, size;
MPI_Comm_rank(comm,&rank);
MPI_Comm_size(comm,&size);

int i, j;
if (rank%2==0) {
    MPI_Bcast(&i,1,MPI_INT,size-1,comm);
} else {
    MPI_Barrier(comm);
}
```

33. Define a collective operation for computing all prefix sums of a distributed array. More precisely, each process contributes an array a of n elements (n may be different for different processes). These arrays together make up a large (virtual) array formed by concatenating the p arrays in rank order (p is the number of processes). Your operation should compute the (inclusive or exclusive) prefix sums on this array. Which collective operation(s) might be convenient as building blocks? Analyze and state the performance of your implementation as a function of n and p and the complexity of the (collective) MPI operations you use.

34. Implement matrix–vector multiplication for row-wise distributed matrices using MPI_Sendrecv on a ring of processes to gather the full input vector at all processes.

35. Implement matrix–vector multiplication for row-wise distributed matrices with the same number of full rows per process with MPI_Allgather as described in Sect. 3.2.30. Perform strong and weak scalability experiments for matrices with m rows and n columns where $p|m$ for different, not too small values of m, n, p.

36. Implement matrix–vector multiplication for row-wise distributed matrices with possibly different numbers of full rows per process with MPI_Allgatherv. See the description in Sect. 3.2.30. Perform strong and weak scalability ex-

periments for matrices with m rows and n columns for different, not too small values of m, n, p.

37. Implement matrix–vector multiplication for column-wise distributed matrices with the same number of full columns per process with MPI_Reduce_scatter_block as described in Sect. 3.2.30. Perform strong and weak scalability experiments for matrices with m rows and n columns where $p|n$ for different, not too small values of m, n, p.

38. Implement matrix–vector multiplication for column-wise distributed matrices with possibly different numbers of full columns per process with MPI_Reduce_scatter. See the description in Sect. 3.2.30. Perform strong and weak scalability experiments for matrices with m rows and n columns for different, not too small values of m, n, p.

39. Consider the three matrix-distributions, row-wise, column-wise, and block-wise, discussed and used in Sect. 3.2.30. Assume that a full matrix needs to be collated at some single, given root process by putting the submatrices from the p processes together. Try to accomplish this with a single MPI_Gather call. Datatypes (MPI_Type_vector and MPI_Type_create_resized) may be useful (and possibly needed) in order to avoid process-local reorganizations of submatrices at either root or non-root processes. It may likewise be that MPI_Gatherv is needed.

40. Complete the implementation of the SUMMA matrix–matrix multiplication described in Sect. 3.2.30. Perform strong and weak scalability experiments for matrices with m rows and n columns for different, not too small values of m, n, p. Your implementation will most likely require that p is a square, and that both $p|m$ and $p|n$. You may assume that $n = m$. The less such restrictions, the better.

41. Devise and implement a distributed memory Breadth-First Search operation in a bulk synchronous way as outlined in Sect. 1.3.9. The input graph, given as a collection of edges stored as adjacency lists (arrays) is distributed in some form over the p MPI processes. The processes explore the graph iteratively in a level-by-level way. In each iteration, each process starts with a set of new vertices that has been reached in the previous iteration. The processes explore their vertices and compute, locally, a collection of vertices that can now be reached for the next iteration and have not been seen so far. At the end of the iteration, these local collections of new vertices are put together and distributed over the processes for the next iterations. The last iteration is reached when no new vertices are discovered. Thus, the total number of iterations equals the largest distance from the given start vertex to another vertex in the graph. The output should be for each vertex the distance of that vertex from the given start vertex and, if possible, a reference to the parent of that vertex in a BFS-tree rooted at the start vertex.

Consider a suitable distribution of the input graph. Either the vertices are distributed roughly evenly over the processes such that a process that has a vertex also has all the edges connected to that vertex. Alternatively, the edges might be distributed roughly evenly across the processes. Use bitmaps to represent seen vertices, and allow each process to maintain a full seen/not seen bitmap for all vertices of the input graph. Use also bitmaps to represent the collections of new vertices considered by the processes in each iteration. Use `MPI_Allreduce` to compute the union of all local bitmap-represented sets. Analyze the complexity of your algorithm in terms of the size of the input graph (n vertices and m edges), the depth of a BFS tree rooted at the given start vertex, and the number of MPI processes. For `MPI_Allreduce`, you can use the estimates from Table 3.3. Benchmark your algorithm with different, randomly generated input graphs for different numbers of MPI processes. What speed-up can you achieve compared to your own, best, sequential BFS implementation?

42. Complete the distributed memory implementation of the Quicksort algorithm discussed in Sect. 3.2.31. Assuming that bad pivots are chosen throughout, how skewed can the resulting output distribution (in terms of numbers of output elements per process) be? What would be the worst-case running time assuming that the final, sequential sorting is done optimally in $O(n \log n)$ time steps.

43. Design and implement a parallel, distributed memory bottom-up mergesort algorithm. The input per process is an unsorted array of n elements and the output per process should be a sorted array of the same size such that the output elements at any one process are smaller than or equal to the output elements of the next (higher ranked) process. The output elements over all processes must of course be a permutation of the input elements over all processes. One approach is to recursively split the input communicator down to communicators with just a single process (as done for Quicksort in the previous exercise). Each process then (merge)sorts its input elements (using a best possible mergesort implementation) after which the processes merge their elements with those of other processes going up the hierarchy of communicators. Use the merging by co-ranking algorithm and implementation of Exercise 20 for the merge steps. You may at first assume that the number of processes p is a power of two. What is the parallel running time of your algorithm as a function of p and n? Conduct strong and weak scaling experiments with your implementation and compute the speed-up and parallel efficiency relative to your best, sequential mergesort (already used for the initial sorting per process; thus, absolute and relative speed-up will coincide). You may be able to improve your implementation by not explicitly creating the communicators as you recursively decrease p and instead stay with only the given input communicator with the p processes. This may in addition make it possible to stay with only one `MPI_Win` (input and output) window. You may be able to generalize this implementation to work well with any number of processes. In that case, what is the parallel running time as a function of n and p? Be as exact as possible. How do the concrete, parallel running time of your improved implementation compare against your first try?

44. Complete an implementation of a distributed counting sort as outlined in Sect. 3.2.31. Each process has input of n integer elements in a range $[0, r[$

stored in an array. The output should be a sorted array segment of n elements of the total input array of pn elements. The elements of some rank $i, i > 0$ must all be equal to or larger than the elements of rank $i - 1$. Your algorithm has to count the number of elements of each key $k \in [0, r]$ over the processes and use this to redistribute the elements. A final, process-local sort or merge may/will be necessary. Which collective operations are you using? What is the estimated running time of your implementation as a function of n nd p? Benchmark and compare against a sequential counting sort for $n = 1\,000\,000$, $n = 10\,000\,000$ and $n = 100\,000\,000$ elements per process. Depending on p, it may not be able to sort sequentially (in a reasonable amount of time).

45. The following code is a sequential implementation of the Floyd–Warshall algorithm discussed in Exercise 60 for Chapter 2 for solving the all-pairs shortest path problem on a weighted graph given by an initial weight matrix $W[n, n]$.

```
void fw_apsp(int *w, int n) {
  int (*W)[n] = (int(*)[n])w;

  int i, j, k;
  for (k=0; k<n; k++) {
    for (i=0; i<n; i++) {
      for (j=0; j<n; j++) {
        if (W[i][j]>W[i][k]+W[k][j]) {
          W[i][j] = W[i][k]+W[k][j];
        }
      }
    }
  }
}
```

The exercise is to give a distributed memory implementation with MPI following the idea of the SUMMA matrix–matrix multiplication algorithm presented in Sect. 3.2.30. More concretely, assume that a square number of MPI processes organized into a Cartesian $\sqrt{p} \times \sqrt{p}$ communicator is available. Create communicators of the processes belonging to the same row and processes belonging to the same column of processes in the Cartesian communicator. Assume further that p divides n, $p|n$, and that the weight matrix is distributed cyclically over the processes as $(n/\sqrt{p}) \times (n/\sqrt{p})$ submatrices. The algorithm perform $\sqrt{p}$ iterations. In iteration k, the kth process in each row and in each column broadcasts its weight matrix to the processes in the row and the column, respectively. The processes can then locally perform $n/\sqrt{p}$ iterations of the Floyd–Warshall update operation (innermost two loops).

Write out this algorithm in detail and state the parallel running time under reasonable assumptions on the broadcast time complexity. Implement your algorithm with MPI, and perform benchmark experiments for a number of larger n values and different numbers of MPI processes. How much speed-up can you achieve compared to your best, sequential implementation of the Floyd–Warshall algorithm?

Appendix A
Proofs and Supplementary Material

A.1 A Frequently Occurring Sum

One of the most frequently occurring (finite) sums in Parallel Computing is the *geometric series* $1 + q + q^2 + q^3 + \cdots + q^n = \sum_{i=0}^{n} q^i$. The geometric series is the sum of the elements of the geometric progression $1, q, q^2, q^3, \ldots, q^n$, where each element of the sequence except the first follow from the previous by multiplying with the common ratio q. For $q = 1$, obviously $\sum_{i=0}^{n} q^i = (n+1)$ (since also $0^0 = 1$). For any other $q, q \neq 1$, it is well-known (and easy to see, even without using induction) that

$$\sum_{i=0}^{n} q^i = \frac{q^{n+1} - 1}{q - 1} \tag{A.1}$$

$$= \frac{1 - q^{n+1}}{1 - q} \ .$$

When $|q| < 1$, the geometric series is convergent, and we can write

$$\sum_{i=0}^{\infty} q^i = \frac{1}{1 - q} \ . \tag{A.2}$$

For instance, with $q = 2$, $\sum_{i=0}^{n} q^i = 2^{n+1} - 1$, and with $q = \frac{1}{2}$, $\sum_{i=0}^{n} q^i = 2 - \frac{1}{2^n}$ (and $\sum_{i=1}^{n} q^i = 1 - \frac{1}{2^n}$). For other elementary sums and series occurring in standard analysis of algorithms, see [33, 47] and other textbooks.

A.2 Logarithms Reminder

The logarithm $\log_b x$ with base b, $b > 0, b \neq 1$ of some $x, x > 0$ is the inverse of exponentiation with base b, that is $x = \log_b b^x$ and $x = b^{\log_b x}$. When clear from context (or not relevant, see the following), the base is left out and the logarithm function is written $\log x$. By convention, $\log^c x$ is notation for $(\log x)^c$ and is not the iterated application of the logarithm function, which is denoted $\log^{(n)} x$ and for integer $n \geq 0$ defined by $\log^{(n)} x = \log \log^{(n-1)} x$ for $n > 0$ and $\log^{(0)} x = x$. For constant $c, c \geq 1$, functions in $O(\log^c x)$ are called "poly-logarithmic" by being polynomials in $(\log x)$.

It follows that $\log_b 1 = 0, \log_b b = 1$. Let $x = b^a$ and $y = b^c$. Then from the laws of exponentiation, $\log_b xy = \log_b(b^a b^c) = \log_b b^{a+c} = a+c = \log_b x + \log_b y$. Similarly, it follows that $\log_b \frac{x}{y} = \log_b x - \log_b y$. Also $\log_b x^d = \log_b (b^a)^d = \log_b b^{ad} = ad = da = d\log_b x$. It now follows that for any other other base $e, e > 0, e \neq 1$, $\log_e x = \log_e b^{\log_b x} = \log_b x \log_e b$, so any two logarithms with different constant bases differ only by a constant factor.

Common logarithm bases in Parallel Computing are $b = 2, b = 2.718281828459 \ldots, b = 10, b = (k+1)$ for some positive integer k. For all of these, $\log_b x$ is in $O(\log n)$. A sometimes useful observation for graphs with n vertices and m arcs is that here $O(\log m) = O(\log n)$ since $m \leq n^2$.

A.3 The Master Theorem

The "Master Theorem", Theorem 9, gives closed form solutions for a range of divide-and-conquer recurrences of the following form, for constants $a \geq 1, b > 1, d \geq 0, e \geq 0$ (the c is omitted to avoid any confusion with constants hidden behind the O) that very often occur in the analysis of (parallel) algorithms:

$$T(n) = aT(n/b) + O(n^d \log^e n)$$
$$T(1) = O(1)$$

The theorem claims a closed-form solution in either of three forms:

1. $T(n) = O(n^d \log^e n)$ if $a/b^d < 1$ (equivalently $b^d/a > 1$),
2. $T(n) = O(n^d \log^{e+1} n)$ if $a/b^d = 1$ (equivalently $b^d/a = 1$), and
3. $T(n) = O(n^{\log_b a})$ if $a/b^d > 1$ (equivalently $b^d/a < 1$).

Let C be a constant at least as large as the leading constant in either of $O(1)$ or $O(n^d \log^e n)$. Then, the recurrence takes the form

$$T(n) \leq aT(n/b) + C(n^d \log^e n) \quad .$$

First, assume $n = b^k$. With this, $\log^e n = (\log b^k)^e = k^e$ and the recurrence takes the form

$$T(b^k) \leq aT(b^k/b) + C(b^{kd} k^e) \quad .$$

A.3 The Master Theorem

Expanding the recurrence for the first few values of k, $k = 1, 2, 3$ yields:

$$T(b) \leq Ca + C(b^d 1^e)$$
$$T(b^2) \leq Ca^2 + Ca(b^d 1^e) + C(b^{2d} 2^e)$$
$$T(b^3) \leq Ca^3 + Ca^2(b^d 1^e) + Ca(b^{2d} 2^e) + C(b^{3d} 3^e)$$

From this, we conjecture that

$$T(b^k) \leq Ca^k(1 + \sum_{i=1}^{k} \left(\frac{b^d}{a}\right)^i i^e) \quad .$$

The claim is easily verified by induction. The base case $T(1) \leq C$ holds, since the sum is void (no summands, per definition 0), by the choice of the constant C. Assuming the claim for $k-1$ yields:

$$T(b^k) \leq aT(b^k/b) + C(b^{kd} k^e)$$
$$= aT(b^{k-1}) + C(b^{kd} k^e)$$
$$= a(Ca^{k-1}(1 + \sum_{i=1}^{k-1} \left(\frac{b^d}{a}\right)^i i^e)) + C(b^{kd} k^e)$$
$$= Ca^k(1 + \sum_{i=1}^{k-1} \left(\frac{b^d}{a}\right)^i i^e)) + Ca^k(\left(\frac{b^d}{a}\right)^k k^e)$$
$$= Ca^k(1 + \sum_{i=1}^{k} \left(\frac{b^d}{a}\right)^i i^e)$$

since $C(b^{kd} k^e) = Ca^k(\left(\frac{b^d}{a}\right)^k k^e)$ by multiplying and dividing again by a^k leading to the last, kth term in sum.

We now distinguish three cases for bounding the sum $\sum_{i=1}^{k} \left(\frac{b^d}{a}\right)^i i^e$ from above.

1. $b^d/a > 1$:

$$\sum_{i=1}^{k} \left(\frac{b^d}{a}\right)^i i^e \leq k^e \sum_{i=1}^{k} \left(\frac{b^d}{a}\right)^i$$
$$= O(k^e \left(\frac{b^d}{a}\right)^{k+1})$$

since the sum is a geometric series and $i^e \leq k^e$ for $i \leq k$ which can be factored out. Therefore,

$$T(b^k) = O\left(a^k \left(\frac{b^d}{a}\right)^{k+1} k^e\right)$$
$$= O\left(b^{kd} \left(\frac{b^d}{a}\right) k^e\right)$$
$$= O(n^d \log^e n) \quad .$$

2. $b^d/a = 1$:

$$\sum_{i=1}^{k} \left(\frac{b^d}{a}\right)^i i^e = \sum_{i=1}^{k} i^e$$
$$\leq k^{e+1}$$

Therefore, by using the bound on $T(b^k)$

$$T(b^k) = O(a^k k^{e+1})$$
$$= O(b^{kd} k^{e+1})$$
$$= O(n^d \log^{e+1} n) \quad .$$

3. $b^d/a < 1$: In this case, we use the fact that an exponential function f^i for $f > 1$ grows faster than the (any) polynomial i^e. We choose a constant $f, f > 1$ with $\left(\frac{b^d}{a}\right) f < 1$. Then, for some constant k', it holds that $i^e < f^i$ for $i \geq k'$.

$$\sum_{i=1}^{k} \left(\frac{b^d}{a}\right)^i i^e \leq \sum_{i=1}^{k'-1} \left(\frac{b^d}{a}\right)^i i^e + \sum_{i=k'}^{k} \left(\frac{b^d}{a}\right)^i i^e$$
$$\leq \sum_{i=1}^{k'-1} \left(\frac{b^d}{a}\right)^i i^e + \sum_{i=k'}^{\infty} \left(\frac{b^d}{a}\right)^i f^i$$
$$= \sum_{i=1}^{k'-1} \left(\frac{b^d}{a}\right)^i i^e + \sum_{i=k'}^{\infty} \left(\left(\frac{b^d}{a}\right) f\right)^i \quad .$$

The first sum is finite. The second sum, which is a geometric series with a quotient smaller than one, is convergent (to a constant). Therefore

$$T(b^k) = O(a^k)$$
$$= O(a^{\log_b n})$$
$$= O(n^{\log_b a}) \quad .$$

When n is not a power of b, it holds that for some k, $b^{k-1} < n < b^k = n'$. Since $T(n)$ is monotone, we have for the three cases

1.

A.3 The Master Theorem

$$\begin{aligned}
T(n) \leq T(n') &= O(n'^d \log^e n') \\
&= O((n'/n)^d n^d \log^e((n'/n)n)) \\
&= O((n'/n)^d n^d (\log^e(n'/n) + \log^e n) \\
&= O(n^d \log^e n)
\end{aligned}$$

since $n'/n < b$ can be upper bounded by the constant b.

2.

$$\begin{aligned}
T(n) \leq T(n') &= O(n'^d \log^{e+1} n') \\
&= O(n^d \log^{e+1} n)
\end{aligned}$$

with the same calculation and argument as in Case 1.

3.

$$\begin{aligned}
T(n) \leq T(n') &= O(n'^{\log_b a}) \\
&= O((n'/n)^{\log_b a} n^{\log_b a}) \\
&= O(b^{\log_b a} n^{\log_b a}) \\
&= O(n^{\log_b a})
\end{aligned}$$

since $n'/n < b$ and also $b^{\log_b a}$ is constant.

The theorem therefore holds for any $n, n \geq 1$. The bounding arguments do not give any useful estimates of the constants incurred by the recurrence; but it can be shown that the bounds are asymptotically tight for recurrences of the form

$$T(n) = aT(n/b) + \Theta(n^d \log^e n)$$
$$T(1) = O(1)$$

The Master Theorem can be improved to give closed-form solutions also for negative values of $e, e < 0$.

References

1. Abolhassan, F., Drefenstedt, R., Keller, J., Paul, W.J., Scheerer, D.: On the physical design of PRAMs. The Computer Journal **36**(8), 756–762 (1993)
2. Abolhassan, F., Keller, J., Paul, W.J.: On the cost-effectiveness of PRAMs. Acta Informatica **36**(6), 463–487 (1999)
3. Aho, A.V., Hopcroft, J.E., Ullman, J.D.: The Design and Analysis of Computer Algorithms. Addison-Wesley (1974)
4. Aho, A.V., Hopcroft, J.E., Ullman, J.D.: Data Structures and Algorithms. Addison-Wesley (1987), reprint of 1983 edition with corrections
5. Ajtai, M., Komlos, J., Szemeredi, E.: An $O(n \log n)$ sorting network. Combinatorica pp. 1–19 (1983)
6. Amdahl, G.M.: Validity of the single processor approach to achieving large scale computing capabilities. In: American Federation of Information Processing Societies (AFIPS) Spring Joint Computer Conference. pp. 483–485 (1967)
7. Anderson, R.J., Miller, G.L.: Deterministic parallel list ranking. Algorithmica **6**, 859–868 (1991)
8. Arora, N.S., Blumofe, R.D., Plaxton, C.G.: Thread scheduling for multiprogrammed multiprocessors. Theory of Computing Systems **34**(2), 115–144 (2001)
9. Axtmann, M., Sanders, P.: Robust massively parallel sorting. In: 19th Workshop on Algorithm Engineering and Experiments (ALENEX). pp. 83–97 (2017)
10. Axtmann, M., Wiebigke, A., Sanders, P.: Lightweight MPI communicators with applications to perfectly balanced quicksort. In: 32nd IEEE International Parallel and Distributed Processing Symposium (IPDPS) (2018)
11. Baddar, S.W.A.H., Batcher, K.E.: Designing Sorting Networks. Springer (2011)
12. Bailey, D.H.: Misleading performance reporting in the supercomputing field. Scientific Programming **1**(2), 141–151 (1992)
13. Batcher, K.E.: Sorting networks and their applications. In: American Federation of Information Processing Societies (AFIPS) Spring Joint Computer Conference. pp. 307–314 (1968)
14. Berlekamp, E.R., Conway, J.H., Guy, R.K.: Winning Ways for Your Mathematical Plays, vol. 4. A.K. Peters, second edn. (2004)
15. Bernstein, A.J.: Analysis of programs for parallel processing. IEEE Trans. Electronic Computers **15**(5), 757–763 (1966)
16. Bilardi, G., Preparata, F.: Horizons of parallel computation. Journal of Parallel and Distributed Computing **27**, 172–182 (1995)
17. Bisseling, R.H.: Parallel Scientific Computation. A Structured Approach using BSP and MPI. Oxford University Press (2004)
18. Blelloch, G.E.: Scans as primitive parallel operations. IEEE Transactions on Computers **38**(11), 1526–1538 (1989)

19. Blelloch, G.E.: Vector Models for Data-Parallel Computing. MIT Press (1990)
20. Blumofe, R.D., Joerg, C.F., Kuszmaul, B.C., Leiserson, C.E., Randall, K.H., Zhou, Y.: Cilk: An efficient multithreaded runtime system. Journal of Parallel and Distributed Computing **37**(1), 55–69 (1996)
21. Blumofe, R.D., Leiserson, C.E.: Scheduling multithreaded computations by work stealing. Journal of the ACM **46**(5), 720–748 (1999)
22. Boehm, H.: Threads cannot be implemented as a library. In: ACM SIGPLAN Conference on Programming Language Design and Implementation (PLDI). pp. 261–268 (2005)
23. Brent, R.P.: The parallel evaluation of general arithmetic expressions. Journal of the ACM **21**(2), 201–206 (1974)
24. Bruck, J., Ho, C.T., Kipnis, S., Upfal, E., Weathersby, D.: Efficient algorithms for all-to-all communications in multiport message-passing systems. IEEE Transactions on Parallel and Distributed Systems **8**(11), 1143–1156 (1997)
25. Bryant, R.E., O'Hallaron, D.R.: Computer Systems. A Programmer's Perspective. Prentice-Hall, third edn. (2015)
26. Chan, E., Heimlich, M., Purkayastha, A., van de Geijn, R.A.: Collective communication: theory, practice, and experience. Concurrency and Computation: Practice and Experience **19**(13), 1749–1783 (2007)
27. Chapman, B., Jost, G., van der Pas, R.: Using OpenMP: Portable Shared Memory Parallel Programming. MIT Press (2008)
28. Codd, E.F.: Cellular Automata. Academic Press (1968)
29. Cole, R.: Parallel merge sort. SIAM Journal on Computing **17**(4), 770–785 (1988)
30. Cole, R.: Correction parallel merge sort. SIAM Journal on Computing **22**(6), 1349 (1993)
31. Cole, R., Vishkin, U.: Deterministic coin tossing and accelerating cascades: Micro and macro techniques for designing parallel algorithms. In: 18th ACM Symposium on Theory of Computing (STOC). pp. 206–219 (1986)
32. Cook, S.A.: A taxonomy of problems with fast parallel algorithms. Information and Control **64**, 2–22 (1985)
33. Cormen, T.H., Leiserson, C.E., Rivest, R.L., Stein, C.: Introduction to Algorithms. MIT Press, fourth edn. (2022)
34. Dean, J., Ghemawat, S.: MapReduce: Simplified data processing on large clusters. Communications of the ACM **51**(1), 107–113 (2008)
35. Dean, J., Ghemawat, S.: Mapreduce: a flexible data processing tool. Communications of the ACM **53**(1), 72–77 (2010)
36. Dongarra, J., Foster, I., Fox, G., Gropp, W., Kennedy, K., Torczon, L., White, A. (eds.): Sourcebook of Parallel Computing. Morgan Kaufmann Publishers (2003)
37. El-Ghazawi, T., Carlson, W., Sterling, T., Yelick, K.: UPC: Distributed Shared Memory Programming. John Wiley & Sons (2005)
38. Faber, V., Lubeck, O.M., White Jr., A.B.: Superlinear speedup of an efficient sequential algorithm is not possible. Parallel Computing **3**, 259–260 (1986)
39. Floyd, R.W.: Algorithm 97: Shortest path. Communications of the ACM **5**(6), 345 (1962)
40. Flynn, M.J.: Some computer organizations and their effectiveness. IEEE Transactions on Computers **C-21**, 948–960 (1072)
41. Forsell, M.: A scalable high-performance computing solution for networks on chips. IEEE Micro **22**(5), 46–55 (2002)
42. Frigo, M., Leiserson, C.E., Prokop, H., Ramachandran, S.: Cache-oblivious algorithms. In: 40th Annual Symposium on Foundations of Computer Science (FOCS). pp. 285–298 (1999)
43. Frigo, M., Leiserson, C.E., Prokop, H., Ramachandran, S.: Cache-oblivious algorithms. ACM Trans. Algorithms **8**(1), 4:1–4:22 (2012)
44. Garey, M.R., Johnson, D.S.: Computers and Intractability: A Guide to the Theory of NP-Completeness. Freeman (1979), with an addendum, 1991
45. van de Geijn, R.A., Watts, J.: SUMMA: scalable universal matrix multiplication algorithm. Concurrency and Computation: Practice and Experience **9**(4), 255–274 (1997)

46. Gorlatch, S.: Send-receive considered harmful: Myths and realities of message passing. ACM Transactions on Programming Languages and Systems **26**(1), 47–56 (2004)
47. Graham, R., Knuth, D.E., Pataschnik, O.: Concrete Mathematics. Addison-Wesley, second edn. (1994)
48. Grama, A., Gupta, A., Kumar, V.: Isoefficiency: Measuring the scalability of parallel algorithms and architectures. IEEE Transactions on Parallel and Distributed Systems **1**(3), 12–21 (1993)
49. Grama, A., Karypis, G., Kumar, V., Gupta, A.: Introduction to Parallel Computing. Addison-Wesley, second edn. (2003)
50. Greenlaw, R., Hoover, H.J., Ruzzo, W.L.: Limits to Parallel Computation: P-Completeness Theory. Topics in Parallel Computation, Oxford University Press (1995)
51. Gropp, W., Hoefler, T., Thakur, R., Lusk, E.: Using Advanced MPI. MIT Press (2014)
52. Gropp, W., Huss-Lederman, S., Lumsdaine, A., Lusk, E., Nitzberg, B., Saphir, W., Snir, M.: MPI – The Complete Reference, vol. 2, The MPI Extensions. MIT Press (1998)
53. Gropp, W., Lusk, E., Skjellum, A.: Using MPI: Portable Parallel Programming with the Message-Passing Interface. MIT Press (1994), second printing, 1995
54. Gropp, W., Lusk, E., Thakur, R.: Using MPI-2: Advanced Features of the Message-Passing Interface. MIT Press (1999)
55. Hagerup, T., Rüb, C.: Optimal merging and sorting on the EREW PRAM. Information Processing Letters **33**, 181–185 (1989)
56. Hardy, G.H., Wright, E.M.: An Introduction to the Theory of Numbers. Oxford University Press, 5th edn. (1979)
57. Helmbold, D.P., McDowell, C.E.: Modeling speedup (n) greater than n. IEEE Transactions on Parallel and Distributed Systems **1**(2), 250–256 (1990)
58. Hennessy, J.L., Patterson, D.A.: Computer Architecture. A Quantitative Approach. Morgan Kaufmann Publishers, sixth edn. (2017)
59. Herlihy, M., Shavit, N.: The Art of Multiprocessor Programming. Morgan Kaufmann Publishers, revised 1st edn. (2012)
60. Hillis, W.D., Guy L. Steele, J.: Data parallel algorithms. Communications of the ACM **29**(12), 1170–1183 (1986)
61. Hoare, C.A.R.: Monitors: An operating system structuring concept. Communications of the ACM **17**(10), 549–557 (1974)
62. Hoare, C.A.R.: Communicating sequential processes. Communications of the ACM **21**(8), 666–677 (1978)
63. Hoare, C.A.R.: Communicating Sequential Processes. Prentice-Hall (1985)
64. JáJá, J.: An Introduction to Parallel Algorithms. Addison-Wesley (1992)
65. Jones, N.D., Laaser, W.T.: Complete problems for deterministic polynomial time. Theoretical Computer Science **3**, 105–117 (1977)
66. Keller, J., Keßler, C.W., Träff, J.L.: Practical PRAM Programming. John Wiley & Sons (2001)
67. Kernighan, B.W., Pike, R.: The Practice of Programming. Addison-Wesley (1999)
68. Kernighan, B.W., Ritchie, D.M.: The C Programming Language. Prentice-Hall, second edn. (1988)
69. Knuth, D.E.: Searching and Sorting, The Art of Computer Programming, vol. 3. Addison-Wesley (1973)
70. Kuszmaul, W., Leiserson, C.E.: Floors and ceilings in divide-and-conquer recurrences. In: 4th Symposium on Simplicity in Algorithms (SOSA). pp. 133–141 (2021)
71. Lamport, L.: How to make a multiprocessor computer that correctly executes multiprocess programs. IEEE Computer **28**(9), 690–691 (1979)
72. Leighton, F.T.: Introduction to Parallel Algorithms and Architechtures: Arrays, Trees, Hypercubes. Morgan Kaufmann Publishers (1992)
73. Leiserson, C.E.: The Cilk++ concurrency platform. The Journal of Supercomputing **51**(3), 244–257 (2010)
74. Martin, M.M.K., Hill, M.D., Sorin, D.J.: Why on-chip cache coherence is here to stay. Communications of the ACM **55**, 78–89 (2012)

75. Mattson, T.G., He, Y.H., Koniges, A.E.: The OpenMP Common Core. Making OpenMP Simple Again. MIT Press (2019)
76. Mattson, T.G., Sanders, B.A., Massingill, B.L.: Patterns for Parallel Programming. Addison-Wesley (2005)
77. McCool, M., Robison, A.D., Reinders, J.: Structured Parallel Programming. Morgan Kaufmann Publishers (2012)
78. Mellor-Crummey, J.M., Scott, M.L.: Algorithms for scalable synchronization on shared-memory multiprocessors. ACM Transactions on Computer Systems **9**(1), 21–65 (1991)
79. Midkiff, S.P.: Automatic Parallelization. An Overview of Fundamental Compiler Techniques. Synthesis Lectures on Computer Architecture, Springer (2012)
80. Milner, R.: Communication and Concurrency. Prentice-Hall (1988)
81. Moore, G.E.: Cramming more components onto integrated circuits. Electronics **38**(8), 114–117 (1965)
82. MPI Forum: MPI: A Message-Passing Interface Standard. Version 3.1 (June 4th 2015), www.mpi-forum.org
83. MPI Forum: MPI: A Message-Passing Interface Standard. Version 4.1 (November 2nd 2023), www.mpi-forum.org
84. Nagarajan, V., Sorin, D.J., Hill, M.D., Wood, D.A.: A Primer on Memory Consistency and Cache Coherence. Synthesis Lectures on Computer Architecture, Morgan & Claypool Publisher, second edn. (2020)
85. Naishlos, D., Nuzman, J., Tseng, C.W., Vishkin, U.: Towards a first vertical prototyping of an extremely fine-grained parallel programming approach. Theory of Computing Systems **36**(5), 521–552 (2003)
86. Papadimitriou, C.H.: Computational Complexity. Addison-Wesley (1994)
87. van der Pas, R., Strotzer, E., Terboven, C.: Using OpenMP – The Next Step. MIT Press (2017)
88. Paterson, M.: Improved sorting networks with $O(\log N)$ depth. Algorithmica **5**(1), 65–92 (1990)
89. Patterson, D.A., Hennessy, J.L.: Computer Organization and Design. The hardware/software interface. Morgan Kaufmann Publishers (2020)
90. Rauber, T., Rünger, G.: Parallel Programming for Multicore and Cluster Systems. Springer, second edn. (2010)
91. Raynal, M.: Concurrent Programming: Algorithms, Principles, and Foundations. Springer (2013)
92. Roughgarden, T.: Algorithms Illuminated. Part 1: The Basics. Soundlikeyourself Publishing (2017)
93. Roughgarden, T. (ed.): Beyond the worst-case analysis of algorithms. Cambridge University Press (2021)
94. Sanders, P., Lamm, S., Hübschle-Schneider, L., Schrade, E., Dachsbacher, C.: Efficient parallel random sampling – vectorized, cache-efficient, and online. ACM Transactions on Mathematical Software **44**(3), 29:1–29:14 (2018)
95. Schardl, T.B., Lee, I.A.: Opencilk: A modular and extensible software infrastructure for fast task-parallel code. In: 28th ACM SIGPLAN Symposium on Principles and Practice of Parallel Programming (PPoPP). pp. 189–203 (2023)
96. Schmidt, B., González-Domínguez, J., Hundt, C., Schlarb, M.: Parallel Programming. Concepts and Practice. Morgan Kaufmann Publishers (2018)
97. Sedgewick, R.: Quicksort with equal keys. SIAM Journal on Computing **6**(2), 240–267 (1977)
98. Sedgewick, R.: Implementing quicksort programs. Communications of the ACM **21**(10), 847–857 (1978), corrigendum *ibidem* **23** (79) 368
99. Sedgewick, R., Flajolet, P.: Analysis of Algorithms. Addison-Wesley, second edn. (2013)
100. Sedgewick, R., Wayne, K.: Algorithms. Addison-Wesley, 4th edn. (2011)
101. Shiloach, Y., Vishkin, U.: Finding the maximum, merging and sorting in a parallel computation model. Journal of Algorithms **2**, 88–102 (1981)

102. Siebert, C., Träff, J.L.: Perfectly load-balanced, stable, synchronization-free parallel merge. Parallel Processing Letters **24**(1) (2014)
103. Snir, M.: On parallel searching. SIAM Journal on Computing **14**(3), 688–708 (1985)
104. Snir, M.: Depth-size trade-offs for parallel prefix computation. Journal of Algorithms **7**(2), 185–201 (1986)
105. Snir, M., Otto, S., Huss-Lederman, S., Walker, D., Dongarra, J.: MPI – The Complete Reference, vol. 1, The MPI Core. MIT Press, second edn. (1998)
106. Snir, M., Otto, S.W., Huss-Lederman, S., Walker, D.W., Dongarra, J.: MPI: The Complete Reference. MIT Press (1996)
107. Strassen, V.: Gaussian elimination is not optimal. Numerische Mathematik **13**, 354–356 (1969)
108. Sutter, H., Larus, J.R.: Software and the concurrency revolution. ACM Queue **3**(7), 54–62 (2005)
109. Tanenbaum, A.S., Austin, T.: Structured Computer Organization. Prentice-Hall, sixth edn. (2012)
110. Tanenbaum, A.S., Wetherall, D.J.: Computer Networks. Pearson Prentice-Hall, 5th edn. (2011)
111. Toffoli, T., Margolus, N.: Cellular Automata Machines: A New Environment for Modeling. MIT Press, second edn. (1987)
112. Torrellas, J., Lam, M.S., Hennessy, J.L.: False sharing and spatial locality in multiprocessor caches. IEEE Transactions on Computers **43**(6), 651–663 (1994)
113. Träff, J.L.: Simplified, stable parallel merging. arXiv:1202.6575 (2012)
114. Träff, J.L.: A note on (parallel) depth- and breadth-first search by arc elimination. arXiv:1305.1222 (2013)
115. Träff, J.L.: Parallel quicksort without pairwise element exchange. arXiv:1804.07494 (2018)
116. Träff, J.L., Hunold, S., Mercier, G., Holmes, D.J.: MPI collective communication through a single set of interfaces: A case for orthogonality. Parallel Computing **107** (2021)
117. Träff, J.L., Vardas, I.: Library development with MPI: Attributes, request objects, group communicator creation, local reductions and datatypes. In: 30th European MPI Users' Group Meeting (EuroMPI) (2023)
118. Valiant, L.G.: A bridging model for parallel computation. Communications of the ACM **33**(8), 103–111 (1990)
119. Valiant, L.G.: General purpose parallel architectures. In: van Leeuwen, J. (ed.) Handbook of Theoretical Computer Science, vol. A, chap. 18, pp. 943–972. Elsevier (1990)
120. Valiant, L.G.: A bridging model for multi-core computing. Journal of Computer and System Sciences **77**(1), 154–166 (2011)
121. Wagar, B.: Hyperquicksort – a fast sorting algorithm for hypercubes. In: Hypercube Multiprocessors. pp. 292–299. SIAM Press (1987)
122. Williams, M., Sanders, P., Dementiev, R.: Engineering multiqueues: Fast relaxed concurrent priority queues. In: 29th Annual European Symposium on Algorithms (ESA). LIPIcs, vol. 204, pp. 81:1–81:17 (2021)
123. Williams, S., Waterman, A., Patterson, D.A.: Roofline: an insightful visual performance model for multicore architectures. Communications of the ACM **52**(4), 65–76 (2009)
124. Zhu, H., Cheng, C.K., Graham, R.L.: On the construction of zero-deficiency parallel prefix circuits with minimum depth. ACM Transactions on Design Automation of Electronic Systems **11**(2), 387–409 (2006)

Index

k-ported, 175
(fully) strict computations, 146

accelerator, 119, 172
access epoch, 240
active synchronization, 240
adaptive routing, 181
algorithm
   cost-optimal, 18
   work-optimal, 19
algorithmic efficiency, 27
all-pairs shortest path problem, 168
all-to-all operation, 249
all-to-all problem, 178
allgather operation, 249
allreduce operation, 63
Amdahl's Law, see Law
Architecture Review Board, 119, 183
arithmetical circuit, 69
array compaction, 53, 64, 69, 117, 134, 144
atomic operation, 98, 114, 115, 118, 129, 139, 140, 236, 240

barrier, 55, 125
barrier operation, 249
barrier synchronization, 66, 121
basic datatype, 206
Bernstein conditions, 44
BFS, 5, 6, 15, 51, 86, 167, 261
bidirectional, 176
bidirectional send-receive, 176
bidirectional telephone, 175
bisection width, 173, 213
Bitonic sequence, 61
block index type, 230
blocking, 71, 215, 248
Breadth-First Search, see BFS
Brent's Theorem, 21, 43

bridging model, 6
broadcast operation, 63, 247, 249
broadcast problem, 176
BSP, 6, 51, 261
bucket sort, 269
buffered send, 226
Bulk Synchronous Parallel, see BSP

cache, 86
   k-way set associative, 87
   cache hit, 87
   cache miss, 87
   capacity miss, 87
   coherent, 92
   cold miss, 87
   compulsory miss, 87
   conflict miss, 87
   directly mapped, 87
   eviction policy, 88
   false sharing, 93, 130
   fully associative, 87
   hit rate, 87
   miss rate, 87
   non-coherent, 92
   replacement policy, 88
   set associative, 87
   spatial locality, 88
   temporal locality, 88
   write allocate, 88
   write back, 88
   write non-allocate, 88
   write-through, 88
cache coherence problem, 92, 98
cache coherence protocol, 92
cache coherence traffic, 92
cache line, 87, 130
cache-aware algorithm, 91
cache-oblivious algorithm, 91

canonical form, 126, 127, 129, 132
Cartesian communicator, 194, 195, 197–199
Cilk, 40, 43, 145, 146
co-rank, 59, 60, 148, 245
collective communication, 54
collective operation, 63, 191, 246–251, 261, 262, 273, 276
collectives, 54, 248
Communicating Sequential Processes, *see* CSP
communication centric, 183
communication deadlock, 206
communication domain, 184, 189
communication epoch, 240
communication round, 213
communication round complexity, 213
communication step, 213
communication window, 237
communicator, 189–191, 195, 197, 199, 222, 246
comparator networks, 62
Compare-And-Swap (CAS), 115, 118, 140, 240
Compare-Exchange (CEX), 115
complexity class
   $\mathcal{NC}$, 75
   $\mathcal{P}$, 75
computational problem, 4, 5, 8, 14, 19
compute-bound, 96
concurrency, 5
Concurrent Computing, 1, 5, 106, 113
concurrent data structures, 112
condition variable, 109, 141
   broadcast, 109
   signal, 109
   wait, 109
congestion, 181
consensus problem, 115
consistent arguments, 246
contention, 181
contiguous type, 230
continuation, 145
cost, 20
cost-optimal, 22, 23, 25–27
cost-optimality, 18
counting sort, 269
critical path, 41
critical section, 105, 138
CSP, 184

DAG, 135, 213
   depth, 41
   span, 41
   work, 40
data distribution, 52

block cyclic, 52, 262
blockwise, 52
column-wise, 53, 263
cyclic, 52
row-wise, 53, 262
data race, 104, 123, 124, 240
data race free, 104, 108, 139
deadlock, 106, 113, 180, 206, 208, 211, 218, 222, 252, 275
deadlock free, 182
deadlock freedom, 180
dense, 6
dependency edges, 39
depth, 72
Depth-First Search, *see* DFS
derived datatype, 220, 221, 229
deterministic routing, 181
DFS, 5, 6, 15, 86
diameter, 172
direct network, 172
Directed Acyclic (task) Graph (DAG), 39
Distributed Computing, 1, 4, 5, 179, 180
distributed graph communicator, 197, 199
distributed object, 199
distribution centric, 183
domain decomposition, 50
dynamic load balancing, 23

efficiency, 26–28, 30, 34, 35
error handlers, 189
Exchange (EXCHG), 115
exclusive prefix sums, *see* prefix sums
exposure epoch, 240
exscan, 62
extent, 221, 230
external memory, 94

Fetch-And-Add (FAA), 114, 140
Fetch-And-Increment (FAI), 114, 140
Fetch-And-Operate (FAO), 115, 240
final task, 40
first touch, 95
FLOPS, 2
flow control, 181
Flynn's taxonomy, 12
fork-join, 40
fully connected network, 173

gather operation, 249
geometric series, 289
GPU, 3, 13, 119, 172
granularity, 23, 88
   coarse grained, 23, 51
   fine rained, 23
graphics processing unit, *see* GPU

Index 303

greedy scheduling, 42, 146

halo, 231
hardware efficiency, 27
hardware multi-threading, 3, 121
High-Performance Computing, *see* HPC
Hillis–Steele algorithm, 69–72
HPC, 1, 2, 27, 92, 183, 238
hypercube network, 175
HyperQuicksort, 269
hyperthreading, 3

immediate operations, 215
in-place, 149
inclusive prefix sums, *see* prefix sums
index type, 230
indirect network, 172
inherently sequential, 76
inter-communicators, 276
interconnect, 172
interconnection network, 171
interleaving, 97
invariant, 7, 51, 59, 60, 68, 70, 97
irregular, 6
irregular collective, 249
iso-efficiency, 27, 30, 35
iso-efficiency function, 27, 75

last level cache, 91
Law
  Amdahl's Law, 24, 26, 42, 50, 51, 60, 112, 114, 262
  Depth Law, 41
  Moore's Law, 2, 86
  Work Law, 20, 26, 41
linear algebra, 5
linear pipeline, 46
linear processor array, 173
linear programming, 76
linear-affine transmission cost model, 178, 182
links, 172
list ranking, 72
list ranking problem, 72
list scheduling, 43
LLC, 91
load balance, 49, 117
load balancing, 23, 51, 58, 60, 117
load imbalance, 23, 44, 126
local completion, 215
local object, 199
Lock, 105, 242
  acquire, 105
  blocking, 108
  contention, 107

deadlock free, 106
fair, 106
lock, 106
nested, 114, 141
readers-writer, 107, 111, 112, 141
recursive, 113, 141
release, 106
spin lock, 108
starvation free, 106
try-lock, 107, 113, 141
unlock, 106
lock-free, 118
lock-freeness, 118
loop dependency, 128
  loop carried anti-dependency, 45
  loop carried dependency, 45
  loop carried flow dependency, 45
  loop carried output dependency, 45
loop of independent iterations, 40
loop schedule
  dynamic, 129
  guided, 129
  static, 128
loop scheduling, 24, 43, 126
lower bound, 18, 20

many-core processor, 3
map-reduce, 55
Master Theorem, 56, 66, 91, 146, 147, 149, 290
master-slave, 50
master-worker, 50, 193, 201
matrix–matrix multiplication, 5, 9, 10, 15, 89–92, 192, 197, 262, 264–266
matrix–vector multiplication, 5, 262, 264
maximum flow problem, 76
memory consistency problem, 98
memory controllers, 95
memory hierarchy, 94
memory-bound, 96
merging, 5, 15, 56–60, 62, 64, 75, 146, 148, 149, 245
merging by co-ranking, 59
merging by ranking, 57
mesh network, 174
message tag, 204, 208, 240
Message-Passing Interface, *see* MPI
MIMD, 12, 99, 171, 182
minimal routing, 181
MISD, 13
monitor, 109
Moore's Law, *see* Law
MPI, 4, 13, 54, 56, 71, 182–187, 215, 236, 246
multi-core processor, 3, 29, 87, 91, 94, 95, 114

multi-ported, 175
multi-stage networks, 175
mutex, 106, 109–111, 242
mutual exclusion, 105, 106, 123, 138, 242
mutual exclusion problem, 105

neighborhood collectives, 198, 273
neighborhoods, 198
network switches, 172
nominal processor performance, 2, 3, 27, 87, 96
non-local completion, 215, 248
non-synchronizing, 248
Non-Uniform Memory Access, see NUMA
nonblocking, 215, 216, 223, 226, 227, 236, 242, 248, 270
NUMA, 12, 95

oblivious, 6, 24, 30, 50, 60, 71, 91, 181, 184
one-ported, 175
one-sided communication, 236, 238–240, 242, 244
opaque, 190, 199
OpenMP, 4, 13, 22, 40, 43, 55, 86, 98, 104, 105, 113, 118, 119, 126, 183, 186, 213
optimistic locking, 139
ordered depth-first search, 76
origin process, 236, 239
overhead, 22, 28
oversubscription, 100, 121, 186
owner computes, 183

packet switching, 181
packets, 180
padding, 94
parallel
    embarrassingly, 24, 52
    pleasantly, 24
    trivially, 24
Parallel Computing, v, vi, 1, 3–6, 11, 13, 15, 16, 22, 24, 25, 27–30, 54, 56, 64, 85, 86, 96, 100, 106, 113, 118, 119, 171–174, 179, 181, 182, 184, 185, 203, 289, 290
parallel data structure, 49
parallel efficiency, 26
Parallel Random Access Machine, see PRAM, 7
parallel region, 120
parallel time, 14, 18, 20, 23, 25, 26, 28, 29, 36, 44, 46, 47, 49, 51, 55, 65, 66, 68, 70, 74, 266
parallel time complexity, 14, 18, 20, 21, 26, 41, 75, 76, 91, 145
parallelism, 21, 26, 28, 32, 42, 64
parallelization, 22, 24, 25

parallelizing compiler, 46
Partitioned Global Address Space, see PGAS
passive synchronization, 241
pattern
    design, 39
    parallel, 39, 63
performance portability, 6
persistent communication, 228
personalized exchange, 249
PGAS, 183
pinning, 86, 186
pipelining, 47, 180, 181
pointer jumping, 78
Posix threads, see pthreads
PRAM, 7, 11, 14, 17, 19–21, 23, 44, 47, 53, 55, 62, 66, 75, 85, 104
    Arbitrary CRCW, 7
    Common CRCW, 7, 9, 11, 45, 71, 73, 105
    CRCW, 7, 54
    CREW, 7, 10, 53, 58–60
    EREW, 7, 10, 60, 62, 72
    Priority CRCW, 7
prefix sums, 5, 15, 55, 56, 58, 62–64, 68–72, 75, 133, 134, 138, 144, 145, 249, 260, 270
prefix sums problem, 62, 64, 65, 72
priority inversion, 114
problem specification, 262
process mapping, 198
processing elements, 3
processor, 3
processor ring, 173, 205
processor-core, 1–3, 85, 91, 100, 106, 108, 182, 186
program order, 97
programming model, 4, 13, 43, 85, 86, 99, 118, 119, 171, 182–184
progress rule, 236
pthreads, 4, 86, 98–100, 104–106, 113, 186, 213, 275

Quicksort, 5, 40, 43, 64, 135, 137, 146, 192, 266, 267, 269

race condition, 45, 102, 104, 119, 123–125, 127, 139, 207, 213, 222, 275
radix sort, 269
RAM, 6, 7, 14, 75, 86
Random Access Machine, see RAM
rank, 57
rank order, 191, 249, 252–254, 257
ready send, 227
recurrence relation, 65
reduction, 15, 63, 133
reduction operation, 249

Index

reduction problem, 62
regular, 6
regular collective, 249
reliable communication, 180, 182, 184, 203
roofline performance model, 96
root process, 249
root task, 40
routing
  centralized, 179
routing algorithm, 179
routing protocol, 179
routing system, 179, 203
row-major order, 48, 89, 232, 233, 265

safe, parallel libraries, 190
scalability, 21
scalability analysis, 28
scan, 55, 62, 64, 133, 249, 258, 260
scan operation, 249
scatter operation, 249
schedule, 41, 126, 128, 130
semaphore, 109
sequential complexity, 14, 57, 63
sequential consistency, 97, 100
serialize, 107
SIMD, 2, 12, 13, 44, 63, 141, 142
SIMT, 13
simulation argument, 17, 20, 75
simultaneous multi-threading, 3
single-ported, 175
SISD, 12
SMP, 86
sorting network, 62
span, 41
sparse, 6
spatial locality, 69
spawning, 145
speed-up, 16, 18, 20, 23–26, 28, 38, 51, 52, 58, 64, 69, 85, 96, 107, 112, 133, 145, 264
  absolute, 16, 22, 27, 30, 32, 33, 35, 70, 267
  linear, 16–18, 20, 27, 28, 75, 76, 96, 138, 146, 266, 269
  perfect, 16, 36, 96
  relative, 21, 22, 42, 267
  scaled, 16, 26
  super-linear, 17, 95, 96
SPMD, 13, 99, 119, 120, 182, 186, 208
stable, 57
stack allocation, 209
start task, 40
start-up latency, 179
static load-balancing, 23
stencil computation, 5, 47–52, 192, 194, 217, 225, 226, 231, 242, 259

stencil rule, 48, 49, 194
store-and-forward, 181
strands, 39
strong scaling, 16
strongly scalable, 28
strongly scaling, 35
structured type, 230
Swap (SWAP), 115
Symmetric MultiProcessing, *see* SMP
synchronization, 22, 23, 26
synchronous send, 226

target process, 236, 239
temporal synchronization, 29
termination detection problem, 50
Test-And-Set (TAS), 114
test-and-test-and-set, 139
thread, 99
thread safe, 102, 117, 122
topological order, 41
topology, 172
torus, 174, 195
torus network, 174
tractable problems, 75
translation look-aside buffer, 91
tree network, 174
type map, 220, 229
type signature, 221, 252

UMA, 11
unidirectional, 175
Unified Parallel C, 183
Uniform Memory Access, *see* UMA
unsafe, 102, 241, 248
unsafe programming, 216, 217, 223, 225, 248, 251
UPC, 183
user-defined datatype, 220, 221, 239

vector computer, 12
vector type, 230

wait-free, 118
wait-freeness, 115, 118
wall clock time, 122, 187
weakly scaling, 27, 28, 35, 231
work, 19, 20, 22, 23, 121
work pool, 49, 50, 118, 129
work sharing construct, 119, 123, 125, 126, 133, 135, 137
work-dealing, 50
work-optimal, 22, 23, 25, 61
work-stealing, 24, 43, 50, 118, 146
write buffer, 95

MIX
Papier aus verantwortungsvollen Quellen
Paper from responsible sources
FSC® C105338

If you have any concerns about our products,
you can contact us on
ProductSafety@springernature.com

In case Publisher is established outside the EU,
the EU authorized representative is:
**Springer Nature Customer Service Center GmbH
Europaplatz 3, 69115 Heidelberg, Germany**

Printed by Libri Plureos GmbH
in Hamburg, Germany